Half Price Books
1835 Forms Drive
Carrollton, TX 75006
OFS OrderID 31295942

Thank you for your order, Bobby Perry!

Thank you for shopping with Half Price Books! Please contact Support@hpb.com. if you have any questions, comments or concerns about your order (113-6859919-4462663)

Visit our stores to sell your books, music, movies games for cash.

SKU	ISBN/UPC	Title & Author/Artist	Shelf ID	Qty	OrderSKU
S382582488	9780070708242	Practice Tests for Communications Licensin... Wilson, Sam,Risse, Joseph A.	27--17--3	1	

SHIPPED STANDARD TO:
Bobby Perry
2800 S MAIN ST APT 314
FINDLAY OH 45840-1051
kcydcn0qnbplbkd@marketplace.amazon.com

ORDER# **113-6859919-4462663**
AmazonMarketplaceUS

Communications Licensing and Certification Examinations

The Complete TAB Reference

Sam Wilson
Joseph A. Risse

TAB Books
Division of McGraw-Hill, Inc.
New York San Francisco Washington, D.C. Auckland Bogotá
Caracas Lisbon London Madrid Mexico City Milan
Montreal New Delhi San Juan Singapore
Sydney Tokyo Toronto

Published by TAB Books, a division of McGraw-Hill, Inc.

pbk 1 2 3 4 5 6 7 8 9 DOH/DOH 9 9 8 7 6 5 4
hc 1 2 3 4 5 6 7 8 9 DOH/DOH 9 9 8 7 6 5 4

Library of Congress Cataloging-in-Publication Data
Wilson, J. A. Sam.
Communications licensing and certification examinations : the complete TAB reference / by Sam Wilson, Joseph A. Risse.
p. cm.
ISBN 0-07-070822-3 ISBN 0-07-070823-1
1. Radio—Examinations, questions, etc. 2. Radio operators--Certification. 3. Electronics—Examinations, questions, etc. 4. Electronic technicians—Certification. I. Risse, Joseph A. II. Title.
TK6554.5.W48 1994
621.384'076—dc20 94-30235
CIP

Acquisitions editor: Roland S. Phelps
Editorial team: Joanne Slike, Executive Editor
Andrew Yoder, Managing Editor
John T. Arthur, Book Editor
Joann Woy, Indexer
Production team: Katherine G. Brown, Director
Susan E. Hansford, Coding
Janice Stottlemyer, Computer Artist
Wanda S. Ditch, Desktop Operator
Lorie White, Proofreading
Design team: Jaclyn J. Boone, Designer
Brian Allison, Associate Designer

0708231
EL1

Acknowledgments

The authors thank Roland Phelps, the TAB acquisitions editor in charge of this book, for his help in getting the book started and finished. The following people were also very helpful in getting the book into production: Joanne Slike, Andrew Yoder, and Kriss Lively-Helman.

In addition, we appreciate the help of Joseph Glynn and William Myers, both of WVIA TV, Pittston, PA. They provided valuable information on the subject of Television Transmission and power measurement.

Thanks also to Ken Muncey of Cocoa Beach, Florida for reviewing the rough draft and for his helpful suggestions.

A special thanks goes to Norma Wilson of Melbourne, FL and Lisa Naipaver of Cleveland, OH. They typed the rough drafts and the manuscript. Working together, they converted our ideas into a readable manuscript.

Electronic Servicing and Technology (ES&T) magazine gave us broad permission to use material from their publication. We appreciate it very much.

Several organizations gave much extra help with study material. We give special thanks to ETA, ISCET, and SBE for sending us a lot of useful material.

Contents

Introduction

Practical applications of electronics started with communications. Throughout the history of electronics, the field of communications has been a leader in new technology. Communications has always been an excellent way to learn about practical electronics.

A number of licenses and certificates are available to people with training in basic electronics and communications. Each is obtained by taking a test in some particular specialty. Below is a list of the available certification and license exams. The abbreviations are explained below.

- GROL
- GMDSS
- Radio Telegraph Certificate
- Radar Endorsement
- Amateur Radio Technician Class License
- Marine Radio Operator Permit
- ETA Associate-Level Certification
- ISCET Associate-Level Certification
- ETA Journeyman Communications Certification
- ISCET Journeyman Communications Certification
- NICET Certification
- SBE Certification
- NARTE

GROL	General Radio Operator License
GMDSS	Global Marine Distress and Safety Service
ETA	Electronics Technicians Association
ISCET	International Society of Electronics Technicians
SBE	Society of Broadcast Engineers
NICET	National Institute for Certification in Engineering Technologies
NARTE	National Association of Radio and Telecommunications Engineers

There is a companion book for this study guide, *The Complete TAB Book of Practice Tests for Communications Licenses and Certification Examinations.* The two books give you an overall review of the material needed for passing required tests for the aforementioned licenses and certifications.

This is not a textbook; rather it is a book designed to give technicians having training and/or experience a chance to evaluate their ability to pass license and certification tests, and to reveal subjects where further study would be useful before taking a test.

Having said that, it is also a useful book for technicians who have recently completed their training in electronics. Because of the wide range of electronics subjects covered, technicians can find areas where they would benefit from further study.

When you answer a question correctly, you reenforce your knowledge. When you cannot answer a question, you have a chance to increase your knowledge.

By showing a willingness to put one's knowledge to a test, technicians demonstrate that they can accept challenges in their work. Also, it demonstrates pride in their knowledge and ability to accept challenge. As an added incentive, for some jobs a license and/or certification is *required* for obtaining employment.

In order to cover the wide range of subject matter necessary, the authors have made a few assumptions. For example, it is assumed that you are preparing to take one or more of the tests required for licensing and/or certification. For some of those tests, but not all of them, it is necessary to have an understanding of basic algebra and trigonometry, but it is not possible to include preparation for those subjects in this book.

It is also impossible to teach Morse code in a book. The amateur Technician license that we prepare you for does not require code. Code is required for the radiotelegraph licenses. So, we are only preparing you for the theory part of the required exam for that license.

Calculations for solved problems in this book are usually carried out to three significant figures.

Very little coverage is given to the review of very basic electricity and electronics in this book. It is assumed you are familiar with Ohm's Law, series and parallel dc and ac circuits, and the theory of operation for transistors and other basic components and circuits. Answers are given for every question used in this book, so it can be used for either classroom work or self study.

To keep in step with most of the license and certification tests, we assume electron flow as the direction for current. If you have been trained in conventional (positive-to-negative) current flow you should practice reversing the direction of current flow when studying for the required tests.

When you add your existing knowledge to the material presented in these books you should not experience any unpleasant surprises when you sit for the actual tests.

1
CHAPTER

Voltage, current, and resistance

The measurements most often made for troubleshooting or evaluating an electronic circuit are of: voltage, current, and resistance.

Voltage

Voltage is not a measure of force. That is why the term *electromotive force* and the abbreviation *EMF* are rapidly disappearing from the technology field. Also, the use of *E* for voltage is on the way out. It is being replaced by *V*. Having said that, remember that you will still see those terms and abbreviations used in books and tests.

It is important to know the exact meanings of all the technical terms you encounter in your work. If you are tracing or analyzing a circuit, it is good practice to consider the voltage as being the force that moves current through a circuit. However, when you measure voltage, you are not measuring a force. Instead, you are measuring a unit of work per unit charge.

To be technically accurate, *voltage* is a measure of work done in moving a unit charge (the coulomb) from a positive point to a less positive point. Because work and energy are numerically equal in this application, you might hear (correctly) that a volt is the energy used in moving a unit charge from a positive point to a less positive point.

The unit charge in electronics is the *coulomb*. It is the sum of all the charges of 6.24×10^{18}, the electrons combined in one location.

Note: You will also see the number as 6.25×10^{18} electrons. Both numbers are correct depending upon how many decimal places are used in its calculation.

In science and technology, *work* is defined as a measure of force exerted through a distance. The simple mathematical relationship is:

$$work = force \times distance$$

The concepts of work and voltage are illustrated in Fig. 1-1.

If voltage was a measure of force, you would always have to know the distance from the more positive point when a measurement was taken. However, the work

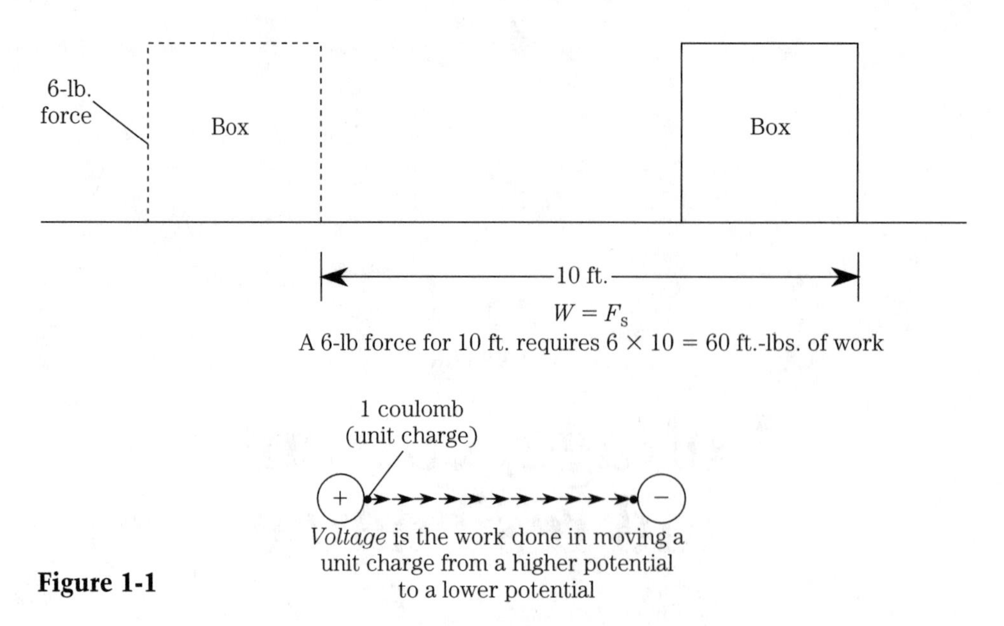
6-lb.
force
Box
Box
10 ft.
$W = F_s$
A 6-lb force for 10 ft. requires 6 × 10 = 60 ft.-lbs. of work
1 coulomb
(unit charge)
+
−
Voltage is the work done in moving a unit charge from a higher potential to a lower potential

Figure 1-1

done in moving a unit charge around a circuit is always the same, regardless of which path is taken. If the distance is increased, the force is decreased, and if the distance is decreased, the force is increased. That point is also illustrated in Fig 1-1.

The overall result is that the force × distance is always the same, regardless of the path taken. That is why voltage is not defined as a force.

This careful definition of voltage becomes very important when making certain calculations. It does not change the way you think of voltage in troubleshooting.

If you are trained as a technician, you learned that the negative voltage of a battery repels electrons and that is what causes current to flow. That is an excellent model and you should stay with it.

Energy is defined as the capacity to do work. There are two kinds of energy that you should be aware of. *Potential energy* is the energy a body has by virtue of its position. A brick sitting on a tree limb has potential energy. *Kinetic energy* is the energy a body has by virtue of its motion. If the brick falls off the tree limb, it has kinetic energy.

Now, this is very important: when the brick hits the ground it must give up its energy. Energy cannot be created or destroyed. So, when the brick hits the ground its kinetic energy is converted to heat energy.

The law that says energy cannot be created or destroyed was not bypassed when the atomic bomb was exploded. What actually happened was the energy was converted from one form into another form.

The understanding of energy is necessary when you review the operation of X-rays and neon lamps.

Methods of generating a voltage

It is not technically accurate to speak of generating a current. Actually, a generated current is defined as a current that flows as a result of a generated voltage.

There are six methods of generating a voltage. Knowing those methods can be helpful in understanding technology. The following section covers the methods along with some practical applications.

Friction (static generators)

You are, no doubt, familiar with some of the examples of how friction can be used to generate a voltage. Rubbing a cat's fur on a dry day, running a comb through your hair on a dry day, and rubbing a piece of paper with the flat side of a pencil. Those are all examples of voltage generated by friction. It is sometimes called the *static method.*

Here are two more examples: sliding across the plastic-covered seats in a car and scuffing your feet across the carpet. In these cases the generated voltage can be sufficient to cause a spark to jump from your finger to a metal surface. When that happens the voltage is over 5000 volts!

You may be surprised by that amount of voltage between you and the metal surface. However, *voltage does not kill*; it is the current that flows as a result of the voltage that kills. The amount of current in the previous examples is so low, and of such short duration, that your life is not endangered.

Generating a voltage with friction in not a highly efficient method. The very fact that friction is needed works against efficiency.

There are some useful static generators. Two examples are the Van de Graff and the Wiemshurst generators. Both are capable of generating many thousands of volts, but they cannot supply any appreciable amount of current to a load resistance. So. they are only used in high-voltage, low-current applications.

One example of how they are used is X-ray tubes. Their theory of operation is illustrated in Fig. 1-2. To understand this application, it is necessary to remember that energy cannot be created or destroyed. However, it can be converted from one form to another.

In the X-ray tube, electrons are emitted by a hot cathode. They are accelerated to a very high speed by the high positive voltage from a static generator. Other sources of high voltage can also be used for this application. The high-speed electrons have a high *kinetic energy*; that is, energy by virtue of their motion.

When the electron beam strikes the highly positive plate in the X-ray tube the kinetic energy of the electrons cannot just disappear, it must be used in some way. In

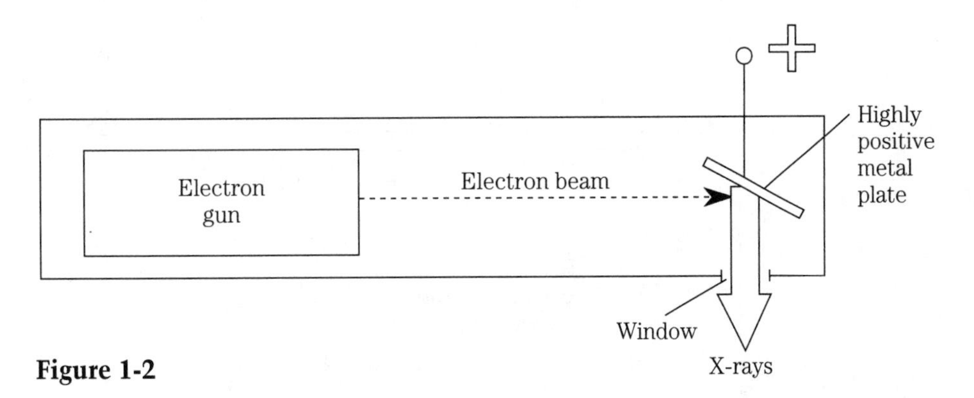

Figure 1-2

this case some is converted to heat and much of it is converted to X-ray energy. A special window in the tube allows the X-rays to be used for various purposes.

From the discussion on X-rays, it is obvious that certain vacuum tubes can be X-ray emitters. Two examples are the cathode-ray tube (CRT) and the vacuum tube diodes in high-voltage rectifiers. Both are used in television receivers. In both of these examples, electrons undergo a high acceleration before striking the anode of the diode or the face of the picture tube (CRT).

At one time, the problem of X-ray exposure was so prevalent that special precautions were necessary to protect television viewers. The high voltage in television receivers is now limited to about 30,000 V in order to limit the electron acceleration. Also, special glass envelopes have been developed to greatly reduce the amount of the X-rays reaching the outside world.

Heat (thermoelectric generators)

Whenever two dissimilar metals that are joined are heated, a voltage is generated across the ends of the conductors. This thermoelectric method of generating a voltage is called the *Seebeck effect*, illustrated in Fig. 1-3. When the metals are properly chosen, they can produce a voltage high enough for some applications. An example of a device that used this principle is called a *thermocouple*. One example of a thermocouple application is illustrated in Fig. 1-3.

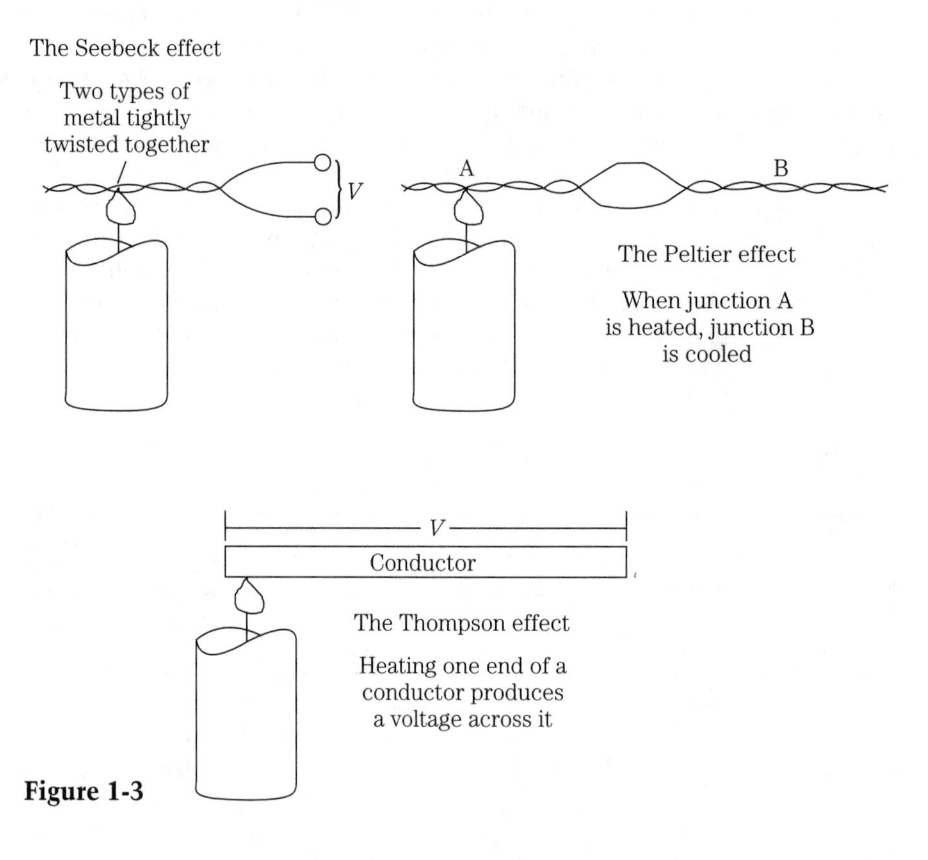
The Seebeck effect
Two types of metal tightly twisted together
V
A
B
The Peltier effect
When junction A is heated, junction B is cooled
V
Conductor
The Thompson effect
Heating one end of a conductor produces a voltage across it

Figure 1-3

Consider the case where the open ends of the conductors are joined together as shown in Fig 1-3. The new junction is cooled when the original junction is heated. This is called the *Peltier effect*.

At one time, this was just a laboratory curiosity. Now, the development of materials has made it possible to use the Peltier effect to advantage in practical applications. A small refrigerator and cooling of some electronic components are examples.

Another method of generating a voltage directly by heat utilizes the Thompson effect (Fig. 1-3). No practical applications of this voltage generator have been made to date.

The Seebeck effect and Thompson effect are called *thermoelectric effects*. When used to produce a voltage, they are called *thermoelectric generators* or *TE generators*.

Pressure (piezoelectric effects)

Certain crystalline materials generate a voltage whenever they are under physical pressure. Examples of these *piezoelectric* materials are: barium titanate, tourmaline, and quartz.

Figure 1-4 illustrates the relationship between piezoelectric devices for accurately controlling the frequency of transmitters. The crystalline material, which might be quartz, is cut into a thin slice. The thin crystal is placed in the circuit of an oscillator. In that circuit, it vibrates at a very specific frequency and generates an accurately controlled frequency.

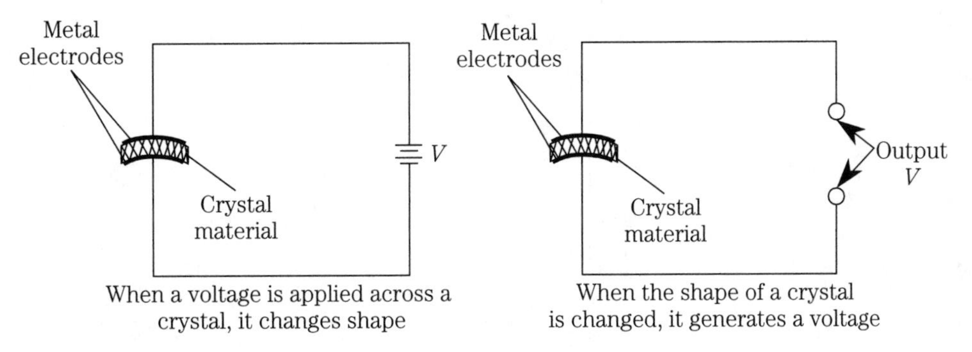
Metal electrodes
V
Crystal material
When a voltage is applied across a crystal, it changes shape
Metal electrodes
Output V
Crystal material
When the shape of a crystal is changed, it generates a voltage

Figure 1-4

In this application, the piezoelectric effect is brought into play. The generated voltage is an ac signal and causes the crystal to vibrate. Because of its size and shape, the vibration frequency is accurately controlled. When the crystal flexes during vibration, it generates a voltage that is used as a signal in the oscillator.

Examples of the use of the piezoelectric effects are illustrated in Fig. 1-5.

The piezoelectric slices are sometimes referred to as *crystals*. Crystals are used in crystal microphones. In this application, sound waves cause the crystal to vibrate and produce an audio voltage. High-frequency speakers, called *tweeters*, generate sound waves when a high-frequency voltage is delivered across their surfaces. Sensitive phonograph pickups produce audio signals when the grooves on the surface of the phonograph record vibrate the crystal connected to the phonograph stylus. The output of the crystal is an audio voltage that corresponds to the audio that made the grooves.

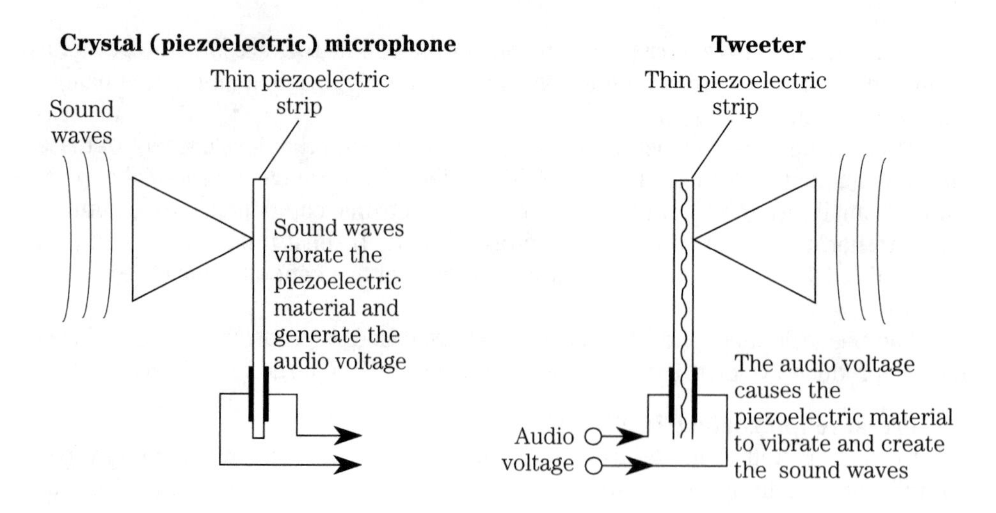

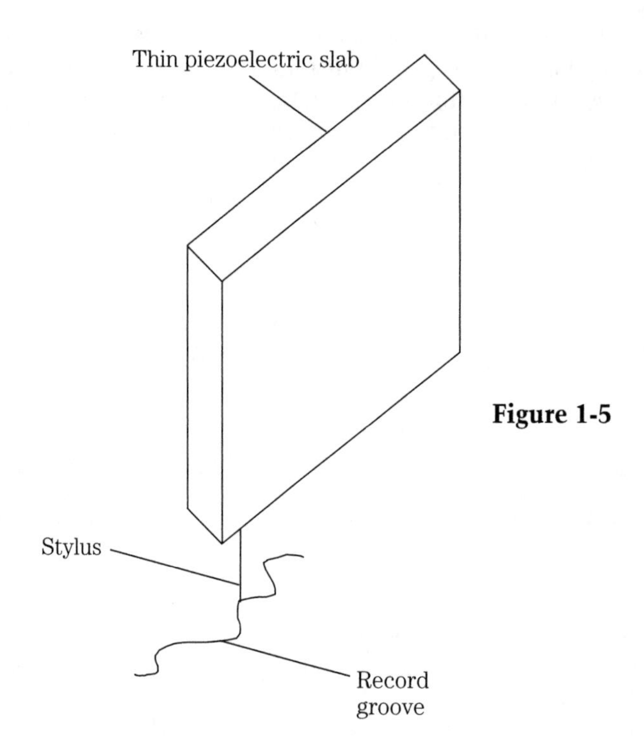

Figure 1-5

Light (optoelectronic devices)

There are two very important optoelectronic phenomena, called *photoelectric effects*, that require attention.

When light strikes the surface of certain materials, such as selenium, they generate a voltage. Photocells are made with this concept. Unfortunately, some companies have marketed photoresistors under the name *photocell*. The difference is that a photocell generates a voltage in response to light, whereas, a *photoresistor*, also known as a *photoconductor*, changes its resistance in response to light. Figure 1-6 shows the symbols for photocells and photoresistors.

When current flows through certain optoelectronic devices, they emit light. An example is the *light-emitting diode (LED)*. Do not confuse this device with the *light-activated diode (LAD)*, which becomes a conducting diode only when exposed to light. Note the difference between the symbols for these devices shown in Fig. 1-6.

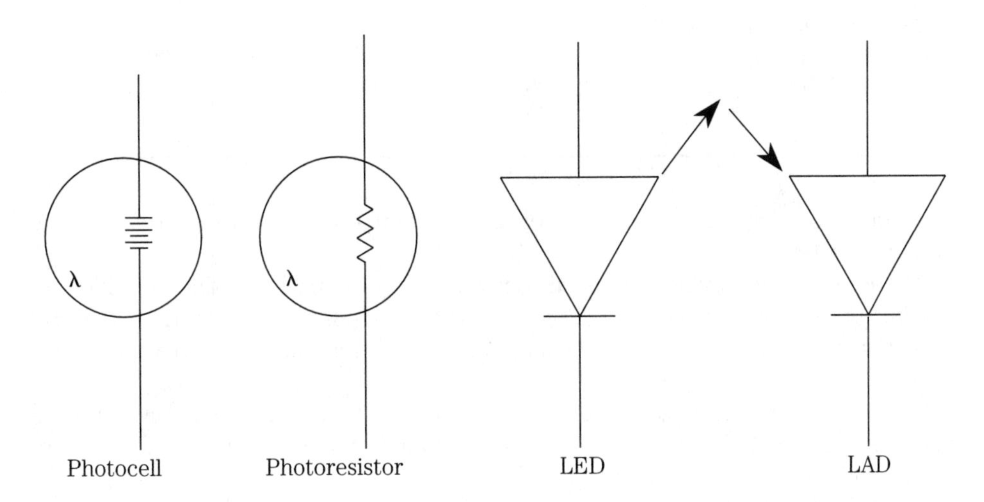

Figure 1-6

You might be asked about one very important optoelectronic effect. The correct name for this effect is *photoconductive effect*. When certain materials are exposed to light, they release internal electrons, making them better conductors of electricity. That is how photoresistors (photoconductors) change their resistance.

The subject of optoelectronic devices will be taken up again in a later chapter when displays are discussed.

Chemical

Whenever two dissimilar metals are immersed in an acid or alkaline solution, there is always a voltage generated. That is the way that batteries are made. Some of the more important examples are given in Table 1-1.

A single positive electrode and negative electrode in the solution, or electrolyte, produces a cell. When two or more cells are connected together, the combination is called a battery. Nontechnical people have occasionally confused the terms cell and battery. However, the terms will be used correctly on any test(s) you decide to take.

All cells and batteries can be classified into two groups: *primary cells*, which cannot be recharged, and *secondary cells*, which can be recharged. The recharging process in secondary cells involves reversing the chemical process used to produce

Table 1-1 Batteries you should know

Name	+ Electrode	– Electrode	Electrolyte
Lead-acid	Porous lead dioxide	Lead alloy and spongy lead	Dilute sulpheric acid
Carbon zinc (Le Clanche)	Carbon rod	Zinc	Ammonium and zinc chlorides
Alakaline	Zinc	Manganese hydroxide	Potassium hydroxide
NICAD	Nickelite hydroxide	Cadmium	Potassium hydroxide

the voltage. Recharging occurs by forcing a current to flow backward through the battery or cell.

Do not be misled by devices used to rejuvenate primary cells. They work by increasing their chemical activity. However, there is no reversal of the chemical process, so the cells are not being recharged. Despite this, there have been some rejuvenators marketed as rechargers.

When cells or batteries are connected in series their voltages add; when connected in parallel their currents add. These connections are shown in Fig. 1-7. Series-parallel combinations are also used.

Care must be taken when connecting cells or batteries in parallel. If their terminal voltages are not identical, one cell or battery tries to recharge the other one (Fig. 1-7). The solution to this problem is to connect diodes in series with each unit. A *diode* is a component that allows current to flow only in one direction. That prevents a reverse current from flowing through any of the cells or batteries.

Your knowledge of the chemical method of generating a voltage makes it easy to understand an undesirable effect that can occur by connecting dissimilar metals to-

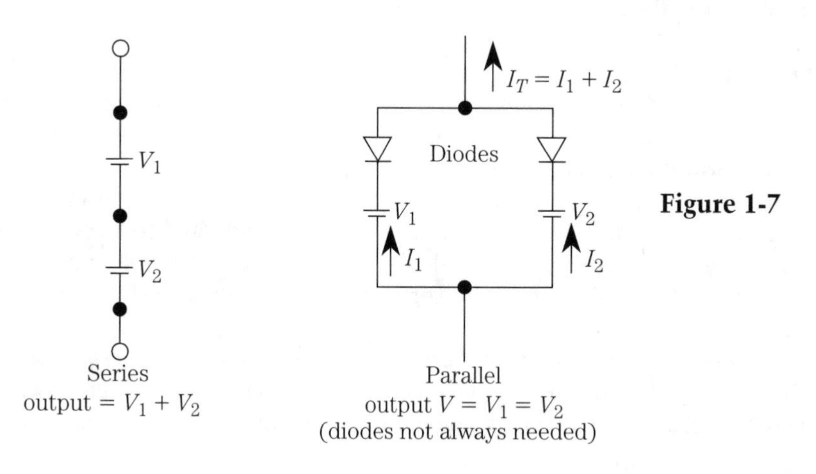

Figure 1-7

gether, called galvanic action. When dissimilar metals are in an acid or alkaline environment, they generate a voltage the same way that a voltage is generated in a battery. That sets up circulating currents that are highly corrosive, and the metals are destroyed by the process.

The electromechanical generator

There are some physical laws and effects that should be understood by technicians. Knowing them is a help in understanding some types of components and circuits.

- If current flows through a resistance, there is always a voltage produced across that resistance.
- If current flows through a resistance, there is always heat produced.
- Regardless of where the current flows, there is always an accompanying magnetic field that surrounds the current.
- Any time a conductor and magnetic field move with relation to each other, there is always a voltage generated in the conductor.

The last of those physical laws and effects is utilized in making generators and alternators. Usually a conductor is moved through a magnetic field, or a magnetic field is moved through a conductor; either way, a voltage is generated.

Because there are standby generators used with transmitters and receivers, it is not unusual for questions to be asked about electric motors and generators on FCC test. Some material on those subjects are given in the programmed review for various chapters.

In the next chapter, under the heading of inductance, the relationships between conductors and magnetic fields is taken up again.

Current

Early experimenters believed that electricity was a fluid. The words "current flow" are a holdover from those days.

Electric current is properly defined as *a flow or motion of charge carriers*. The charge carriers are called *electrons* and *holes*.

For many years, current was defined only as a flow of free electrons in a circuit. Then, the transistor came into being. Its operation depends upon the use of P-type semiconductor material and there aren't enough free electrons in that material to explain current on the basis of free electron flow. So, it is more descriptive to talk about the motion of charge carriers.

The original standard of measurement for current was based upon the amount of silver deposited during silver plating over a given amount of time. Andre Ampere, a French scientist, knew that a magnetic field always accompanies a flow of current. He also knew that the magnetic field circles the current and its strength is directly proportional to the amount of current. Saying it another way, when the current increases the strength the magnetic field also increases.

Using that information, Ampere devised a scientific method of measuring current by relating the magnetic attraction or repulsion between two parallel current-carrying wires in space. The force is exerted because of the attraction or repulsion of the magnetic fields that always accompany a flow of current. When the currents in the two wires are flowing in the same direction, the wires are attracted to each other. When the currents are in the opposite directions, the wires repel each other.

Every electron carries a negative charge. When 6.25×10^{18} electrons are combined in one location—that location is said to have a charge of one coulomb. The coulomb is sometimes called a unit charge.

When one coulomb of charge passes a point in a wire in one second, the current is one ampere.

one ampere = 1 coulomb per second

Resistance and conductance

Resistance is a measure of the opposition to current flowing through a circuit or component. Voltage (V), current (I), and resistance (R) are related by Ohm's Law, which is given mathematically as:

$$I = \frac{V}{R}, \text{ or } V = IR, \text{ or } R = \frac{V}{I}$$

If any one of these equations is known, the other two can be derived mathematically.

One way to define resistance has been to compare it with the opposition offered by a column of mercury in a tube with a given length, cross section and temperature. Today, resistance is referenced to a standard resistor at the Bureau of Standards.

Conductance is a measure of the ease with which a current flows through a circuit or component. The relationship between resistance (R) and conductance (G) is given by the equations:

$$G = \frac{1}{R} \text{ and } R = \frac{1}{G}$$

Originally the unit of conductance was called the *mho*, but conductance is now measured in *siemens*. Conductance is used as a parameter to define the performance of some amplifying devices.

Resistors

A *resistor* is a component used to introduce resistance into a circuit. There are three basic uses of resistors:

- limit current
- introduce a voltage drop
- produce heat

Whenever current flows through a resistor, there is *always* an accompanying magnetic field surrounding the current; and there is *always* heat generated.

Two examples of how a resistor can be used to limit current through a component are shown in Fig. 1-8. In each case, the amount of resistance needed can be computed by using ohm's law.

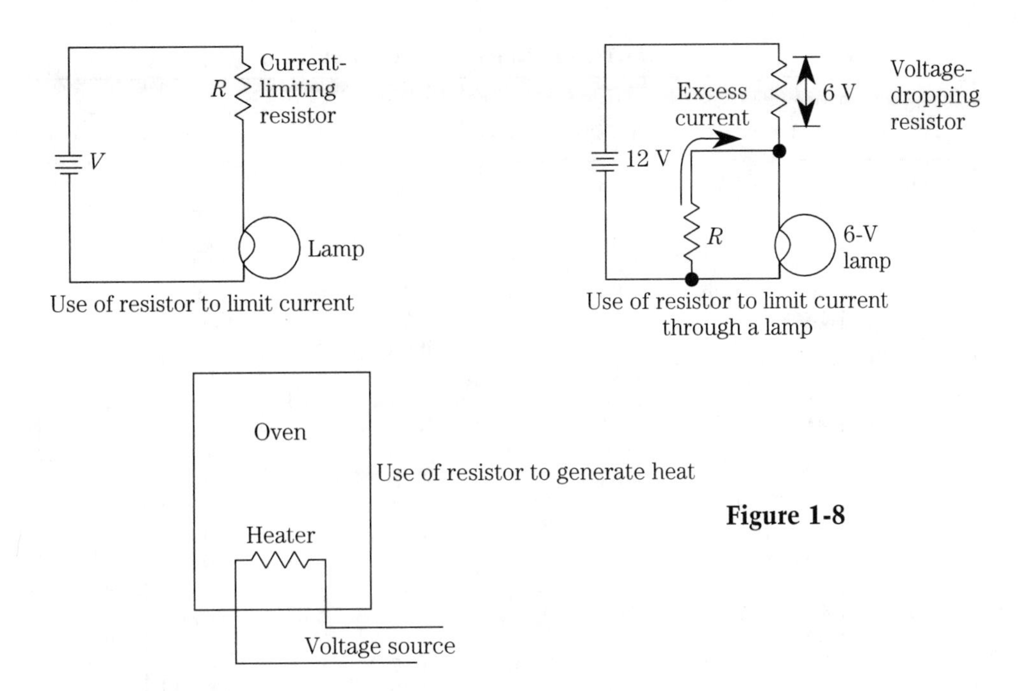

Figure 1-8

Figure 1-8 also shows how a resistor can be used to introduce a voltage drop. The same illustration shows an application of a resistor for generating heat.

All resistors fall into one of two categories: *linear* and *nonlinear*. The use of resistors to produce heat is shown in Fig. 1-8.

Linear resistors

Linear resistors with a fixed value of resistance are rated by the following:

- material used to make the resistor
- resistance value
- reliability
- tolerance
- safe power dissipation
- voltage rating

Examples of linear resistors

The most popular resistors are made with a carbon composition. They are inexpensive and reliable as long as they are operated within the manufacturer's specifications. In some cases, only the resistance value and tolerance are given.

Instructions in the companion volume to this book, *Practice Tests for Communications Licensing and Certifications Examinations: The Complete TAB Reference* show methods of reading the resistance value according to the color code. Test questions related to color codes that are frequently missed are those for 100 Ω, 10 Ω, and resistance less than 10 Ω. Resistors used for surface mounting might have a special code as shown in Fig. 1-9.

Standard Resistance Decade Value

E 48 (±2%)	E 96 (±1%)	E 48 (±2%)	E 96 (±1%)	E 48 (±2%)	E 96 (±1%)	E 48 (±2%)	E 96 (±1%)	E 24 (±5%)
1.00	1.00	1.78	1.78	3.16	3.16	5.62	5.62	1.0
	1.02		1.82		3.24		5.76	1.1
1.05	1.05	1.87	1.87	3.32	3.32	5.96	5.90	1.2
	1.07		1.91		3.40		6.04	1.3
1.10	1.10	1.96	1.96	3.48	3.48	6.19	6.19	1.5
	1.13		2.00		3.57		6.34	1.6
1.15	1.15	2.05	2.05	3.65	3.65	6.49	6.49	1.8
	1.18		2.10		3.74		6.65	2.0
1.21	1.21	2.15	2.15	3.83	3.83	6.81	6.81	2.2
	1.24		2.21		3.92		6.98	2.4
1.27	1.27	2.26	2.26	4.02	4.02	7.15	7.15	2.7
	1.30		2.32		4.12		7.32	3.0
1.33	1.33	1.33	2.37	2.37	4.22	7.50	7.50	3.3
	1.37		2.43		4.32		7.68	3.6
1.40	1.40	249	2.49	4.42	4.42	7.87	7.87	3.9
	1.43		2.55		4.53		8.06	4.3
1.47	1.47	2.61	2.61	4.64	4.64	8.25	8.25	4.7
	1.50		2.67		4.75		8.45	5.1
1.54	1.54	2.74	2.74	4.87	4.87	8.66	8.66	5.6
	1.58		2.80		4.99		8.87	6.2
1.62	1.62	2.87	2.87	5.11	5.11	9.09	9.09	6.8
	1.65		2.94		5.23		9.31	7.5
1.69	1.69	3.01	3.01	5.36	5.36	9.53	9.53	8.2
	1.74		3.00		5.49		9.76	9.1

Figure 1-9

The tolerance of a resistor refers to the range of allowable values of resistance the resistor can have. Popular tolerances are 10% and 5%. Resistors with a tolerance of 20% are no longer being made, but many are still in use. They have no fourth band. Be sure that you know how to read color codes before taking any license or certification test!

The reliability color code (in the fifth band) tells how many resistors can be out of tolerance in a purchased group. It is not of general use to technicians.

Notice that the safe power dissipation is given by the power rating. For a carbon composition resistor, it is indicated by its physical size. The usual practice is to use a resistor with twice the power rating of the actual power dissipation that can be expected in an application. For example, if a resistor is expected to dissipate 0.5 W, it is a general practice to use a 1-W resistor. This practice may vary from one company to another.

Resistors are manufactured in standard, off-the-shelf values. Values in this chart are designated by the Electronic Industries Association (EIA). Any of the values shown can be multiplied by an even multiple of 10, but that is not necessarily true for values beyond 22 MΩ. For example, the color code chart shows that 3.9 Ω is a preferred value. That means the following values are preferred values: 3.9, 39, 390, 3900 (3.9 kΩ), 390 kΩ, and 3900 kΩ (3.9 MΩ).

Film resistors are manufactured by a thick film or thin film process. They are made with a close tolerance, that is, with a low numerical value of tolerance. These resistors are often made with a ceramic coating. Two types are readily available: carbon film and metal film. The carbon types are made in a wider range of resistance values, up to 200 MΩ, but the film types can be made with a closer tolerance, as low as 0.01%.

When replacing a resistor in a circuit, it is not a good idea to substitute one type for another. Some film resistors are designated as *flame retardant.* If one of those types is replaced with one that is not flame retardant and a fire occurs, the technician might be held responsible. Also, the replacement resistor can have a wider tolerance than needed by the circuit. That can make the circuit operation unsatisfactory.

Wirewound resistors are made by winding a resistance wire on a ceramic core. They have high wattage ratings. Many of these have the resistance value marked on their surface. However, some are marked with a manufacturer's code and it is necessary to refer to a catalog to get the rated resistance.

Winding a resistance wire in a spiral onto a ceramic core will produce a resistor that has inductance. In cases where that inductance is objectionable, the noninductive wire-wound types are used. Both types are shown in Fig. 1-10.

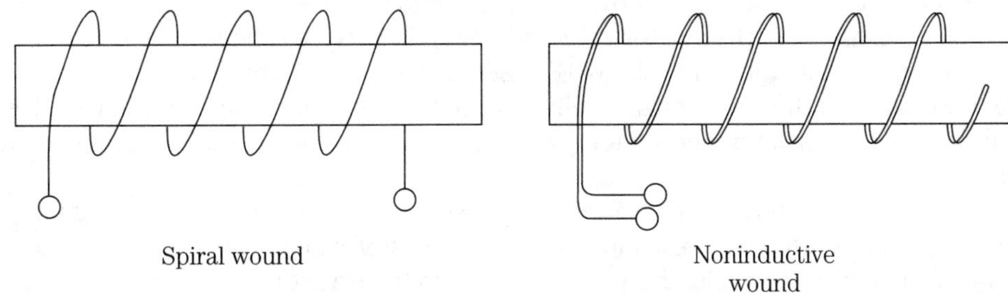

Figure 1-10

When series-parallel resistors are sold in *integrated packages*, they are referred to as an *array*. Figure 1-11 shows an example. One advantage of this construction is that it allows the use of mass production techniques and reduces the amount of board space needed for that many resistors.

Nonlinear resistors

The resistors discussed so far are linear types. The resistance of a linear resistor complies with Ohm's Law. If you double the voltage across a linear resistor, the current through it will also double, assuming the resistor is being operated within its rated power dissipation.

We turn our attention now to the subject of nonlinear resistors. Their resistance values do not conform to Ohm's Law. They are used extensively for making measurements.

Nonlinear resistors are also used in control circuitry. An example of that application is maintaining the temperature of transmitter crystals to a fixed range of values.

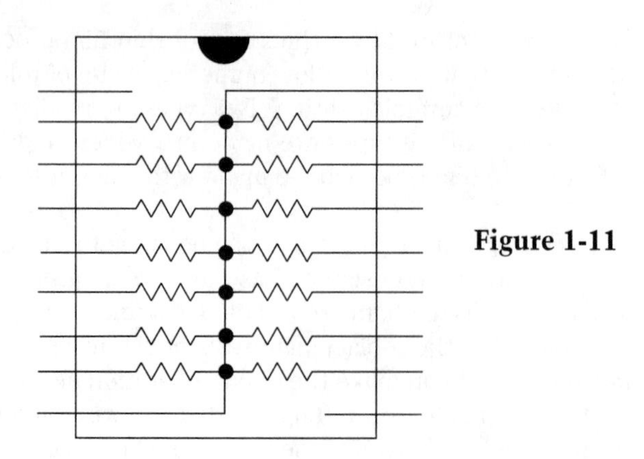

Figure 1-11

When a measurement is taken, the thing being measured is called a *measurand.* A *sensor* is used to convert the measurand to an electric signal. Sensors are also called transducers.

It is not precisely accurate to say that a sensor or transducer "converts" energy from one form to another. By the correct definition, a sensor or transducer uses the energy of one system to control the energy of another system.

As an example, a thermistor (studied next) has a wide range of resistance values determined by temperature. When used as a sensor, its resistance value is directly related to the temperature being measured. However, you shouldn't say the thermistor converts temperature to resistance if you want to be technically accurate.

Sensors (or tranducers) are used extensively in control systems. If the temperature is being sensed, the resistance of the thermistor will not be correct. Its resistance can be used in an electronic control system to correct the temperature.

You will see sensors and transducers being described as "components that convert energy from one form to another." That is a very useful model of their behavior.

Temperature-sensitive resistors

Bolometers are important temperature-sensitive resistors used in communications systems. The two types of bolometers are *thermistors* and *barretters.*

Thermistors undergo a wide change in resistance value for a relatively narrow range of temperature values. An example of a thermistor characteristic curve is shown in Fig. 1-12. The symbol for a thermistor is shown in the same illustration.

The characteristic curve shows that there is a decrease in resistance when there is an increase in temperature. This type is said to have a *negative temperature coefficient (NTC).* Thermistors with *positive temperature coefficients (PTC)* are also available. With a positive temperature coefficient, there is an increase in resistance when the temperature increases.

Barretters are made with different materials than thermistors and they have positive temperature coefficients. As with thermistors, they have a wide change in resistance value for a relatively narrow change in temperature.

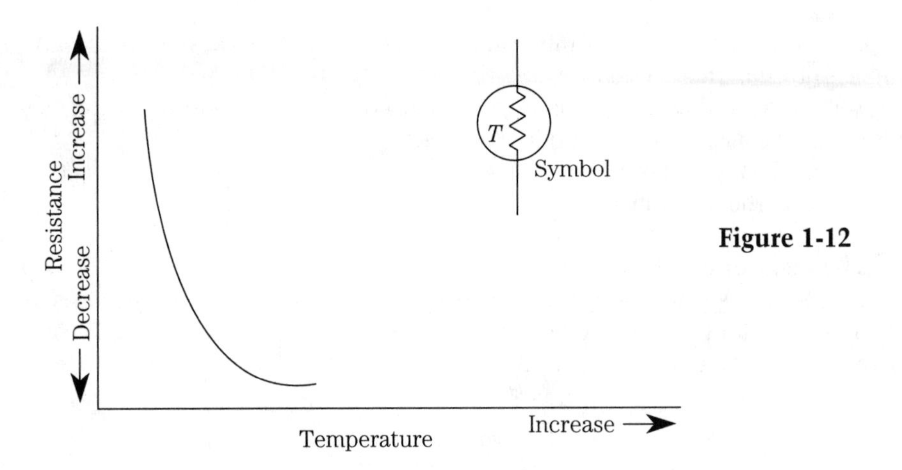

Figure 1-12

The voltage-dependent resistor (VDR)

Figure 1-13 shows a characteristic curve and symbol for a *Voltage-dependent resistor (VDR)*. The VDR is another example of a nonlinear resistor. Its resistance decreases with an increase in voltage across it. This device is often used to protect a component or circuit from an overvoltage.

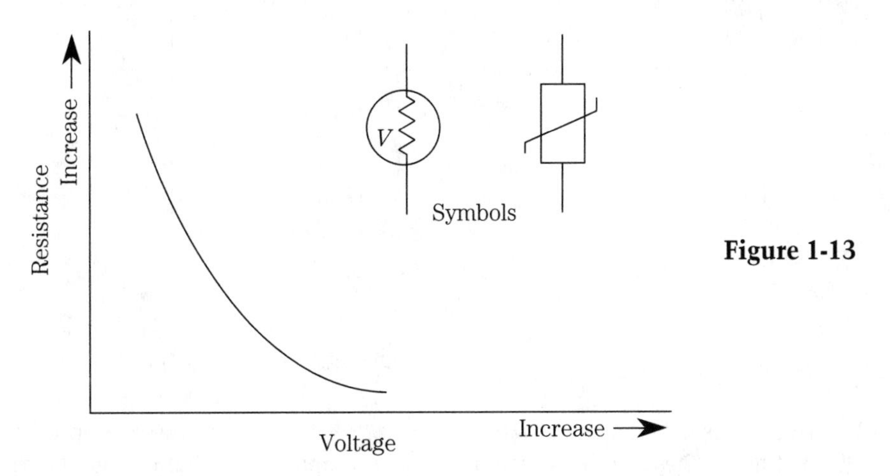

Figure 1-13

For example, a lightning strike near a power transmission line can produce a very high-voltage, short-period pulse, called a *transient* or *spike*, on the transmission line. The amplitude of that pulse can be so high that it can get through a power supply and get into semiconductor circuitry. Some semiconductors are intolerant of those overvoltages. Transient voltages can quickly destroy a complete circuit board.

Some applications of linear and nonlinear resistors

Most applications of resistors are obvious when you keep in mind the three basic uses of resistors in circuits: limit current, introduce heat, or produce a voltage drop. In a few cases a resistor might serve more than one of those purposes. An example is given in this section.

Keep in mind that a resistor can also be used as a transducer or as a voltage divider. That will also be explained in this section.

The applications described in this section have been used as the basis of questions in basic theory exams.

The resistor as a fuse

There are cases where a resistor is used to serve two different purposes in one circuit. Figure 1-14 shows an example of a circuit where the resistor (R) is used as a current limiter; and in some cases it may also serve as a fuse.

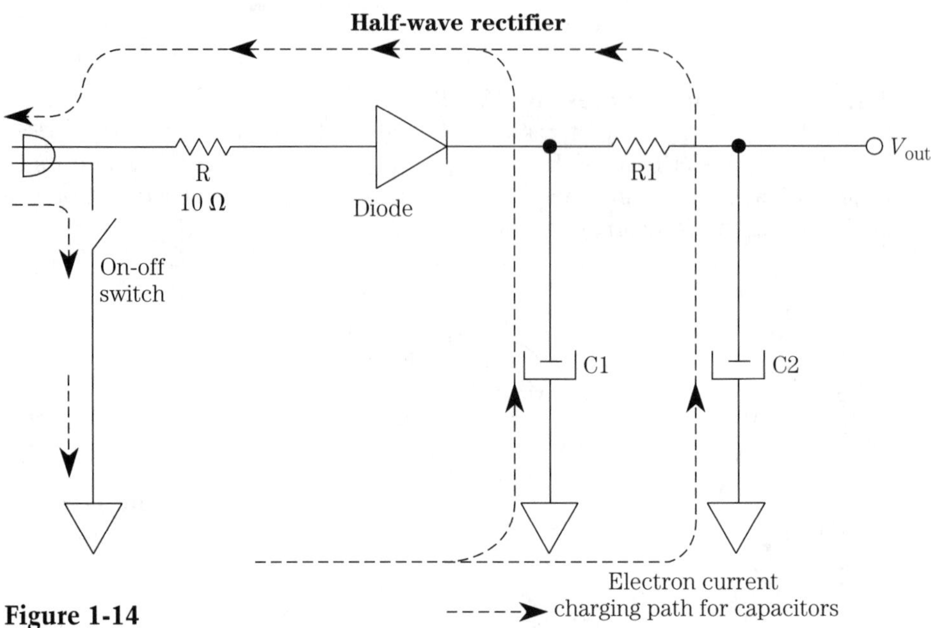

Figure 1-14

Resistor (R) is called a *surge-limiting resistor*. When the power supply is first energized, it limits the charging current of the two capacitors. That charging current can be very high. Unless it is limited, it can damage semiconductor diodes and power supply output circuits.

Once the supply is in operation and the capacitors have been charged, the low resistance of *R* is not significant as far as the operation and output voltage of the power supply is concerned. However, if there is a serious overload in the supply output circuit, such as a short circuit, a high current will flow through R. If the power rating of R is properly chosen, it will burn out and shut down the supply before the diode and circuits are destroyed. In that application, the resistor is acting like a fuse.

A fuse is a resistance chosen to burn out when the current through it exceeds a predetermined value. The basic purpose of the fuse is to generate enough heat (when its current rating is exceeded) to destroy its internal resistor.

Not all surge-limiting resistors are designed to act as fuses. However, if that resistor must be replaced it is a good practice to use the exact resistance and power rating for the replacement.

The crystal oven

Figure 1-15 shows a system that uses a linear resistor as a heater and a nonlinear resistor as a sensor. This type of crystal temperature control is used in some transmitters.

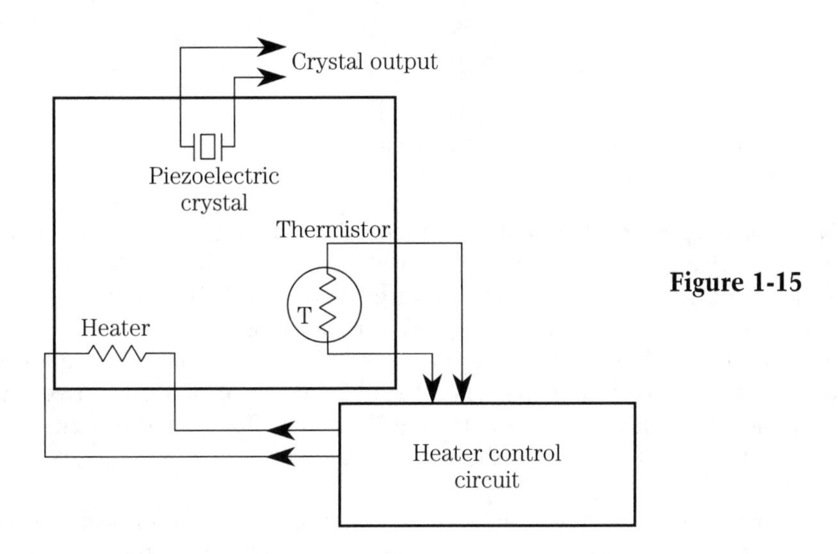

Figure 1-15

The crystal is a piezoelectric type. It vibrates at a precise frequency or combination of frequencies when it is properly excited.

A crystal is used as a standard fixed-frequency source in a transmitter circuit. Its purpose is to establish the transmitter carrier frequency. That frequency must be very accurate to ensure that the transmitter complies with the regulations of the Federal Communications Commission.

The vibrating frequency of the crystal can be affected by temperature, so it is sometimes placed in a small temperature-stabilized *crystal oven*, heated by a resistive element.

The heater circuit for the crystal oven is an excellent example of a *closed loop control*. The heat of the oven is sensed by a thermistor. The resistance of that thermistor is used in a control circuit to adjust the current through the heater.

Suppose, for example, that the oven temperature is too high. The thermistor current delivered to the control circuit will cause it to reduce the heater current and allow the oven to cool. On the other hand, if the temperature of the oven is too low, the thermistor current will cause the control circuit to increase the heater current and raise the oven temperature.

Variable resistors

Variable resistors (also called *adjustable resistors*) can be set to any value of resistance within the range stated by the manufacturer. Figure 1-16 compares the symbols for fixed-value resistors (also called *fixed resistors*) and variable resistors.

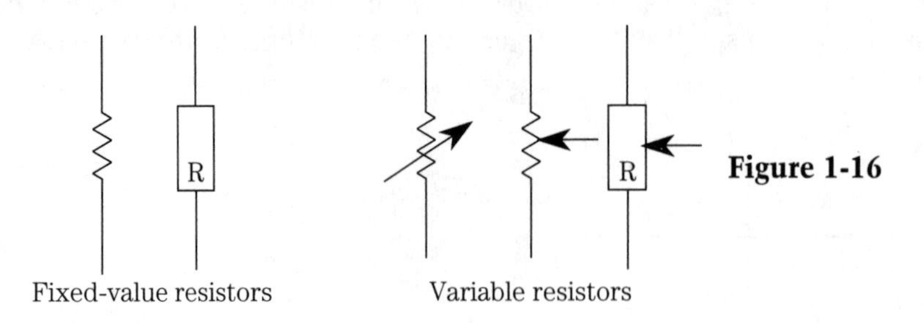

Figure 1-16

An important feature of variable resistors is their taper. It is an indication of how its resistance changes; with the rotation of its shaft. For example, a *linear taper* means that the resistance varies directly with the number of degrees of shaft rotation angle throughout the rotation range. That is shown in Fig. 1-17.

Human sense organs have a *logarithmic response* to all stimuli. At a low magnitude of brightness or sound there is a high change in response for a very small change in the perceived magnitude of the light or sound. So, the eye and the ear have response curves that are roughly the same as the curve marked "reverse" on the curve in Fig. 1-17.

In order to make changes in the output sound or picture brightness, audio volume controls or brightness controls must appear to be linear to the one operating

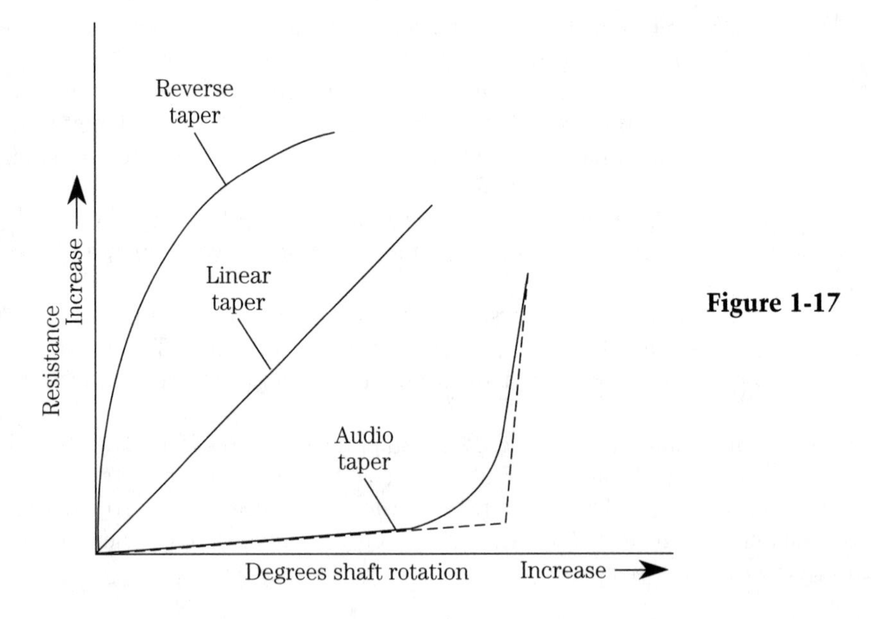

Figure 1-17

the controls. The *audio taper* (also called *logarithmic taper)* is used in receivers to change the magnitude of the output sound or brightness signal in such a way that the resulting change in the intensity or brightness appears to be linear to the observer. In other words, changing the control by a small angle should change the observed sound or brightness a small amount, and changing the control by turning it a greater angle should produce a greater change in volume or brightness.

Combining the reverse and audio curves in Fig. 1-17 will result in the approximate linear response shown in the illustration. That means when using an audio taper on an audio volume control or brightness control, there *appears* to be an even amount of change in audio volume or brightness for a given angle of shaft rotation throughout the range of adjustment.

Saying it another way, a rotation of 15° at the low end of the adjustment will produce the same amount of change in audio or brightness to the observer as a 15° rotation at the high end of the adjustment.

Variable resistors with the response curves shown in Fig. 1-17 are readily available. The broken line shows the actual taper for many audio volume and brightness controls. With only two slopes, the variable resistor is easier to make; therefore, it is less expensive. The human ear or eye cannot perceive the difference between the two audio tapers.

In addition to the taper, variable resistors are rated by the maximum power they can safely dissipate. As a general rule, wirewound types are made with higher power ratings than those made of a carbon composition.

Rheostats and potentiometers

Variable resistors can be connected in either of two ways, as is shown with the *rheostat* and *potentiometer* (Fig. 1-18). Although they are often referred to as *potentiometers* or "*pots*," it is possible to use most variable resistors in either configuration. An exception is that some large variable resistors are made with only two terminals, and they are specifically designed to be used as rheostats.

When the variable resistor is connected for controlling current, it is called a *rheostat*. Two examples are shown in the illustration. A disadvantage of the one on the left is that loss of contact at the moveable arm will result in an open circuit. That can be a destructive situation in some applications. If the arm loses contact in the connection on the right, the maximum resistance is inserted into the circuit.

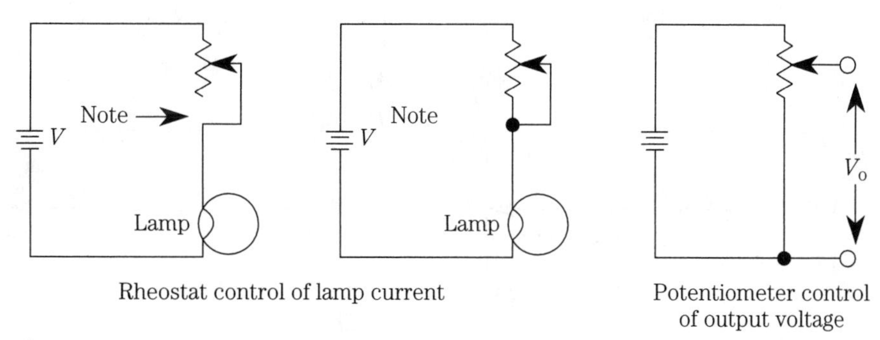

Figure 1-18

Adjusting the rheostat in either circuit will control the current through the lamp. That, in turn, will control the lamp brightness.

When a variable resistor is connected so as to control output voltage (V_0), it is called a potentiometer. That connection is also shown in Fig. 1-18.

Instead of a smooth variation in resistance with rotation, there are also variable resistors designed to change resistance in discrete steps. They are used extensively as attenuators and pads (discussed in a later chapter).

Of course, if you change the voltage across a resistor the current through the resistor will also change. However, the output of the potentiometer is connected to a point where voltage, not current, change is desired.

In addition to the voltage adjustment described above, the word potentiometer is also used to describe an instrument that measures voltage. That meaning has no relationship to the connection of a variable resistor.

Wire resistance

When technicians analyze circuits, they usually disregard the resistance of connecting wires; however, there are some isolated cases where wire resistance can affect circuit performance. An obvious case is where there is a very long wire run.

The resistance of a wire depends upon its length and its cross-sectional area. Think of a piece of wire as a cylinder, as shown in Fig. 1-19. The volume (V of any cylinder is given by the equation:

$$V = \textit{area of base} \times \textit{height}$$

or in the case of the piece of wire,

$$V = \pi D^2 \times \text{length } (l)$$

where (πD^2) is the area of the cross-section of the wire (Fig. 1-19).

The resistance (R) of a piece of wire is directly proportional to its length (l) and inversely proportional to its cross-sectional area (A). That proportion can be written as follows:

$$R \propto l/\pi D^2$$

Read $\propto$ as "is proportional to."

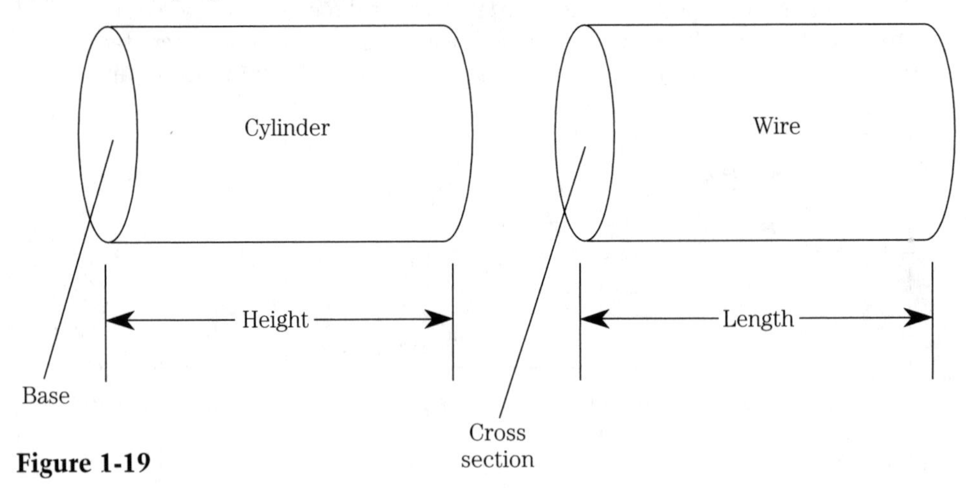

Figure 1-19

When parameters are proportional, as in this equation, they can always be made equal by introducing a constant of proportionality. In this case, it is ρ (rho).

$$R = (\rho)\ l/\rho D^2$$

In this important equation ρ, the constant of proportionality, is called the *resistivity* of the wire material. It is also called *specific resistance* of that material. This constant is dependent upon the type of material used to make the wire. Resistivity values are given in circular mils per foot (circular mils/foot). At the end of this chapter, circular mils are covered. The importance of this equation is in the fact that it clearly shows which factors affect wire resistance.

Figure 1-20 shows a soldering gun that is sometimes sold in stores. It is not a good idea to use this type of soldering device on modern electronic systems.

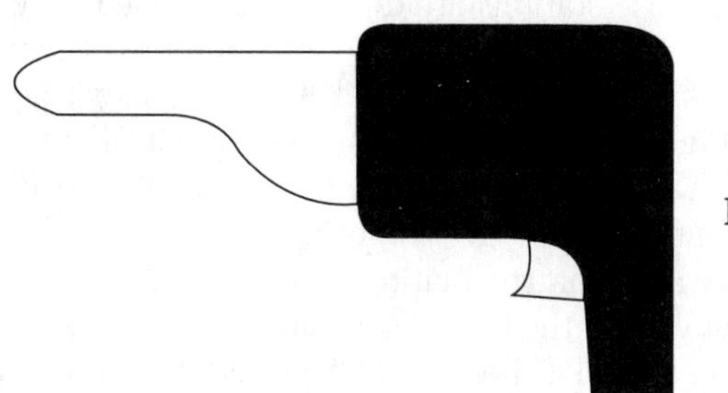

Figure 1-20

This gun can destroy some types of components in modern systems. In some soldering guns, the secondary is floating. In other words, there is no ground connection to the secondary. Suppose a negative or positive electrostatic charge accumulates on that secondary. An "electrostatic charge" means an excess or deficiency of electrons. There is no way for a charge to get off the secondary.

Assume there is an electrostatic charge on the secondary. It can represent a very high voltage. The instant that soldering tip is touched to the wire connection of some kinds of modern electronic components, that component will be destroyed! Before you solder with any type of soldering device, make sure it is not capable of storing an electrostatic charge.

There is another problem to be aware of regarding the soldering gun shown in Fig. 1-20. The heat generated by that type is usually too high for working in the delicate modern components.

Very strong electromagnetic fields surround the tip when it is being used for soldering. To understand why that can be a problem, review some very important laws and effects.

Every time there is an electric circuit flow, there is always a magnetic field around the current. That magnetic field is always at a right angle (90°) with the current. In other words, the magnetic field surrounds the current. The strength of the magnetic field directly depends upon the amount of current.

The ac current flowing in the tip increases and decreases with time. That means the electromagnetic field is expanding and contracting with the current variations.

Faraday's Law says that *every time a moving magnetic field cuts through a conductor a voltage is induced.* So, the varying magnetic field around the tip is capable of inducing voltages in the conductors near the point being soldered. Those induced voltages can be destructive in some cases—even though the equipment is off.

That brings up another important rule about soldering, never solder in equipment that is energized. Always make sure the equipment is off before you start soldering. That is also a rule for human safety. The tip of the soldering iron is a conductor of electricity. During the soldering procedure, that tip can easily cause a destructive short circuit (high-current flow) in energized equipment.

High-wattage soldering irons must never be used on modern equipment. A 25-W iron is sufficient for most work today.

Soldering stations are available. They make the job easier and enable you to do professional work. Some of those stations are designed to sense the temperature of the tip and regulate that temperature. They are a very good investment.

During the soldering procedure, the tip of the soldering iron will accumulate dirt and impurities. You cannot do a professional job of soldering if the tip is dirty. In the early days of electronics, it was common practice to file the tip. Never do that with modern soldering equipment! Many of the soldering irons sold today have plated tips. If you file the tip, you will file off the plating and that will make it useless.

It is very important that you tin the tip periodically. That makes it possible to deliver the heat directly to the materials being soldered; reducing the amount of time the iron is held on the work. That is very important because too much heat can destroy many of today's electronic components.

The procedure for tinning the iron is simple. The first step is to make sure the tip is clean. Many of the soldering irons sold today have a sponge in a holder. That sponge is for use in cleaning the tip by rubbing it against the sponge. The sponge can be lightly damp, but never wet!

It is a bad practice to use a piece of cloth to rub the tip clean. That method can cause specks of hot solder to fly in all directions. Some companies require the use of safety glasses during soldering. It is a good safety practice.

In addition to tinning the iron, it is important to tin the wires and terminals, and anything else, being soldered. That permits you to make a good solder connection in a short period of time, and it aids in making the connection.

Heat can travel along the conducting leads of a component and destroy that component. To help protect the component when its leads are being soldered, a *heatsink* should be used. Its purpose is to carry heat away from a component. As shown in Fig. 1-21, the heatsink connected between the soldering point and the component being soldered.

You can buy heatsinks at parts suppliers, but do not use alligator clips as heatsinks! Remember, the purpose of a heatsink is to carry heat away from the lead so the heat does not reach the component. Heatsinks are made of good thermally conductive material; alligator clips are not.

Soldering is as much an art as it is a science. Novices are inclined to use too much solder. According to the manuals on the subject, you should be able to see the outlines of the wires and terminals after the soldering is completed. Time spent practicing is time well spent.

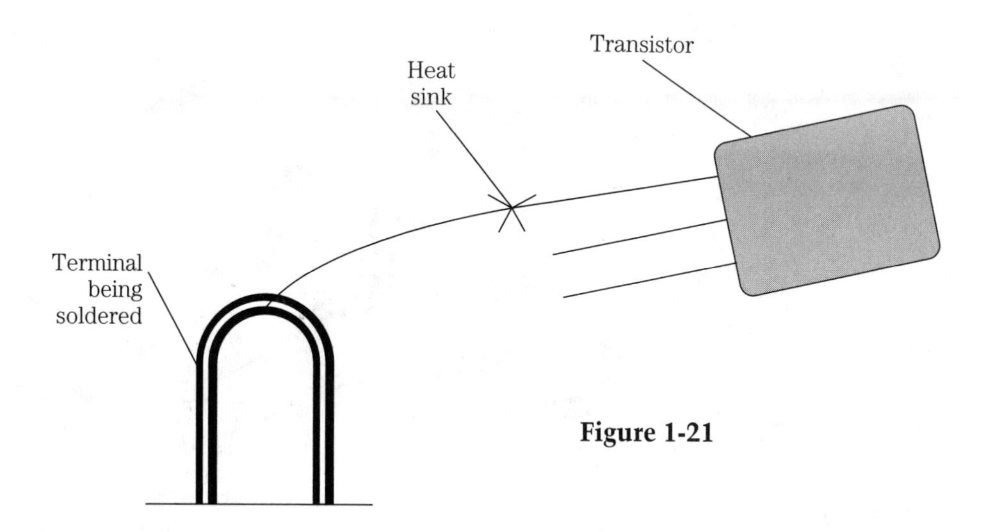

Figure 1-21

A large glob of solder makes it impossible to see if the materials are well soldered. *Cold solder joints* have solder, but no connection. The solder of a cold solder connection does not fuse onto the materials being soldered.

The soldered connection should appear smooth and it should have a shine. A dull grey appearance is not acceptable because it often hides a cold solder joint.

The type of solder used for most electronics work is called *60/40*. That means it is made with 60% tin and 40% lead. Remember that!

Most solder passes from a solid to a paste and then to a liquid when it is heated to the melting point.

Another important type of solder is called 63/37. (63% tin and 37% lead) It melts at a lower temperature than 60/40 and it goes directly from a solid to a liquid. That reduces the amount of time heat must be applied. 63/37 solder is strongly recommended for soldering very small and sensitive components.

A method of fabricating circuits, called *surface mount*, requires the use of 63/37 solder for parts placement. Surface mount means all of the very small components are mounted and soldered on top of the board as opposed to running the connecting leads through the board and soldering on the bottom. Figure 1-22 shows an example of a surface mount component.

It is very important that you do not handle surface mount components. Handling can easily cause their destruction. Those components are to be picked and placed by using tweezers. If you are going to work with surface mount equipment, be sure to get the proper tools.

Here are two very important rules to follow when making a solder connection:

- It is important that the wires and connectors do not move during the period when the solder is cooling.
- Do not blow on the solder joint to hasten cooling. That cools the surface too fast and it can result in a poor connection. Also, it can cause high-resistance joints in equipment designed for use at ultra-high frequencies.

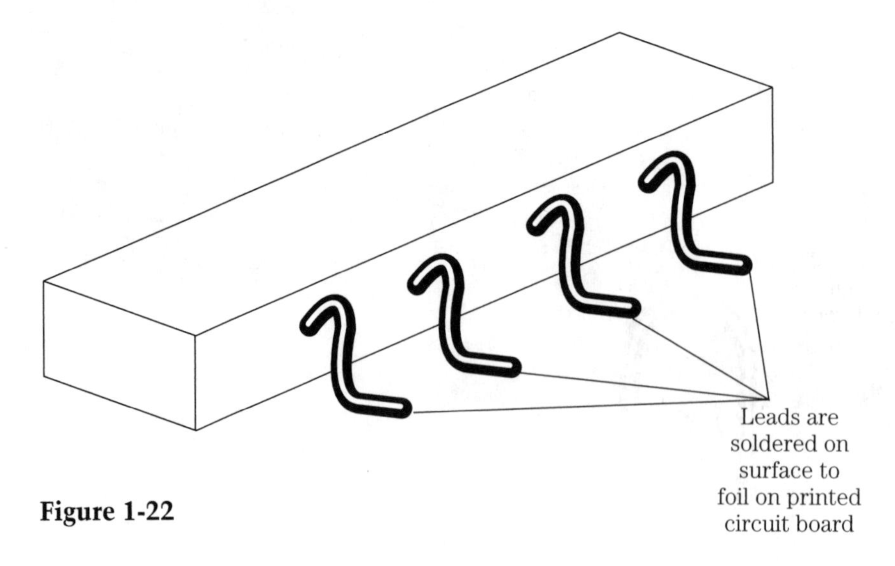
Leads are
soldered on
surface to
foil on printed
circuit board

Figure 1-22

Most solders used by technicians have a rosin flux in the core. The flux helps to clean the surfaces being soldered. That, in turn, makes it easier for soldering to occur. You should be aware of the following kinds of fluxes:

- Rosin flux is very active during soldering. After soldering, the residue is not corrosive and it is easily removed with an alcohol swab. This kind of flux is most often used.
- Organic fluxes are made with mild organic acids. They are susceptible to thermal decomposition and that can be an advantage in some applications because it limits corrosion.
- Acid flux (also known as chloride flux) must never be used for soldering electronic equipment. It is very corrosive.

In the early days of electronics, it was felt that soldering should not occur until after a very extensive mechanical connection was made (Fig. 1-23). Research shows that the tight mechanical connection shown in that illustration should never be used. One reason is that it is nearly impossible to replace a component soldered that way. When you try to unwrap the wire, it is necessary to apply heat for too much time. Also, the terminal can be destroyed by the excessive heat and mechanical stress.

Figure 1-23 also shows the proper way to attach a wire to a terminal in preparation for soldering. If the wire has been tinned it can be mechanically affixed to the terminal by lightly squeezing it with a pair of long-nosed pliers. It is sometimes necessary to apply a small amount of tension to the wire during soldering.

When unsoldering, and sometimes during soldering, it is necessary to remove excess melted solder. There are two ways to do that quickly and effectively. One method is to use a solder sucker (Fig. 1-24). It is used to pull the excess melted solder away from the area. Usually, the solder is solidified before it reaches the bulb and it is easily removed from the bulb.

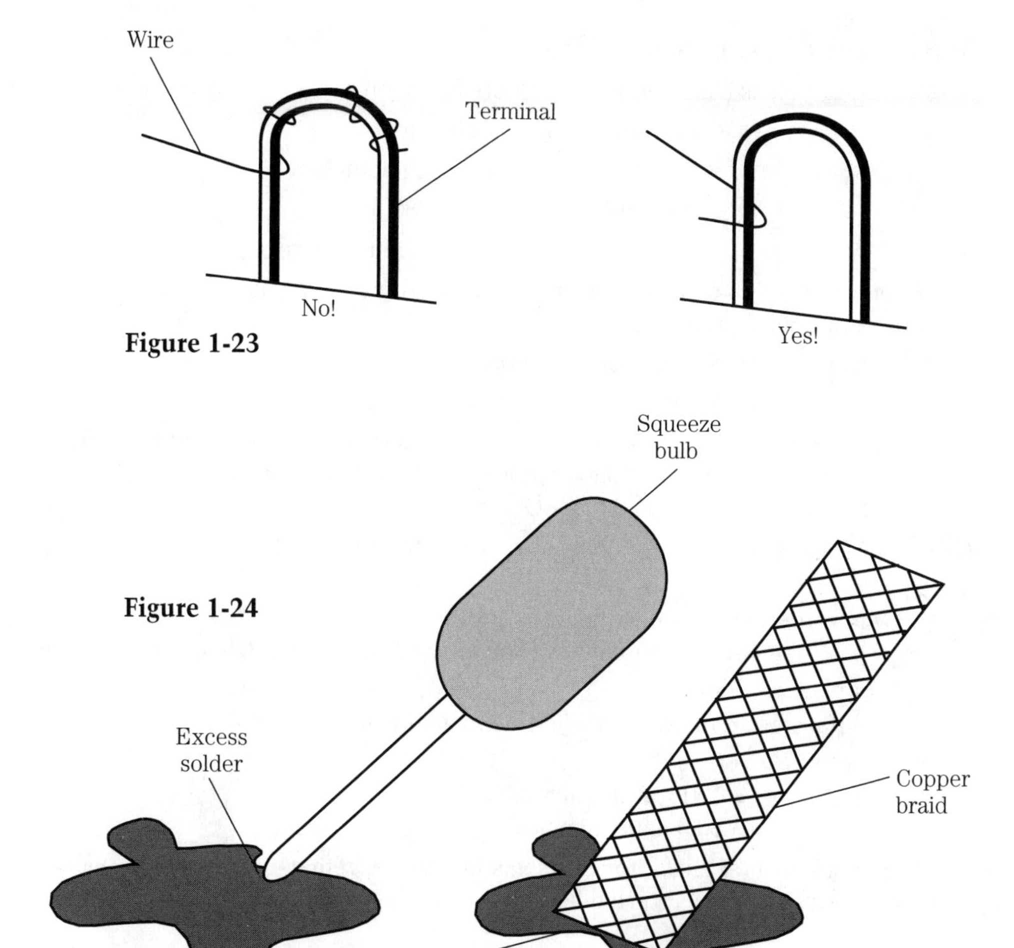

Figure 1-23

Figure 1-24

The second method of removing unwanted excess solder is also shown in Fig. 1-24. Removal is achieved by using a braid of copper wire. Although it is preferred by many technicians, this method is not recognized in the FCC pool of GROL questions.

The braid is called a *solder wick*. It is inserted into the unwanted melted solder. That solder flows up the wick and away from the area. The wick, with the unwanted solder, is cut away and more solder wick is unwound from the spool for the next operation.

The last step in the soldering process is to clean the wires and terminals. The preferred method is to use an alcohol swab to remove excess flux. All excess solder should be removed.

Conversions

There are some questions on basic conversions that you are expected to be able to answer. Equations for conversions are given. You should know those equations.

Statute miles vs. nautical miles

A nautical mile is based upon one minute of arc on a great circle of the earth.

One nautical mile = 6076 feet
= 1.852 kilometers

One statute (land) mile = 5280 feet
= 1.609 kilometers

When you need to convert from statute miles to nautical miles, or from nautical miles to statute miles, remember this number: *1.15*.

$$\textit{Statute miles} = \textit{nautical miles} \times 1.15$$

$$\textit{Nautical miles} = \textit{statute miles} \div 1.15$$

Do not expect your answers to be exactly equal to one of the choices in a multiple choice test. Just pick the choice that is closest to your answer. Here are two examples taken from the FCC pool of GROL questions:

Example 1.73 nautical miles equals how many statute miles?

Solution

Statute miles = 1.73 nautical miles × 1.15
= 1.99 statute miles (The answer given in the GROL exam book is 2.0)

Example 100 statute miles equals how many nautical miles?

Solution

Nautical miles = 100 statute miles ÷ 1.15
= 86.9 nautical miles

As a quick mental check for problems like these, remember that there will be more statute miles than nautical miles.

Knots

One *knot* is equal to one nautical mile per hour. When speaking of speed it is not correct to say "knots per hour." Use the same conversion number (1.15) that was used to convert between nautical miles and statute miles.

$$\textit{Miles per hour} = \textit{knots} \times 1.15$$

$$\textit{Knots} = \textit{statute miles per hour} \div 1.15$$

Example 10 miles per hour equals how many knots?

Solution

Knots = *miles per hour* ÷ 1.15
= 10 ÷ 1.15
= 8.695 knots

When working problems like this, remember that there will be more miles per hour than knots.

Phonetic alphabet

In a noisy radio communication, it is advisable to use the phonetic alphabet for spelling words and for identifications. It has been adopted by international agreement. There are questions in the GROL exam book on this alphabet. Take time to learn it.

ALPHA	BRAVO	CHARLIE
DELTA	ECHO	FOXTROT
GOLF	HOTEL	INDIA
JULIET	KILO	LIMA
MIKE	NOVEMBER	OSCAR
PAPA	QUEBEC	ROMEO
SIERRA	TANGO	UNIFORM
VICTOR	WHISKY	X-RAY
YANKEE	ZULU	

Radians vs. degrees

For most scientific work, angular measurement is in radians rather than degrees. Refer to Fig. 1-25. If you take the radius of a circle and lay it along the circumference, it represents on radian of arc. The angle at the center, called "x" in the illustration, represents on radian.

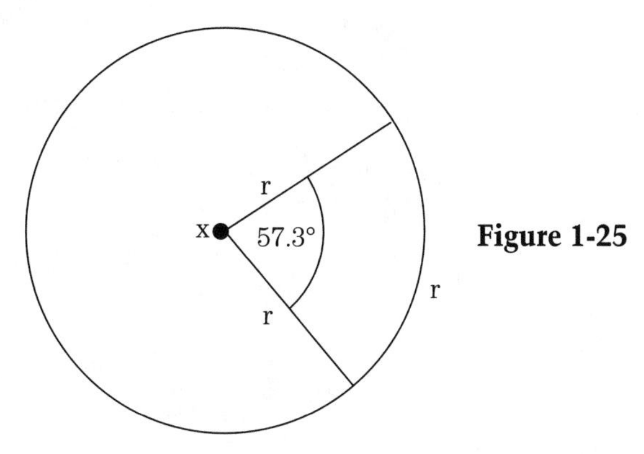

Figure 1-25

If you continue to lay the radius along the circumference, you find you can do that π times in one half circle (180°). Therefore:

$$\pi \text{ radians} = 180 \text{ degrees}$$

Divide both sides of that equation by π and you get:

$$1 \text{ radian} = 57.3 \text{ degrees}$$

The key number to remember for conversion between radians and degrees is 57.3!

$$Degrees = radians \times 57.3 \text{ and}$$

$$Radians = degrees \div 57.3$$

As a check, there will always be more degrees than radians.

It is common practice to measure angular velocity in revolutions per minute (RPM). However, in scientific work, it is expressed in radians per minute or radians per second. Remember that there are 2π radians per revolution.

$$\textit{Revolutions per minute} = \textit{radians per minute} \div 2\pi \text{ and}$$

$$\textit{Radians per minute} = \textit{revolutions per minute (RPM)}/2\pi$$

The division sign can be read as "per." For example, revolutions/minute is revolutions per minute (RPM).

Temperature conversion

It is not known for sure what happens when the temperature reaches absolute zero (–273.16°C). Most of the literature agrees that molecular motion stops; however, because that temperature has never been reached, it is still an educated guess.

The temperature at absolute zero is called *zero degrees Kelvin*. The Kelvin scale starts at absolute zero. The following equation is used for converting a celsius temperature to the corresponding Kelvin temperature given here:

$$\textit{Kelvin degrees} = \textit{celsius degrees} + 273.16 \text{ or}$$

$$°K = °C + 273.16$$

The Rankin scale also starts at zero degrees Kelvin. It is based upon the Fahrenheit scale.

$$°R = °F + 459.7$$

You are more likely to be asked to convert between celsius and Fahrenheit temperatures:

$$°C = \frac{5\,(°F - 32)}{9} = \frac{5\,(°F) - 160}{9}$$

$$°F = 9\left(\frac{°C}{5}\right) + 32 = \frac{(9)\,(°C) + 160}{5}$$

Example Convert 320°F to °C.

$$°C = \frac{(5)\,(320°) - 160}{9} = 160°\text{C}$$

Example Convert 60°C to °F.

$$°F = \frac{(9)(60°) + 160}{5} = 140°\text{F}$$

Programmed review

In some cases, there are questions in the programmed section about subjects that were not discussed in the text. You will probably be able to answer those questions as a result of your experience and training. If not, consider those questions and answers to part of your training for taking FCC and other exams.

Start with Block No. 1. Pick the answer you believe is correct. Go to the next block and check your answer. All answers are in italics. There is only one choice for each block. There might be some material covered in this section that was not covered in the chapter.

1. When the third digit on a color-coded resistor is silver it means to multiply the first two color-coded digits by:

   A. 0.001. B. 0.01. C. 0.1. D. 1.0.

   The correct answer is B.

2. Two resistors are connected in parallel. Their values are 39 kΩ and 68 kΩ. The parallel resistance value should be:

   A. greater than 68 kΩ.
   B. less than 39 kΩ.
   C. equal to 68 kΩ to 39 kΩ.
   D. none of these choices is correct.

   The correct answer is B. The value is 24.8K.

3. Which of the following is a resistor that changes resistance with the amount of light it receives?

   A. LDR. B. CDR. C. RER. D. COR.

   The correct answer is A. (Light Dependent Resistor)

4. Ohm's law cannot be used unless it is applied to a component or circuit that is:

   A. unilateral.
   B. exponential.
   C. trivalent.
   D. linear.

   The correct answer is D.

5. Number 28 wire has a cross-sectional area of about 160 circular mils. What is its diameter?

   A. 12.65 mils. B. 126.5 mils.

   The correct answer is A. (The diameter is the square root of the circular mil area.)

6. The reciprocal of resistance is:

   A. conductance. B. admittance.

   The correct answer is A.

7. 133 radians = __________

   A. 7620+ degrees. B. 2.32 degrees.

   The correct answer is A.

8. What is the value of Ri in the voltage divider shown in Fig. 1-26.

   A. 43.0. B. 16.1.

   The correct answer is B.

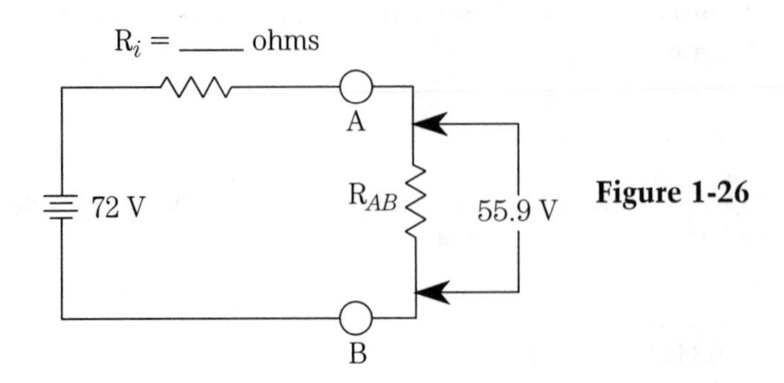

Figure 1-26

9. Name two kinds of bolometers.

   The correct answer is thermistor and barretter.

10. Express an angle of 77 degrees in radians.

    The correct answer is 1.34 radians.

11. What is the unit of measurement for conductance?

    The correct answer is siemen. At one time it was called the mho (ohm spelled backward).

12. Wire A has a resistance of 0.001 Ω per inch. Another piece of wire, made of the same material, has a diameter twice the diameter of wire A. What is its resistance per inch?

    A. 0.025 Ω.
    B. 0.0025 Ω.
    C. 0.00025 Ω.
    D. 0.0005 Ω.

    The correct answer is C. When the diameter is double the resistance is reduced to one-fourth.

13. When you hold a flame to a certain type of resistor it will not start to burn. The resistor is a type known as:

    A. flame proof.
    B. fire retardant.
    C. type 137XT.
    D. none of the choices is correct.

    The correct answer is A.

14. To replace the intensity (brightness) variable resistor control on an oscilloscope use:

    A. a video taper.
    B. an audio taper.
    C. a linear taper.
    D. a reverse taper.

    The correct answer is B.

15. When you hold a match to a certain type of resistor it will start to burn. However, if you remove the flame will go out by itself. The resistor is a type known as:

    A. flame proof.
    B. flame retardant.
    C. type 137XT.
    D. none of the choices is correct.

    The correct answer is B.

16. Which of the following is more popular?

    A. Carbon composition.
    B. Wirewound.

    The correct answer is A.

17. Instead of using a single 5K, 1-W resistor a technician uses two 1OK, 1-W resistors in parallel. This will give:

    A. a wider tolerance.
    B. a higher resistance.
    C. less than the original power rating.
    D. more than the original power rating.

    The correct answer is D.

18. Which of the following is an acceptable method of solder removal from holes in a printed board?

    A. Compressed air.
    B. Toothpick.
    C. Soldering iron and a suction device.
    D. Power drill.

    The correct answer is C.

19. What is the purpose of flux?

    A. Removes oxides from surfaces to be joined.
    B. Prevents oxidation during soldering.
    C. Acid cleans printed circuit connections.
    D. Both A and B.

    The correct answer is D.

20. What is the photoconductive effect?
    A. The conversion of photon energy to electromotive energy.
    B. The increased conductivity of an illuminated semiconductor junction.
    C. The conversion of electromotive energy to photon energy.
    D. The decreased conductivity of an illuminated semiconductor junction.

    The correct answer is B.

You have now completed the Program Review.

Quiz

1. What is the resistance of ten 100-Ω resistors in parallel?

2. Four 37-kΩ resistors are connected in series. What is the resistance of the combination?

3. The following resistance values are connected in parallel: 25 kΩ, 27 kΩ, 33 kΩ, and 3900 Ω. What is the equivalent resistance of the parallel combination?

4. Refer to the circuit marked "A" in Fig. 1-27. What is the resistance of the circuit ($R_{A\text{–}B}$)?

5. Refer to the circuit marked "B" in Fig. 1-27. What is the range of resistance values for the circuit?

   Maximum $R_{A\text{–}B}$ = ________________

   Minimum $R_{A\text{–}B}$ = ________________

6. What is the voltage across the output terminals ($V_{A\text{–}B}$) in the circuit marked "C" in Fig. 1-27.

7. Assume that 10 V is applied to the circuit marked "D" in Fig. 1-27. How much current is flowing through R_L?

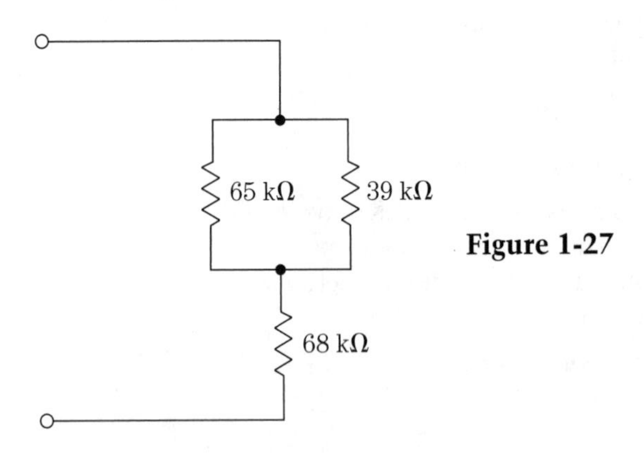

Figure 1-27

8. Return to the circuit marked "D" in Fig. 1-27. How much applied voltage is needed to get a current of one ampere through R_L?

9. What is the value of maximum power dissipated by R_L in the circuit marked "E" in Fig. 1-27?

10. The color-coded bands on the resistor of Fig. 1-28 are marked with letters. Write the color codes for the following resistors.

100 kΩ ±10%	a.________	b.________	c.________	d.________
10 kΩ ±5%	a.________	b.________	c.________	d.________
1 kΩ ±2%	a.________	b.________	c.________	d.________

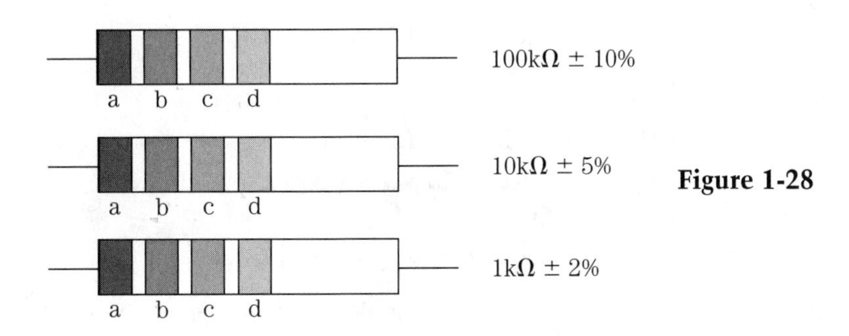

Figure 1-28

11. A cell that can be recharged is called a ______________ cell.

12. A variable resistor connected to control current is called a ______________.

13. A variable resistor connected to control voltage is called a ______________.

14. What is the name of a nonlinear resistor used to sense temperature? __________

15. What is the name of a nonlinear resistor that changes its resistance when the voltage across it changes?

16. A cell that cannot be recharged is called a ______________ cell.

17. Which type of device generates a voltage when under physical pressure?

    ____

18. Name six methods of generating a voltage.

    ______________ ______________

    ______________ ______________

    ______________ ______________

19. Five 2-V cells are connected in series. The output voltage of this battery is:

    ____

20. Five 2-V cells are connected in parallel. The output voltage of this battery is:

    ____

Answers to quiz

1. When all of the parallel resistors have the same value, you can divide that value by the number of branches.

$$100/10 = 10\ \Omega \text{ (answer)}$$

2. When resistors in series have the same value, you can multiply that value by the number of resistors.

$$4 \times 37\ \text{k}\Omega = 148\ \text{k}\Omega \text{ (answer)}$$

3. Change 3900 to 3.9 kΩ.

$$1/R_{eq} = 1/R_1 + 1/R_2 + 1/R_3 + 1/R_4$$
$$1/R_{eq} = 1/25\ \text{k}\Omega + 1/27\ \text{k}\Omega + 1/33\ \text{k}\Omega + 1/3.9\ \text{k}\Omega$$
$$1/R_{eq} = 0.364 \times 10^{-3}$$
$$R_{eq} = 1/(0.364 \times 10^{-2}) = 2.75\ \text{k}\Omega \text{ (answer)}$$

Note: It is not necessary to carry the k multiplier throughout the solution. Use 25, 27, 33, and 3.9, then add k at the end.

4. First, find the equivalent resistance of the parallel combination.

$$R_{eq} = [(65 \times 39)/(65 + 39)]\ \text{k}\Omega = 24.4\ \text{k}\Omega$$

or, use this equation:

$$1/R_{eq} = 1/65\ \text{k}\Omega + 1/39\ \text{k}\Omega$$

Add the parallel resistance to 68 kΩ – $R_{A\text{–}B} = 92.4$ kΩ (answer)

5. The highest value occurs when the 0- to 100-kΩ resistor is set at 100 kΩ. The two resistors in the right-hand branch combine to give 93 kΩ.

$$R_{A\text{–}B} = [(100 \times 93)/100 + 93)]\ \text{k}\Omega + 50\ \text{k}\Omega = 98.2\ \text{k}\Omega \text{ (answer)}$$

The lowest value occurs when the 0- to 100-kΩ resistor is set at 0 Ω.

$$R_{A\text{–}B} = 0 + 50\ \text{k}\Omega = 50\ \text{k}\Omega \text{ (answer)}$$

6. Using the proportional method: $V_{A\text{–}B} = 10 \times (75/100) = 7.5$ V. The 100-Ω resistor does not enter into the calculation because there is no current flowing through it. So, there is no voltage across it.

7. The voltage source (V) is "looking into" a resistance of:

$$R_T = 20 + (10 \times 15)/(10 + 15) = 26\ \Omega$$

Therefore, the voltage source must supply a total current of: I = V/R = 13/26 = 0.5 A. That is the total current supplied by the source. Using that current, you can find the voltage across the parallel combination. That is the same as the voltage across the 10-Ω resistor. Use Ohm's Law to find the current through the 10-Ω resistor.

Another way is to use the reciprocal method:

$$I_{IO} = I_T[(15)/(10 + 15)] = 0.5 \times 0.6 = 0.3\text{A (answer)}$$

8. If one ampere is flowing through R_L, then the voltage across R_L (call it V_L) can be determined by Ohm's Law:

$$V = I \times R_L = 1 \times 10\ \Omega = 10\ \text{V}$$

That is also the voltage across the 15-Ω resistor, so, the current through that resistor is:

$$I_{15} = 10\ \text{V}/15\ \Omega = 0.667\ \text{A}$$

The combined currents in the parallel resistors equals the total current flowing into the junction:

$$I_T = 1\text{A} + 0.667\text{A} = 1.667 \text{ amperes}$$

That total current flows through the series 20-Ω resistor, so, the voltage across that resistor is:

$$V = IR = 1.667\text{A} \times 20\ \Omega = 33.34\ \text{V}$$

The voltage across the 20-Ω resistor added to the voltage across the parallel branch equals the applied voltage required for 1A to flow through R_L.

$$\textit{Required voltage} = 33.34\ \text{V} + 10\ \text{V} = 43.34\ \text{V (answer)}$$

9. The maximum power transfer theorem says that the maximum power will be delivered to R_L when $R_L = R_i$. Obviously, that will be one-half the total power dissipated by the circuit.

$$\textit{Total power} = V^2/R_T = 10^2/20 = 5\ \text{W}.$$

10. 100 kΩ ±10%. Brown, black, yellow, silver.
    10 Ω ±5%. Brown, black, black, gold.
    1 Ω ±2%. Brown, black, gold, red.

11. Primary.

12. Rheostat.

13. Potentiometer.

14. Thermistor.

15. VDR (voltage-dependent resistor).

16. Primary.

17. Piezoelectric.

18. Chemical, electromechanical.
    Pressure, light.
    Friction, heat.

19. 10 V.

20. 2 V. Diodes might be needed for isolation.

2
CHAPTER
Components (ac)

The two most important passive, two-terminal components in ac circuits are capacitors and inductors. Their characteristics and applications are covered in this chapter.

Experience with CET testing has shown that capacitors and inductors are sometimes misunderstood by technicians. For that reason, a complete chapter is devoted to a better understanding of those components.

Capacitors

The purpose of a capacitor is to store energy. The energy is stored in the dielectric in the form of an electric field. The charge of a capacitor is stored in the plates.

Capacitors are used for only one purpose: to store energy. However, there are many applications of this purpose. In some cases, those applications are so frequently encountered that they are included as purposes.

A few examples of capacitor applications are: ac voltage dividers, parametric amplifiers, tuning components in RF circuits, and time-delay elements.

Remember that none of the applications are possible unless the capacitor can store energy.

Capacitor construction

In its simplest form, a capacitor is made with two parallel plates that are separated by an insulator (Fig. 2-1). The insulator is called the *dielectric.* Because the dielectric plays such an important part of the capacitor action, capacitors are often identified by their type of dielectric. So, there are mica capacitors, air-dielectric capacitors, mylar capacitors, etc.

Any time that two conductors are separated by a dielectric, a capacitor is formed. For that reason, you should always be alert to capacitance between two wires, or between a wire and a metal surface. Those kinds of capacitance can cause problems.

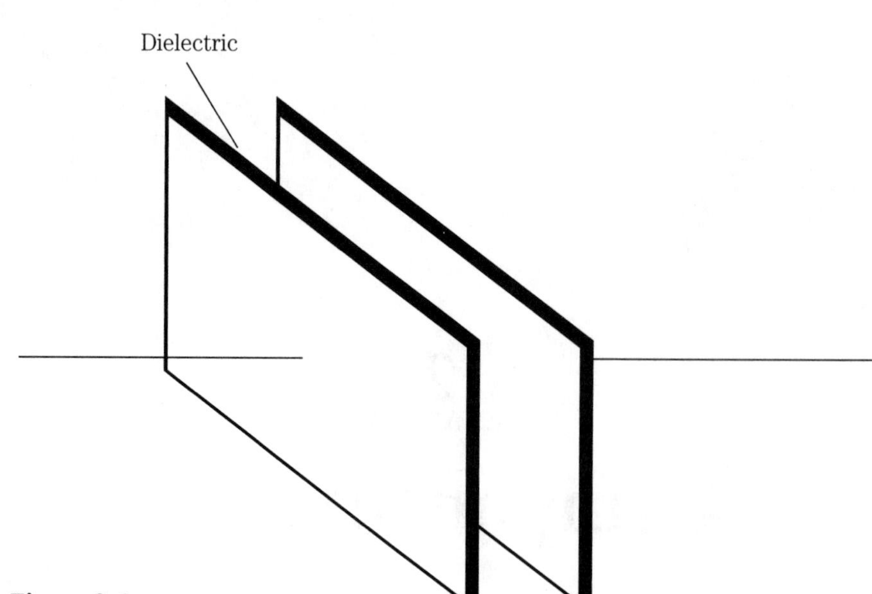

Figure 2-1

Capacitance

Capacitance is a measure of how much energy can be stored in a capacitor. You will sometimes see the term *capacity* (which is an archaic term) used instead of capacitance.

The amount of capacitance in a capacitor depends on:

- the area of the plates that faces each other
- the distance between the plates
- the type of dielectric material

The amount of capacitance is directly proportional to the area of the plates that face each other. In other words, the greater the area (A), the higher the capacitance.

The amount of capacitance is inversely proportional to the distance between the plates (d). The shorter that distance, the higher the capacitance value (Fig. 2-2).

A proportion can be written for the capacitance from the above statements:

$$C \propto (A/d)$$

where $\propto$ is a symbol that means "is proportional to."

Any time a proportion can be written for parameters they can be made equal by introducing a constant of proportionality. In this case, it is called the *dielectric constant* and it is often represented by the symbol ε_r.

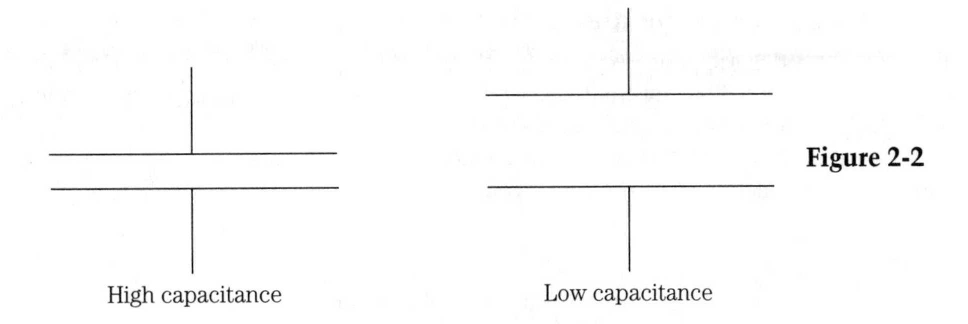

Figure 2-2

With the constant introduced, the equation becomes:

$$C = \varepsilon_r(A/d)$$

The symbol ε_r means relative dielectric constant. It is the dielectric constant compared with the dielectric constant of a vacuum. For example, Table 2-1 shows that distilled water has a dielectric constant that is 80 times greater than that of a vacuum.

Table 2-1

Dielectric	Value of ε_r	Dielectric	Value of ε_r
Vacuum	1.0	Paraffin	2.35
Air	1.00058	Quartz	4.3 to 5.1
Mica	6 to 7	Distilled water	80
Glass	5 to 7		

For all practical purposes, the relative dielectric constant of air is considered to be 1.0.

Observe that in the equation for capacitance there is no parameter that compares C with the type of material that is used for the plates. Short of trying to use an insulating material for the plates, different conductors work equally well.

A few relative dielectric constant values are given in Table 2-1.

You can think of the relative dielectric constant as being a capacitance multiplier. For example, assume that a certain air-dielectric capacitor has a capacitance of 50 pF. If the space between the plates is filled with distilled water, the capacitance is 80 times greater; that is, the capacitance becomes 4000 pF (0.004 μF).

Experiments that describe capacitor operation

The experiments described in this section have been performed many times in school laboratories. Each demonstrates an important feature of capacitors and their operation.

The importance of the dielectric

Figure 2-3 describes an experiment that was first proposed by Ben Franklin. Three pails are used. Two are made of conducting metal and the third, shown shaded in Fig. 2-3, is made of an insulating material.

The pails are assembled to form a capacitor with the two pails serving as the plates and the third used as the dielectric. A voltage of about 100,000 V is applied

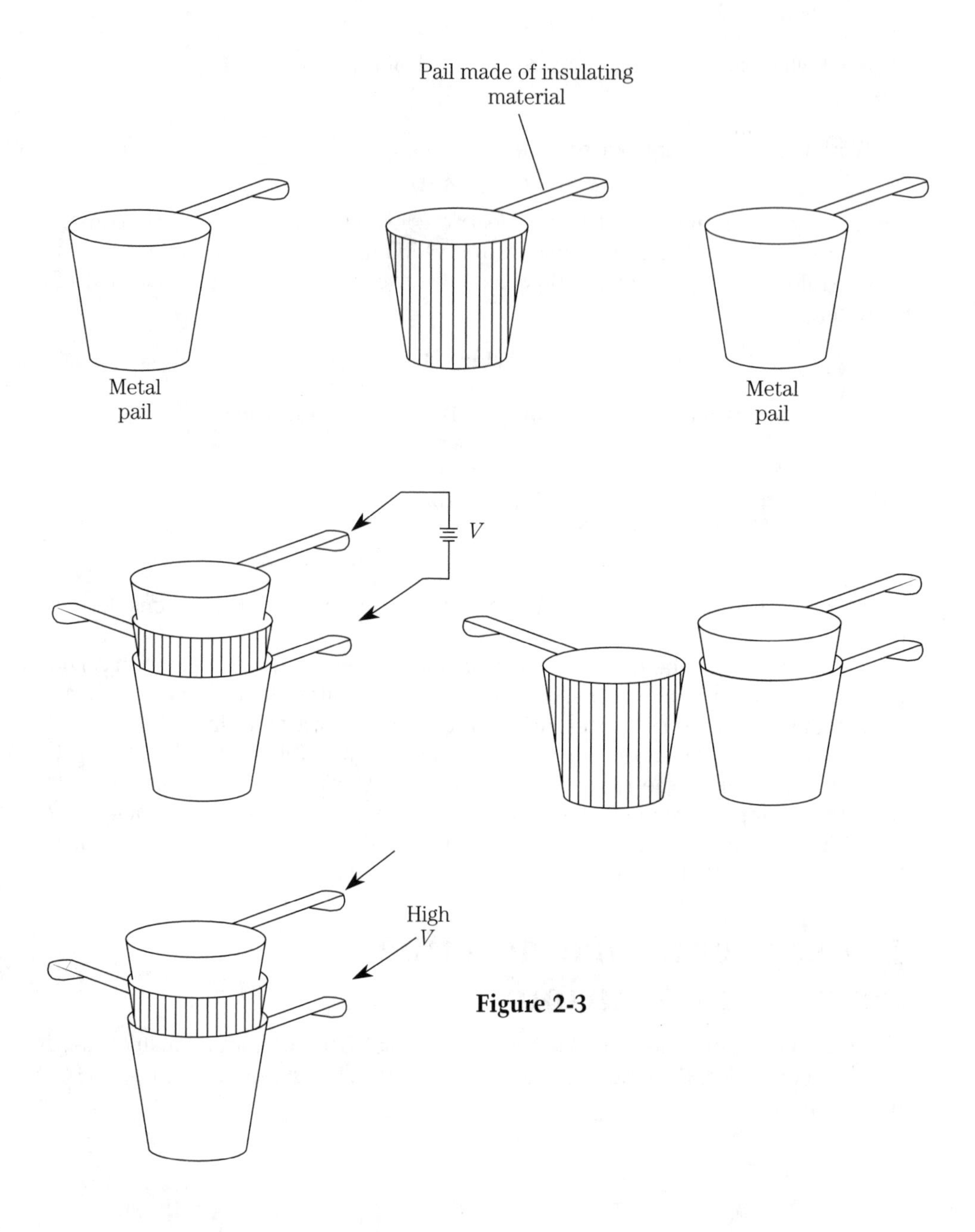

Figure 2-3

across the capacitor and it becomes charged. Then, the charged capacitor is disassembled very carefully. A long insulating stick is used to lift each pail.

The two metal pails are assembled as shown in the third illustration. After a short period of time, the pails are carefully assembled again into their original configuration, as shown in the fourth illustration. A very large discharge spark can be drawn from the metal plates after the capacitor has been reassembled.

An erroneous idea about capacitor operation is that a capacitor is charged by forcing electrons into one plate and drawing them out of the other plate. If a capacitor was charged by establishing a difference in the number of electrons on the plates, then it would have been discharged when the plates were assembled as shown in the third illustration!

The point made here is very important: *the energy is stored in the dielectric!* When the pails are separated, the electrons in the metal pails repel each other. That causes them to be distributed throughout the pails. When the pails are touched together, they do not cross from one pail to the other.

The relationship between capacitance and voltage

Figure 2-4 shows an experiment with a charged capacitor. It works best if distilled water is used as a dielectric.

The capacitor is charged by a dc voltage (V_1). The voltage source is removed and the plates are moved apart. When this is done, the voltage across the capacitor increases! The higher voltage is called V_2 in the illustration.

The explanation is quite simple. Voltage is a measure of work per unit charge, not a measure of force. Because the voltages on the plates have opposite polarities,

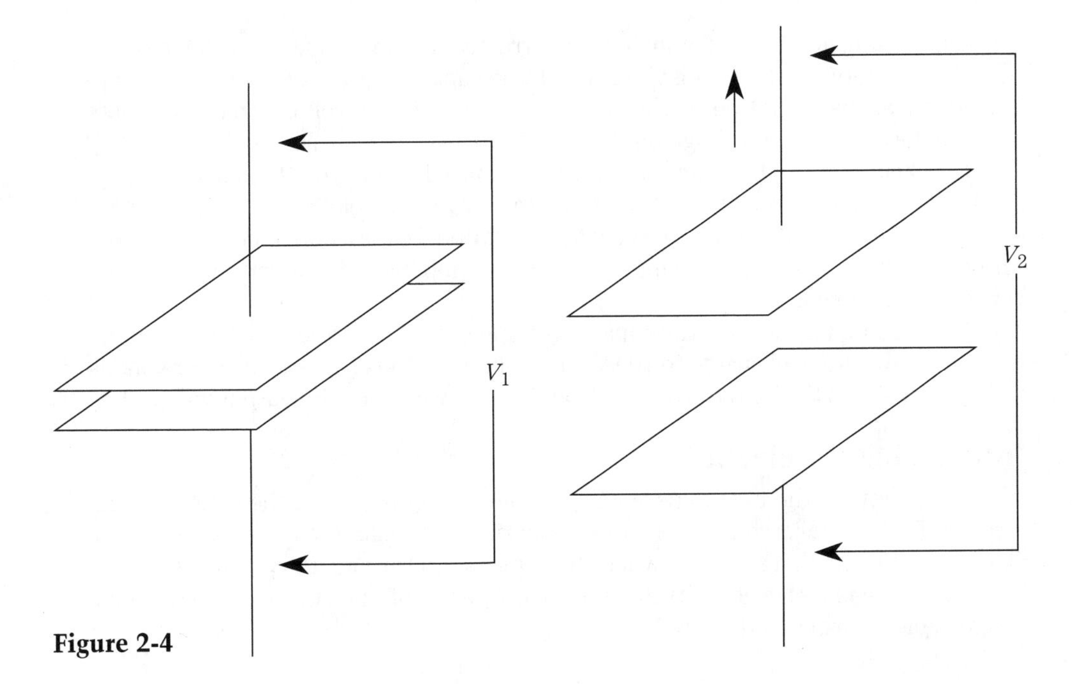

Figure 2-4

a force must be used to move the plates apart (remember that opposite charges attract). The force exerted through a distance indicates that work is being done on the system.

$$Work = Force \times Distance$$

The work done in moving the plates shows up as an increase in voltage across the capacitor. In fact, the work shows up as an increase in energy stored. Work and energy are directly related. *Energy* is defined as the ability to do work.

Capacitors do not "block" dc

Figure 2-5 is an illustration taken from an early CET test. Capacitor C is uncharged. The question was: "What is the voltage across the output terminals?"

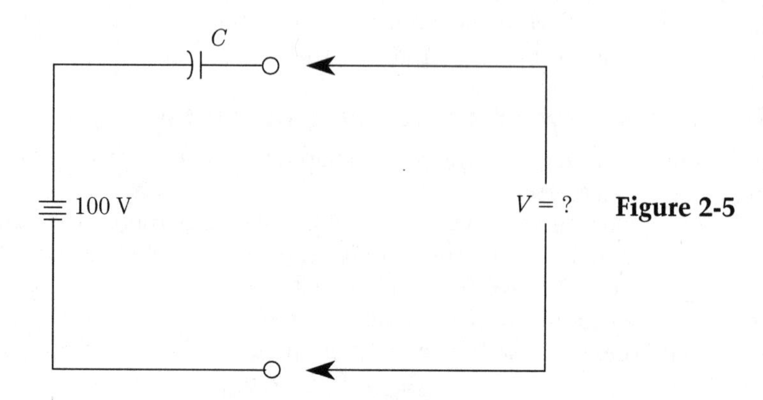

Figure 2-5

A surprising number of technicians incorrectly answered that the voltage is 0 V. Actually, the output voltage must be 100 V. If the capacitor is not charged, then there is an identical charge (0 V) on each plate. In other words, the voltage on the left plate must be the same as the voltage on the right plate.

It is likely that the high number of technicians who believed the output voltage should be 0 V were oversold on the model that says "capacitors block dc and pass ac." That, of course, is not true in the way it is stated, but is a convenient way to describe the action of a coupling capacitor between amplifiers. However, that coupling capacitor is charged.

If the output terminals of the capacitor in Fig. 2-5 are shorted together for a moment, the capacitor will charge to 100 V. The polarity of the charged capacitor voltage will be opposite to the battery voltage. That, in turn, will make the output voltage 0 V.

Introducing the electret

To understand the next experiment, it is necessary to review the concept of an *electret*. Probably everyone has had the experience of charging a comb with static electricity. It occurs on a dry day when the comb is run through thick hair. The comb becomes charged with electricity and for a short period of time, it can be used to pick up small bits of paper and straw.

The comb becomes an electret when it is charged. So, an electret is a dielectric material that has been charged with positive or negative electricity. In this case, the charge doesn't last very long.

Permanent electrets are now being made that will hold their charge for 99 years (according to the manufacturers). They are used in high-frequency microphones and speakers, and in electrostatic recordings.

The experiment of Fig. 2-6 uses a permanent electret. At the start of the experiment, the capacitor is not charged and there is no connection to its terminals. When the charged electret is inserted between the plates, the capacitor has a voltage across its terminals. In other words, the capacitor is charged.

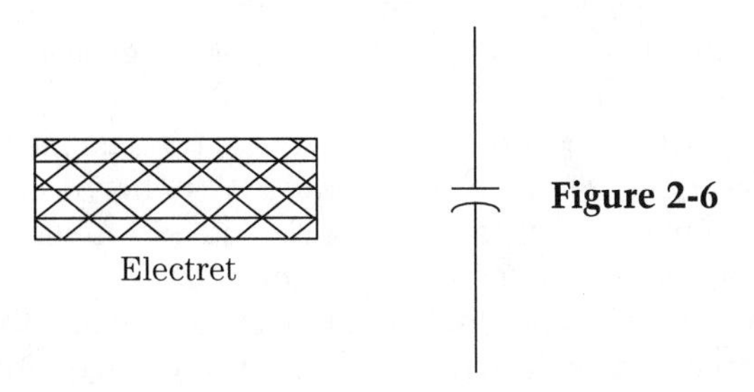

Figure 2-6

The capacitor is charged without the use of an external voltage to move electrons into and out of the capacitor plates.

When a capacitor is charged in a circuit, the dielectric becomes electrified like the electret. That might only be for a short moment, but the fact is that the capacitor is charged by the action in the dielectric, not just by the motion of electrons on the plates.

When electrons are moved into and out of the plates, there is an electrostatic field between the plates. That electric field is what charges the dielectric. As covered in Table 2-1, the amount of charge the capacitor can hold depends upon the type of dielectric used in the capacitor construction.

With the descriptions of the experiments just given, it is obvious that the models used for describing capacitor action should not be taken literally. They serve an important purpose for beginning students. However, you should always be aware of the fact that they are models. In order to understand the operation of certain circuits and systems you must use the correct theory of capacitor operation.

Capacitor voltage ratings

The experiments just described demonstrate the importance of the dielectric in capacitor operation. Actually, the dielectric serves two purposes. The one already mentioned is that it stores energy.

To understand the second purpose, remember that a charged capacitor has a voltage across its terminals. The dielectric prevents that voltage from producing a current from one plate to the other. If a conductor was between the plates a current would flow easily from one plate to the other and no energy would be stored.

There is no such thing as a perfect insulator. Given a high enough voltage, there is always a point at which the insulator will begin to conduct. For that reason, capacitors must be rated by the amount of voltage that can be placed across it without having an insulation breakdown.

When you buy capacitors, there are two ratings you need to specify: the capacitance value and the voltage rating. In some applications, the temperature coefficient is also very important. It tells how much capacitance change will occur when the temperature changes.

Usually, the voltage *rating* of the capacitor will be higher than the actual voltage that is expected to be placed across it. That extra voltage rating is called the *safety factor.*

The tolerance for capacitors can also be shown with a color code. Tolerance for capacitors is calculated the same way as for resistors. It gives the range of values of capacitance for each capacitor.

Electrolytic capacitors, covered in the next section, might have a tolerance of +100%, –50%. That range of values in tolerance is unheard of in resistors. Other types of capacitors can be purchased with ±2% and ±1% tolerances. However, the closer the tolerance, the more expensive the capacitor.

If a capacitor is charged and discharged by an ac voltage, the peak voltage occurs only two instants of time during a full cycle. The voltage rating is based on that peak value. The peak voltage that should be allowed across a capacitor when the applied voltage is ac or pulsed dc is called the *working voltage*, and it is sometimes referred to simply as its *voltage rating*.

An electrolytic filter capacitor in a dc circuit has a continuous voltage across its terminals, so the "working voltage, dc" (WVDC) rating is for a continuous dc voltage. The type of dielectric is the determining factor of both the capacitance value and voltage rating.

Types of capacitors with fixed values of capacitance

In all of the capacitors covered in this section, the type of capacitor is also the type of material used as a dielectric. Although technicians do not normally design capacitors, they should have a general idea about which types are available and the way they are used.

In ceramic capacitors, the ceramic dielectric is plated on two sides with a metal. Those metal platings are the capacitor plates. Figure 2-7 shows a typical ceramic capacitor. It is obvious why they are often referred to as *disc capacitors.*

Mica capacitors are made with a *planar* construction, rather than as a rolled construction as shown in Fig. 2-8. That can be very important in some applications. Whereas the rolled version has a built-in undesirable inductance because of its coiled construction, the planar construction has very little inductance.

Silver-mica capacitors are made with a silver plating to form the capacitor plates onto the mica dielectric. They have the same advantages of the planar construction as the mica type, but it is more difficult to make them with a close tolerance.

Glass capacitors are made the same way as the mica types, but are cheaper. That is an important advantage because of the high cost of mica.

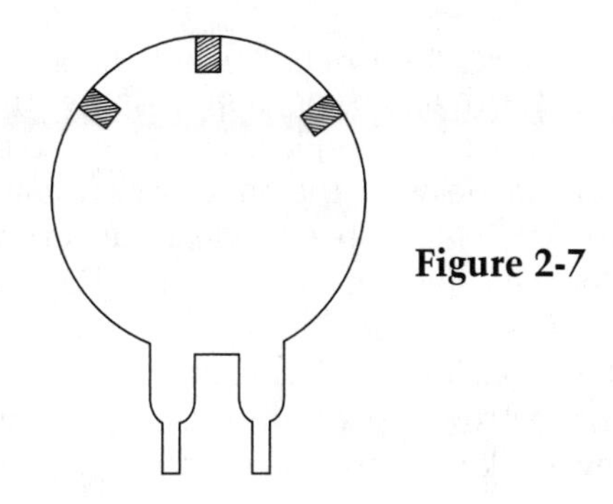

Figure 2-7

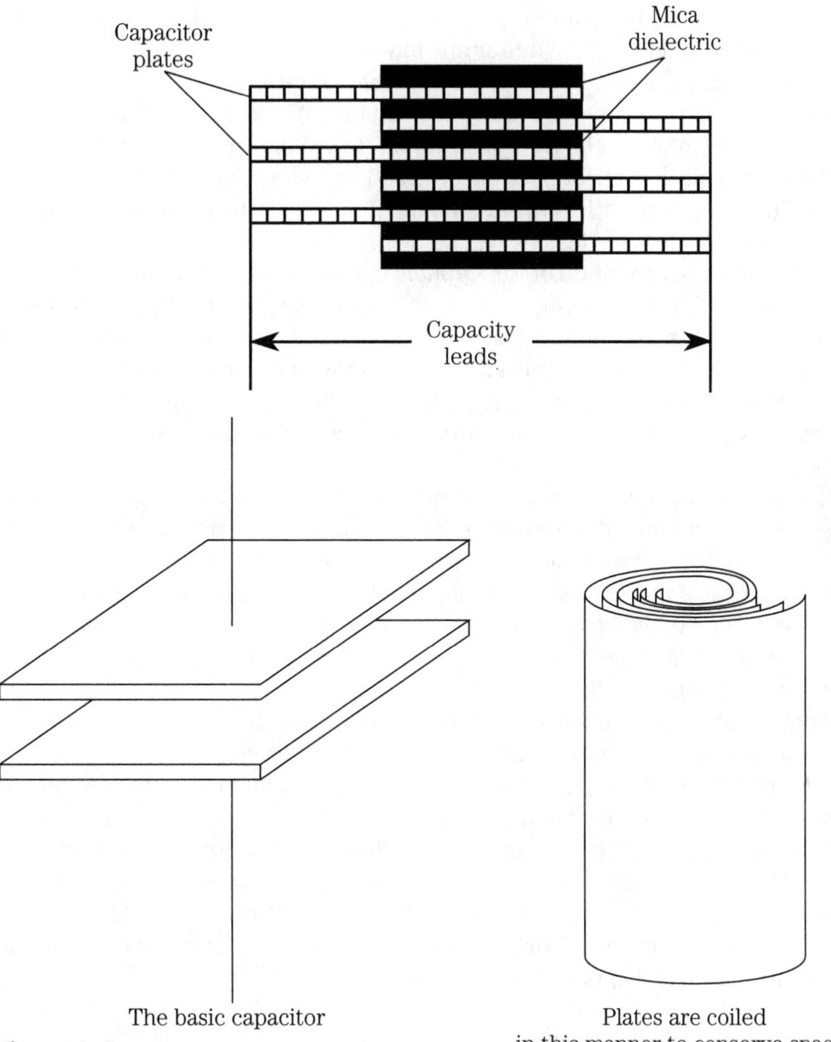

Figure 2-8

Paper capacitors are made with the rolled construction shown in Fig. 2-8. The paper is often impregnated with a poisonous material, so they should not be taken apart. This type of capacitor has been replaced by plastic types.

Plastic capacitors are made with the same rolled construction as the paper types. However, they are much less subject to leakage currents between the plates. This is especially true for the capacitors as they age. The type of plastic is used for their name; e.g.: mylar and teflon.

Remember that the capacitance of a capacitor increases as the plates are brought closer together. Electrolytic capacitors use a very thin chemically coated dielectric. Therefore, they are known for their high-capacitance characteristic. Two types of electrolytic capacitors are available: aluminum and tantalum.

Aluminum types are less expensive, however they cannot be made with close tolerances. Tantalum types can be made with a closer tolerance, and they are usually more expensive than aluminum types. They cannot withstand over-voltages as well as the aluminum electrolytics. Tantalums have a longer shelf life than the aluminum types and can be made with higher working voltage ratings.

It is important to remember that electrolytic capacitors are polarized. In other words, you must always observe the proper dc voltage polarity when connecting them into a circuit. If you connect one with a positive voltage delivered to its negative terminal, and a negative voltage polarity delivered to its positive terminal, it might explode! At the very least it will be destroyed.

An aluminum-type electrolytic capacitor will deteriorate if unused for a long period of time. If you are going to put one into service after it has been stored for more than a year, it is a good idea to form the dielectric before using it in a circuit. That is done by starting with a low dc voltage across it at first. A good point to start at is 20% of its WVDC rating, then increase the voltage by 20% every 10 minutes or so. The procedure reestablishes the oxide coating used for the dielectric.

Electrolytic capacitors are usually rated by their Equivalent Series Resistance (ESR), which combines the leakage resistance and the series terminal resistance. Leakage resistance is the resistance of the dielectric. There is no such thing as a perfect dielectric, so some leakage current flows between the plates of a capacitor, but it is so low it can be disregarded in most capacitors. However, in electrolytic capacitors it is an important measurement that determines whether or not the capacitor will be useful in some applications.

Many types of capacitors made today are encased in a plastic form. That prevents destructive moisture from getting into the construction.

It is a good idea to use manufacturers catalogs to obtain in-depth training in capacitors available, and their characteristics.

Failure of all types of capacitors occurs when the dielectric breaks down. An arc between the plates, through the dielectric, is usually destructive. Also, high leakage currents through the dielectric will make most capacitors useless. An exception to that is the electrolytic capacitors, which have a characteristic leakage that can be tolerated in their usual applications.

Nonpolarized electrolytic capacitors

Not all electrolytic capacitors are polarized. There is a type, called the *nonpolarized electrolytic capacitor*, that can be used in circuits where the voltage can be positive or negative! The concept is illustrated in Fig. 2-9. Electron currents for charging the capacitors are shown with arrows.

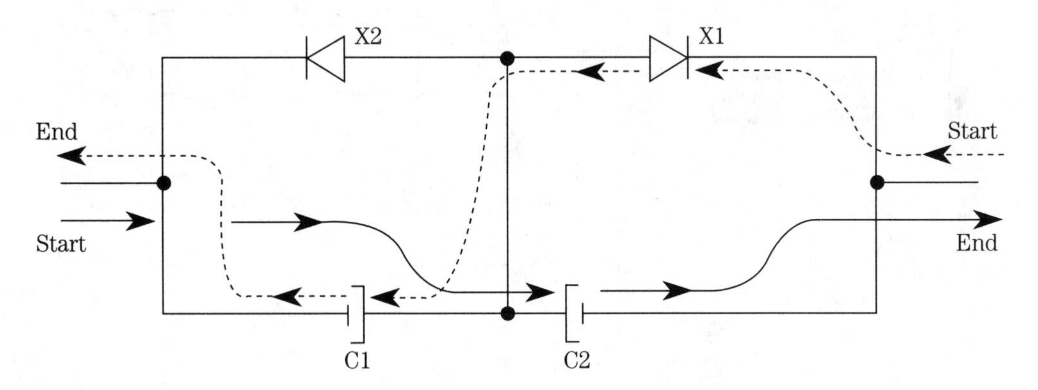

Figure 2-9

The charging current for C1 is shown with broken arrows. Notice that it flows through diode X1. A diode is a unilateral device, meaning it can only conduct current in one direction. Therefore, it is not possible to charge C1 or C2 in the reverse direction.

The charging current for C2 flows through diode X2, and is shown with the solid arrows. Nonpolarized electrolytic capacitors are sold in a tubular package. However, if the one you need is not available you could make your own in an emergency. Use Fig. 2-9 as a guide. Make sure the capacitors and the diodes are identical components.

Variable capacitors

The equation for C shows that there are three things that determine the capacitance of a capacitor: area of plates facing each other (A), distance between the plates (d), and type of dielectric (k). A variable capacitor can be made by varying any of those parameters. Figure 2-10 gives examples.

Trimmer capacitors have a narrow range of capacitance. They are used to slightly modify the capacitance of capacitors that have a wider capacitance range. As shown in Fig. 2-10, one type of trimmer operates by varying the distance between the plates with a screwdriver adjustment.

A variable capacitor is usually used to tune table-model radios. It operates by varying the area of the plates facing each other. A shaft changes the position of the moveable plate and that is how the radio is tuned. A smaller version of this type uses a plastic separator between the plates, making it possible to have a higher capacitance in a smaller package. The plastic also prevents the plates from shorting together.

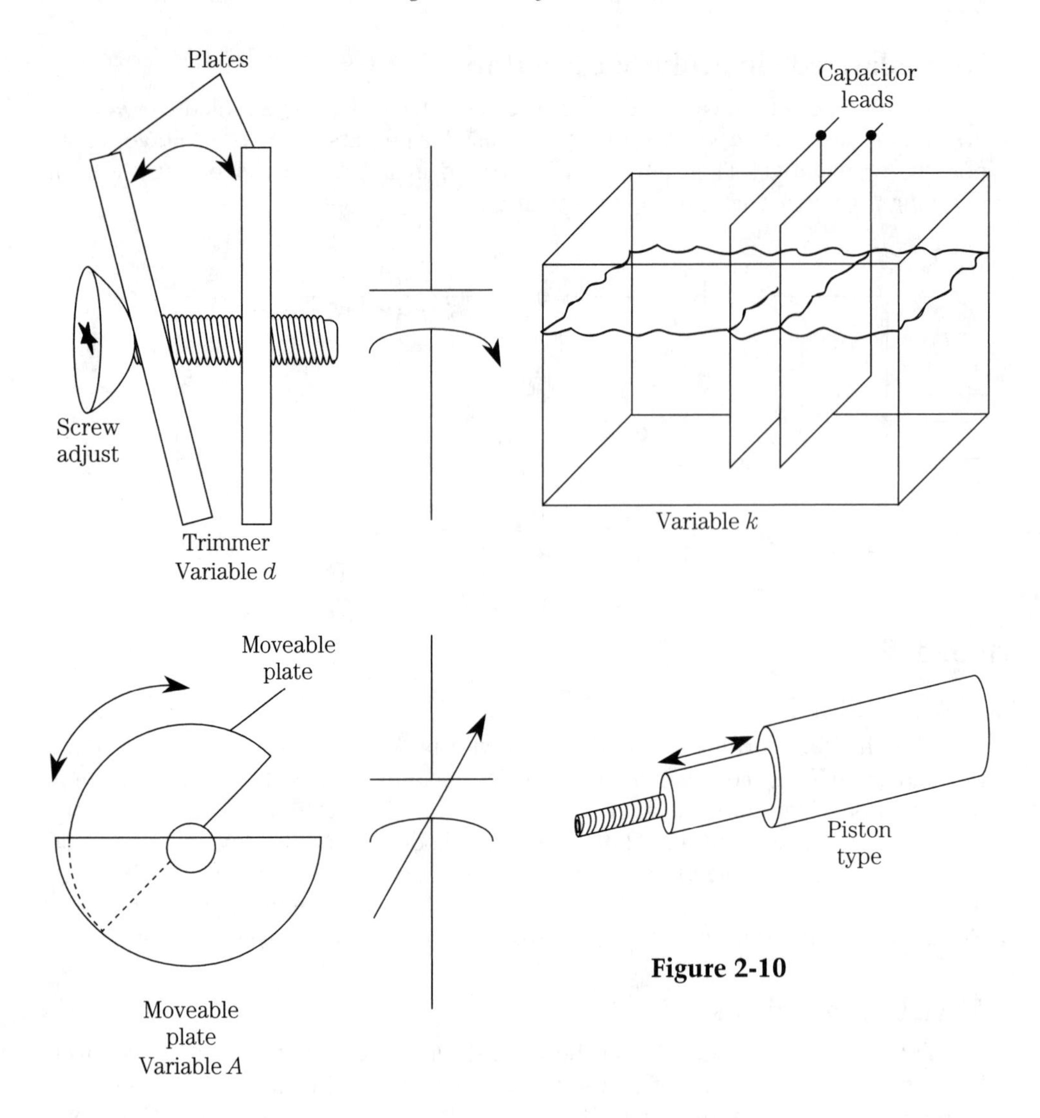

Figure 2-10

The piston type also operates by varying the area of the plates. In this case, the plates are shaped as cylinders.

The variable dielectric type varies capacitance by changing the dielectric material. For the example shown, when the tank in Fig. 2-10 is full, the dielectric fluid between the plates gives the capacitor maximum capacitance. As the fluid drops in the tank, the capacitance decreases because air has a lower dielectric constant. This application uses the capacitor as a sensor. It permits an electronic circuit to monitor the amount of fluid in the tank.

Inductors

An inductor is a component that stores energy in the form of a magnetic field. That is all that it does. However, as with capacitors, there are many applications of inductors and some of those applications are used so frequently that they are sometimes thought to be the purpose of the inductor.

Inductors can be made by coiling a conductor. An example is shown in Fig. 2-11. (Not all inductors are made that way.) Inductors follow some very specific laws. Knowing those laws help in understanding applications.

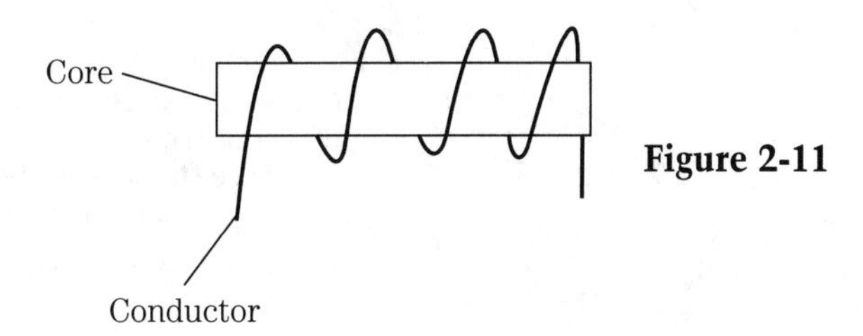

Figure 2-11

There is a popular model for analyzing inductors in circuits, and it works well. Here it is compared with the corresponding popular model for capacitor behavior in circuits:

- A capacitor is a component that opposes any change in voltage across its terminals.
- An inductor is a component that opposes any change in current flowing through it.

The key words in those models are "change in." The laws that govern the behavior of inductors in circuits will be discussed next.

Field direction

Whenever an electric current flows, regardless of whether it flows through a conductor or through space, there is always an accompanying magnetic field surrounding the current. The direction and magnitude of that magnetic field depends upon the direction and amount of current flowing.

Figure 2-12 illustrates the magnetic field around a current-carrying conductor. Note the direction of the field lines. This direction can be determined by the left-hand rule, as shown in the illustration.

Remember that electron flow is used in this book because that has been the way it has been presented in the tests. If the time comes when conventional current is used on the tests, the illustrations will still work, but you will have to use the right hand to determine the direction of the magnetic field around a current.

The direction of the field lines around the current are based upon the same rule for the direction of field lines around a permanent magnet. It is the direction that a unit north pole would move if placed in the field. Magnetic fields around currents, inductors, and other electromagnetic devices are all the same, and all follow the same rules.

Faraday's Law and Lenz' Law

Whenever there is relative motion between a conductor and a magnetic field, there is always a voltage induced across the conductor. *Relative* is a key word in this

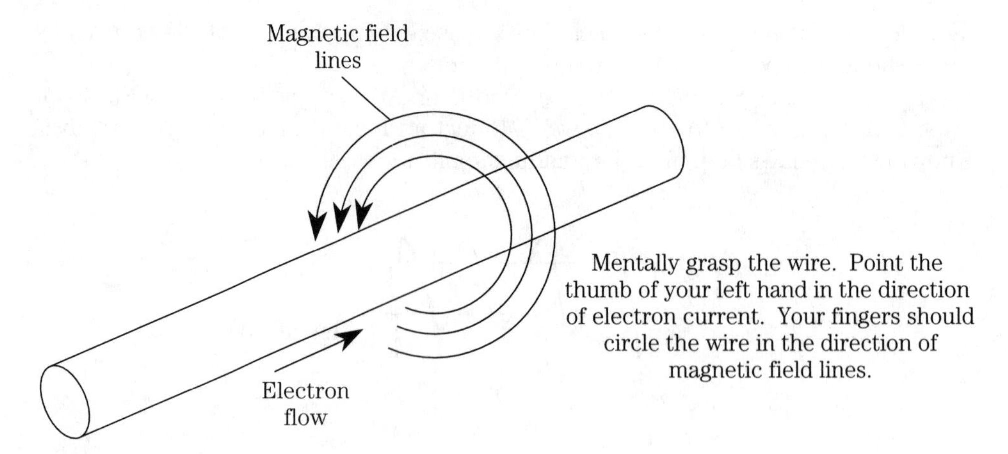
Magnetic field lines
Mentally grasp the wire. Point the thumb of your left hand in the direction of electron current. Your fingers should circle the wire in the direction of magnetic field lines.
Electron flow

Figure 2-12

statement. It doesn't matter whether the magnetic field is stationary and the conductor is moving through it, or if the conductor is stationary and the magnetic field is moving through it. In fact, they both can be moving as long as they are not moving in the same direction and at the same speed.

According to Faraday's Law, the amount of induced voltage depends upon two things:

- the number (N) of conductors involved; and,
- the amount of relative motion between the conductor and the field.

Written as an equation:

$$v = (N)(d\phi/dt)$$

The expression $d\phi/dt$ needs some explanation; it is a mathematical way of saying "the rate of change of magnetic field lines with respect to time." As shown in Fig. 2-13, the maximum rate of change in flux occurs when the motion of the conductor and/or the motion of the field are perpendicular. If they cross at an angle that is less than 90°, the rate of change of flux is less than for 90°.

The equation for Faraday's Law is sometimes written as:

$$v = (L)(di/dt)$$

This version of the equation is based upon the fact that the inductance (L) varies directly to the number of turns of wire. Also, the rate of change of current (di/dt) varies directly as rate of change of the flux that it produces. The term *magnetic flux* is used to mean magnetic field lines.

There is one other very important addition to the equation for Faraday's Law. It is usually written with a negative sign as shown here:

$$v = -(N)(d\phi/dt) \text{ and}$$

$$v = -(L)(di/dt)$$

The negative sign is because of Lenz' Law. An induced current is defined as the current that results from an induced voltage. The negative sign on the equations is a

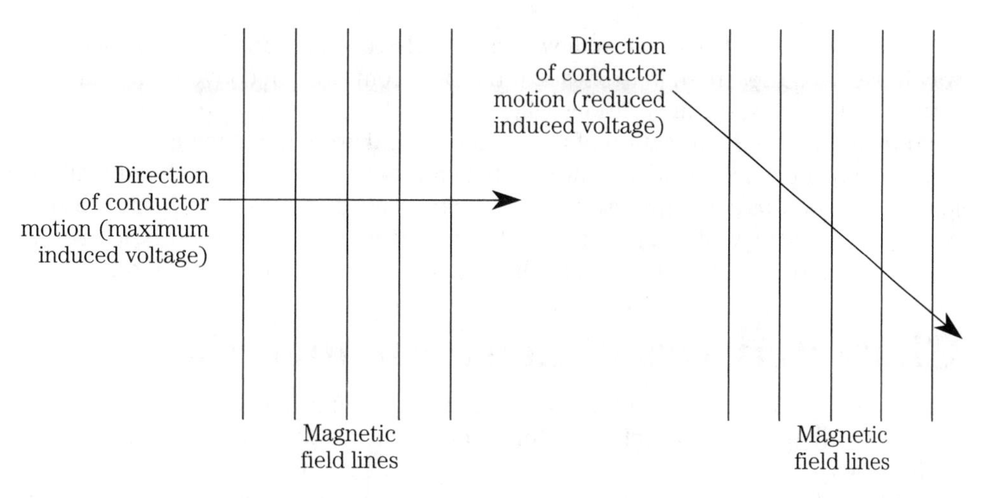

Figure 2-13

reminder that *the induced voltage produces an induced current that opposes the change in current that produced it.* That is a statement of Lenz' Law.

The arrows usually shown on magnetic flux lines seem to indicate that those lines move from the north pole to the south pole. However, the field lines do not move! The arrows are only meant to show the direction that a unit north pole would move if it was placed on a field line.

Whenever an electric current flows, regardless of whether it flows through a conductor or through space, there is always an accompanying magnetic field surrounding the current. The direction and magnitude of that magnetic field depends upon the direction and amount of current flowing.

Figure 2-14 illustrates the magnetic field around current-carrying conductors. Notice the direction of the field lines. This direction can be determined by the left-hand rule, as shown in Fig. 2-16.

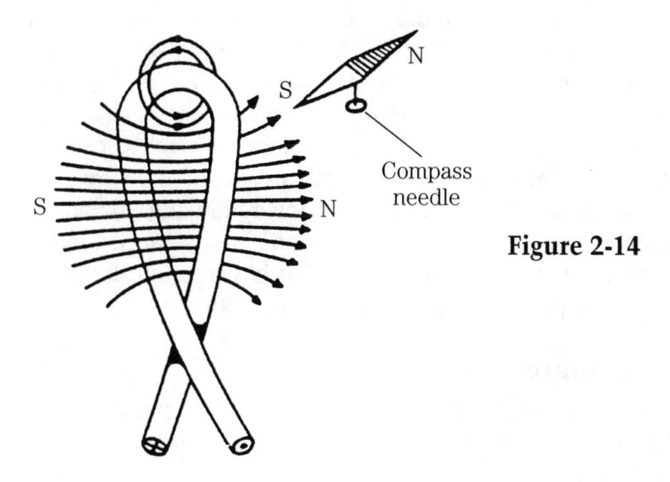

Figure 2-14

Keep it in mind that electron flow is used in this book because that has been the way it has been presented in the tests. If the time comes when conventional current is used on the tests, the illustrations will still work, but you will have to use the right hand to determine the direction of the magnetic field around a current.

The direction of the field lines around the current are based upon the same rule as for the direction of field lines around a permanent magnet. It is the direction a unit north pole would move if placed in the field. Magnetic fields around currents, inductors, and other electromagnetic devices are all the same, and all follow the same rules.

Characteristics of magnetic materials

Reluctance is a measure of how difficult it is to establish magnetic flux lines in a material. A material with a high reluctance does not readily permit the establishment of flux lines.

Permeability is the opposite of reluctance. It is a measure of the ease with which flux lines can be established in a material. For example, a material with a high permeability will readily conduct magnetic flux lines. Permeability is often represented by the Greek letter mu (μ).

From these definitions, it is obvious that a low-reluctance material has a high permeability.

All materials can be divided into one of three categories: ferromagnetic, paramagnetic, and diamagnetic.

Ferromagnetic materials

Whenever you see the term *ferrous*, or one of its related terms, think of the word iron. Originally, ferromagnetic materials were made of, or contained, iron, which is a high-permeability material. Today, there are ferromagnetic materials that do not contain any iron. An example is *alnico*, which is made of *al*uminum, *ni*ckel and *co*balt.

Some ferromagnetic materials can be permanently magnetized. Alnico, for example, will retain its magnetism for a very long time.

The expression "permanently magnetized" requires some modification. No material has ever been found that cannot be demagnetized. One way to demagnetize a material is to heat it to a high temperature. At some point, called its Curie temperature, every material will lose its "permanent magnetism." That is why devices that utilize permanent magnets, such as ammeters, should be protected from high temperatures.

Another way to demagnetize a permanent magnet is to expose it to a strong varying magnetic field. The magnetic field around an alternating current is an example.

The term *degaussing* is used to mean demagnetizing. In some cases, degaussing is necessary. For example, it is used to demagnetize a magnetic tape when it is desired to remove its stored information. That is why recorded tapes must be protected from high temperatures and ac magnetic fields.

Paramagnetic materials

Paramagnetic materials do not exhibit any appreciable amount of magnetic properties. Examples are mostly plastics, glass, and wood.

Avoid saying that paramagnetic materials have no magnetic properties at all. Under the right circumstances practically all materials have some slight magnetic properties.

One condition where this is important is in the use of very sensitive magnetic measuring instruments. It is difficult to make such instruments because they must be encased in a material that does not disturb the measurement by its magnetic properties. Such materials as Bakelite and wood—once thought to have absolutely no magnetic properties—are too magnetic for use in the sensitive measuring instruments.

Diamagnetic materials

Ferromagnetic materials are *attracted* to a magnetic field. Diamagnetic materials are *repelled* by a magnetic field, and will move away from the source of magnetic flux if free to do so. Copper is an example of a diamagnetic material.

Air-core coils

Remember that there is a magnetic field around a wire that carries a current. If the wire is coiled into a single turn (as shown in Fig. 2-14), the magnetic field lines combine to give the coil a total magnetic field. If the wire is carrying one ampere of current, the magnetomotive force of the coil field is said to be a 1-A turn.

Magnetomotive force can be thought of as being a measure of the magnetizing force of the coil's magnetic field.

A number of turns can be combined, as shown in Fig. 2-15, to give a greater combined magnetic field. However, as shown in that illustration, not all of the field lines emerge at the end of the coil, so the total magnetic field of the coil is not equal to the sum of the fields around each turn.

The coil can be wound around a soft-iron core to increase the total number of lines at the ends of the coil. Figure 2-16 shows the effect of the soft-iron core. Note that nearly all of the flux lines follow the core so that the electromagnet formed this way is more intense than that of the air-core coil.

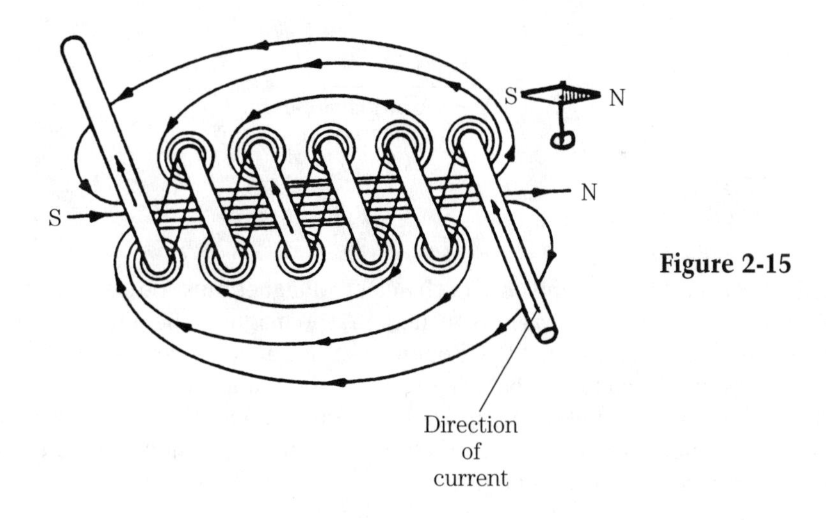

Figure 2-15

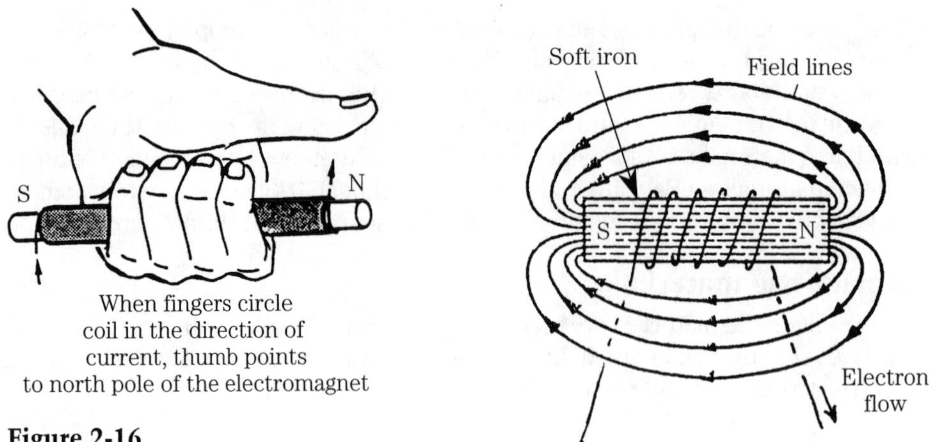
S
N
When fingers circle
coil in the direction of
current, thumb points
to north pole of the electromagnet
Soft iron
Field lines
S
N
Electron
flow

Figure 2-16

The coil in Fig. 2-16 is called a *solenoid* and it is often used to obtain a strong magnetic field. In the example of Fig. 2-17, it is used to produce a magnetic field in a relay. When a dc current flows in the coil, the strong magnetic field attracts the armature of the relay. That, in turn, closes the contacts of the relay.

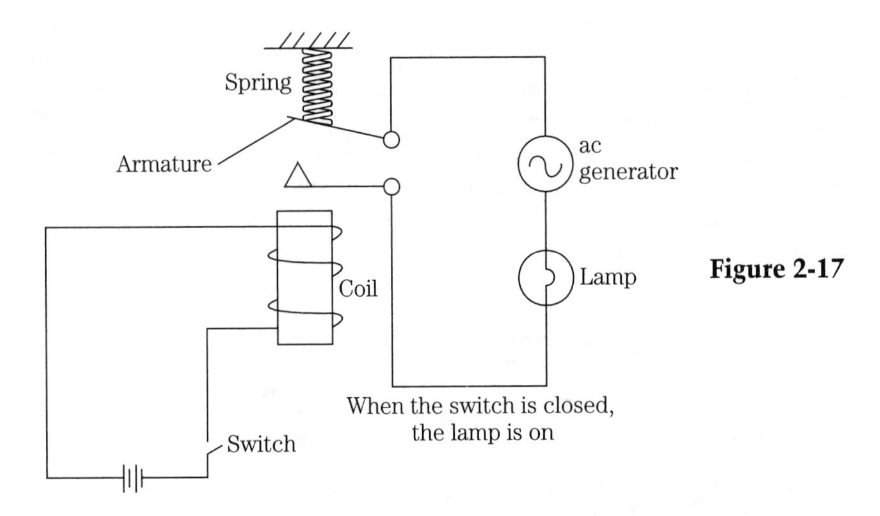
Spring
Armature
ac
generator
Coil
Lamp
When the switch is closed,
the lamp is on
Switch

Figure 2-17

In applications like this, the core of the electromagnet must be a soft-iron material. That way, when the dc current is turned off, the magnetic field is removed, and the spring returns the contacts to the open position. In some versions of the relay, there are more switch contacts that are operated simultaneously.

The advantages of using the relay are that a current can be used to operate the switch, the switching job can be performed from a remote position, and a number of contacts can be closed at the same time.

RC and RL dc circuits

Components, R, L, and C, are usually used in combinations that make circuits. They are called *two-terminal, bilateral, passive components*. The meaning of two-terminal is obvious. In a bilateral component, current can flow with equal ease in either direction. A component that does not generate a voltage is called *passive*. If it generates a voltage it is an active component.

RC time constants

Figure 2-18 illustrates the concept of RC time constants. At the start, it is assumed that the capacitor is uncharged and the switch is in the center position.

When the switch is changed to the charge (C) position, the RC series combination is connected directly across the battery. At the instant the switch is closed, the capacitor begins to charge. The instant that the switch is closed is usually considered to be time zero and is marked t_0 on the graph.

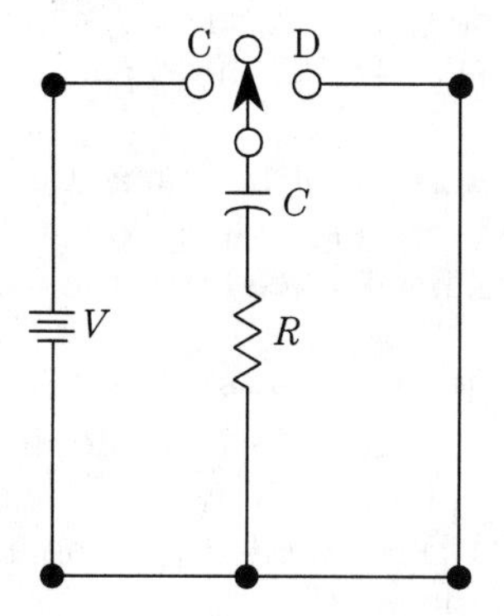

Figure 2-18

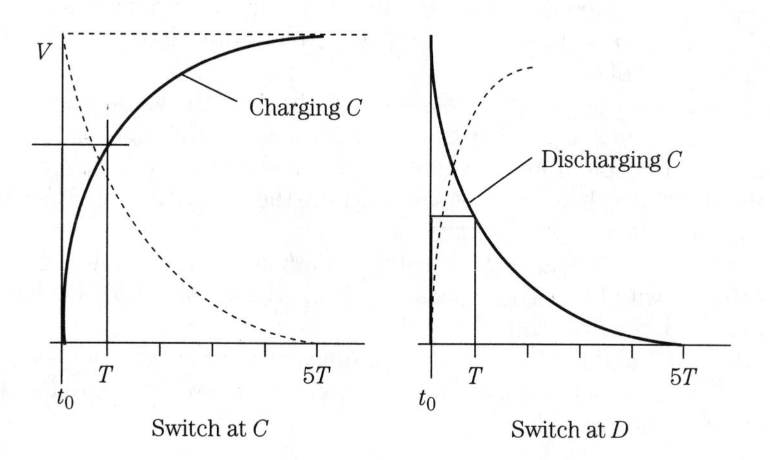

At t_0, the voltage across the resistor is equal to the battery voltage. Remember: the capacitor is uncharged at the initial condition. If 0 Ω resistance was in the circuit, the voltage across the capacitor would be equal to the battery voltage at $t = 0$.

So, when $t = 0$, the capacitor starts to charge. The graph of the charging voltage across the capacitor is shown in Fig. 2-18. Notice that the rate of charging is rapid at first, but as the voltage across the capacitor nears V the rate of charging is nearly zero.

The curve showing the voltage across the capacitor is called the *time constant curve.* That curve has the same shape regardless of the capacitance of C or the resistance of R. The only thing that changes when the values of R and C are changed is the value of T along the x-axis.

The meaning of T is very important; it is the time constant of the RC combination. Mathematically:

$$\text{Time constant } (T) = RC$$

Do not confuse T with t. The time at any instant after the switch is closed is abbreviated t, the instantaneous value of time.

The curve shows that at one time constant the capacitor has charged to about 63% of the maximum charge (V). At $5T$, the capacitor is *considered* to be fully charged.

It can be shown mathematically that the capacitor never actually reaches the voltage of the battery. However, forever is a long, long time and there is no practical value to that concept. So, it has been agreed that the capacitor is considered to be fully charged when $t = 5T$.

At $t = 0$, all of the battery voltage is across the resistor. The voltage across the resistor will decrease as the voltage across the capacitor increases. The sum of the instantaneous voltages across C and R will always equal the battery voltage.

The curve for the switch at D is exactly the same as the curve for the voltage across R during the charging period of C.

That is an important point because it means that one set of curves can be drawn for both positions of the switch, and for the voltage across both C and R. That set of curves is shown in Fig. 2-19. They are called universal time constant curves because they can be used to represent the voltage across a charging or a discharging capacitor. Likewise, they can be used for the voltage across R during the charge period or the discharge period of C.

During discharge the voltage across the capacitor and the voltage across the resistor at one time constant is illustrated by lines on the charging curve in Fig. 2-19. Observe that at one time constant the voltage across the capacitor is about 63.2% of the charging voltage. At that same instant, the voltage across R has dropped to about 36. 8% of the charging voltage.

The curve for the discharging capacitor (switch at D) shows how the capacitor voltage decreases with time. The capacitor is discharged to about 37% of its original charged voltage at one time constant.

The values of T and $5T$ on the universal time constant curve can be very accurately determined mathematically. The actual equation for the charging and discharging capacitor are given here:

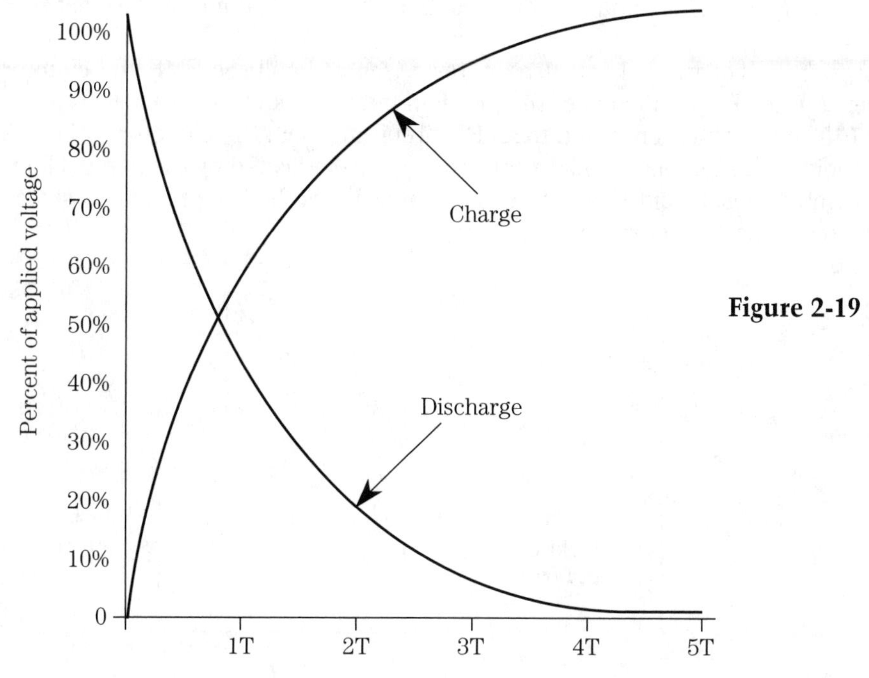

Figure 2-19

For the *charging capacitor*:

$$v_c = V[1 - \varepsilon^{-t/RC}]$$

and, for the *discharging capacitor*:

$$v_c = V[\varepsilon^{-t/RC}]$$

In both equations, v is the instantaneous voltage at any instant t. The values of R and C are in ohms and farads. The applied voltage is V.

Epsilon (ε) is a constant equal to 2.71828 . . ., which is an irrational number. In other words, its value could be calculated to an infinite number of decimal places and there would be no repeating combinations.

The equations can be used for determining the voltage across the resistor by subtracting the values of v from the applied voltage V.

When you let $t = RC$ in the charging capacitor equation you get 0.632V, or, 63.2% of the applied voltage. When you let $t = RC$ in the discharging equation, you get v = 0. 3678V, or, about 36.8% of the original voltage across the capacitor. That explains the values for T for both the charging and discharging conditions.

RL time constants

The concept of RL time constants is not as simple as it is for the RC case. The thing that complicates the RL picture is inductive kickback, sometimes called *counter voltage*. At one time, it was called *counter electromotive force* or *counter EMF*. Those terms were dropped when the expression electromotive force went out of favor because it gave the impression that voltage is a force. As mentioned earlier in this chapter, voltage is *not* a unit of force, but rather, it is a unit of work per unit charge.

However, on license and certification tests, you will still see EMF and counter EMF being used. To get an idea of the problem inductive kickback causes, refer to Fig. 2-20. The switch is connected to an RL circuit in which the inductor is the coil of a solenoid. The solenoid coil has a soft-iron core as indicated by the parallel lines on the symbol. A solenoid is an electromechanical switch that provides a strong pulling force when it is activated.

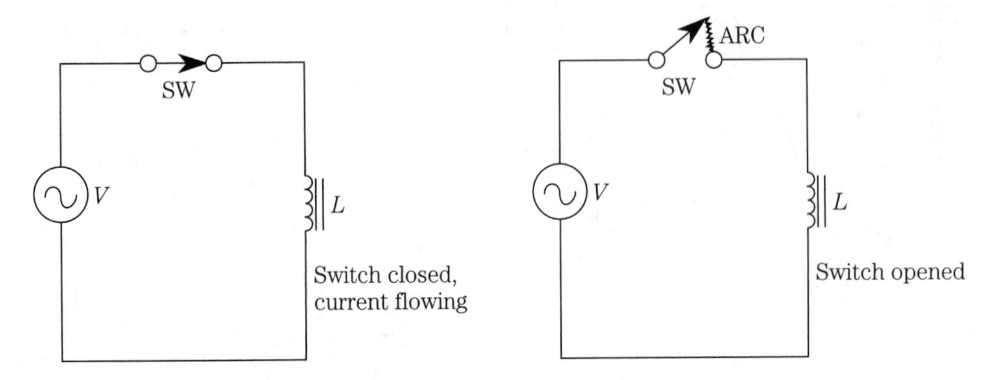

Figure 2-20

You now have two uses of the word *solenoid*. It is a coil wound on a soft iron material, and it is an electromechanical switch.

Assume the switch (SW) has been closed and a magnetic field has been stored around the coil. When the switch is opened to stop the current from flowing in the coil, the inductive kickback acts to keep the current flowing. The voltage associated with that kickback is so high that it produces a heavy arc across the switch contacts. That arc is so strong that it can destroy the switch contacts unless some method of preventing it is used.

The same thing happens in the older automotive ignition systems (Fig. 2-21). The contact points in the distributor interrupt current through the so-called coil. (Although it is called a coil, it is actually a form of transformer). The inductive kickback that occurs with switching the primary circuit on and off would normally destroy the distributor points, except that a "condenser," actually a capacitor, is connected across the points. The condenser stores the inductive kickback voltage and prevents the arc from occurring, then it discharges harmlessly through the circuit.

There are many examples of electronic circuits where some provision has to be taken to prevent the effects of inductive kickback. One method is to connect a capacitor across the endangered component, as just described. Another popular method is to connect a voltage-dependent resistor (VDR) across the coil. When the high inductive kickback voltage occurs the VDR resistance becomes very low. That has the same effect as connecting a very low resistance across the coil, effectively short circuiting the coil counter voltage.

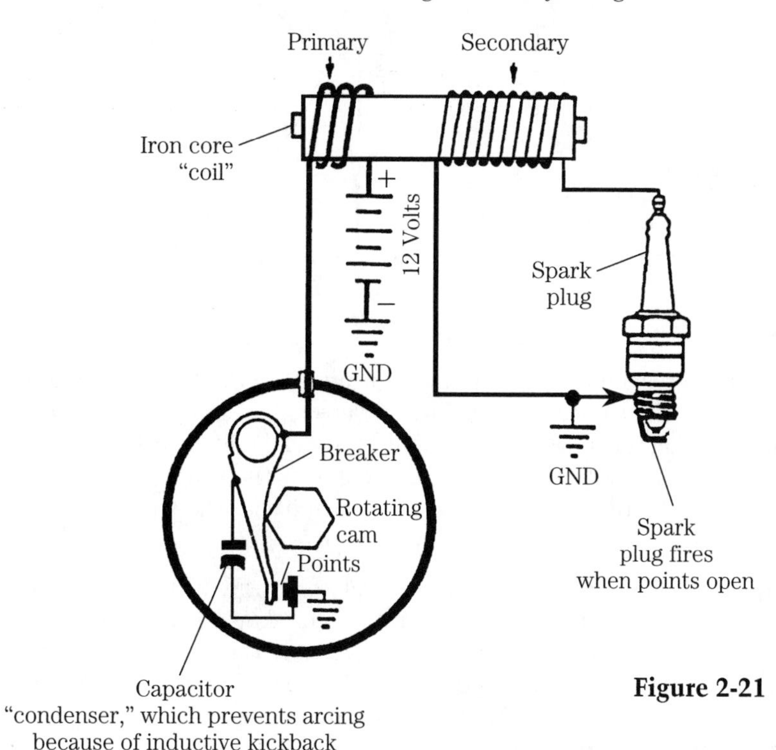
Magnetic field collapses
when points open. Rapid change of
flux induces high secondary voltage
Primary
Secondary
Iron core
"coil"
+
12 Volts
–
GND
Spark
plug
Breaker
Rotating
cam
Points
GND
Spark
plug fires
when points open
Capacitor
"condenser," which prevents arcing
because of inductive kickback

Figure 2-21

Figure 2-22 shows a circuit that is similar to the one used for studying RC time constants. From what has been said about inductive kickback you can see that it could happen here. However, for now we will disregard its effects and discuss only the charge and discharge voltages and currents.

Actually, if the inductance and resistance are properly chosen, the circuit can be operated with a minimal kickback effect, so the circuit has some practical application.

Assume there has been no current through the coil for some time and the switch is in position N. When the switch is turned to position C, there is an inrush of current. That will cause the countervoltage of the coil to oppose the rapid increase of circuit current. Therefore, the current builds along the time constant curve shown in the illustration. The equation for the current (i) at any instant of time (t), given here, is very similar to the voltage equation in the RC circuit.

$$i = I[1 - \varepsilon^{(-t/(L/R))}]$$

where I is the maximum possible current and is equal to V/R. L and R are the inductance and resistance in the circuit. Epsilon is the same as in the RC time constant circuit, that is, 2.71828. Compare this equation with the one for voltage rise in the RC circuit.

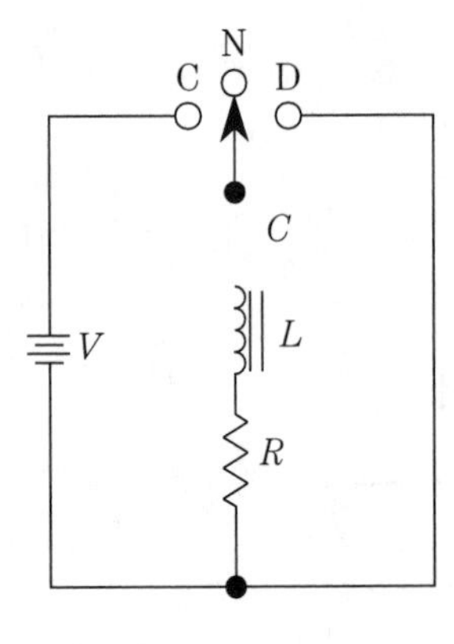

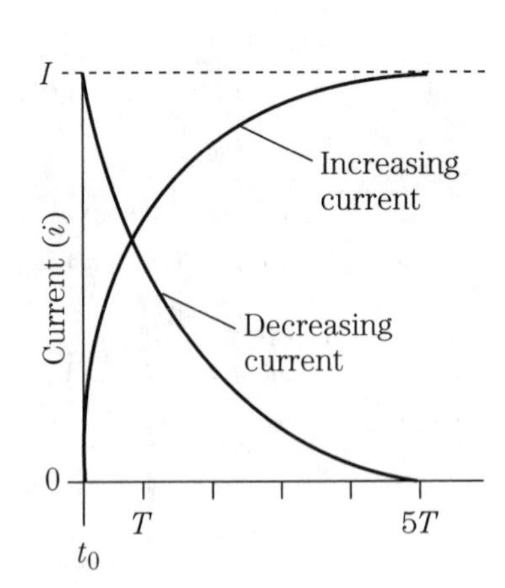

Figure 2-22

If the current has reached the maximum value and the switch is turned to the D position, the magnetic field around the coil will collapse. That produces an inductive kickback in most applications. (Remember that kickback is disregarded in this discussion.) The decay in current follows the same curve as the voltage across the capacitor in the discharge circuit. The decaying coil current has the equation:

$$i = I\ [\varepsilon \text{ exponent } -t/(L/R)]$$

In both equations the exponent has an L/R value. Consider, first, the equation for increasing current. If you let t in the numerator equal L/R, the exponent of ε becomes equal to –1. Solving the equation with that exponent gives an instantaneous current that is about 63% of the total possible current (I). The maximum possible current is assumed to be reached in five time constants.

The overall result is that the time constant of an RL circuit is:

$$T = L/R$$

You get the same result by making t in the discharging current equation equal to L/R. As before, that makes the exponent of epsilon equal to –1. The result is that the discharge current will drop to about 37% of the maximum current in a period of one time constant.

Make a special note of the fact that the equation for RL time constant shows that when the resistance (R) decreases, the time constant (T) increases. Also, the time constant increases when the inductance (L) is increased.

The equations for v in RC circuits and i in RL circuits are easily manipulated with a scientific calculator. However, they are seldom, if ever, needed for passing certification or licensing exams. You *are* likely to encounter RC and RL time constant problems requiring the $T = RC$ and $T = L/R$ equations.

Solving GROL time constant problems

So far, you have looked at time constant problems using a mathematical approach. Now, the solution of time constant problems is extended by using a piecewise graphical approach. This method makes it possible to solve all of the time constant problems in the FCC GROL exam and many other types of exams. The approach gives a very good estimation for all time constant problems, and exact values for the FCC exams.

The graphical method usually requires the use of a universal time constant curve as a starting point. But, what to do if you can't lay your hands on that curve?

The answer is that you can make one by using graphical coordinates and a calculator. You can make the required coordinates with two pieces of paper. If you have graph paper, the procedure will be easier.

Refer to Fig. 2-23. The horizontal axis is used for time constants and the vertical axis represents the percent value of the applied dc voltage. For convenience, you can say the vertical axis marked for 100 V, however it can also represent 100% of any voltage.

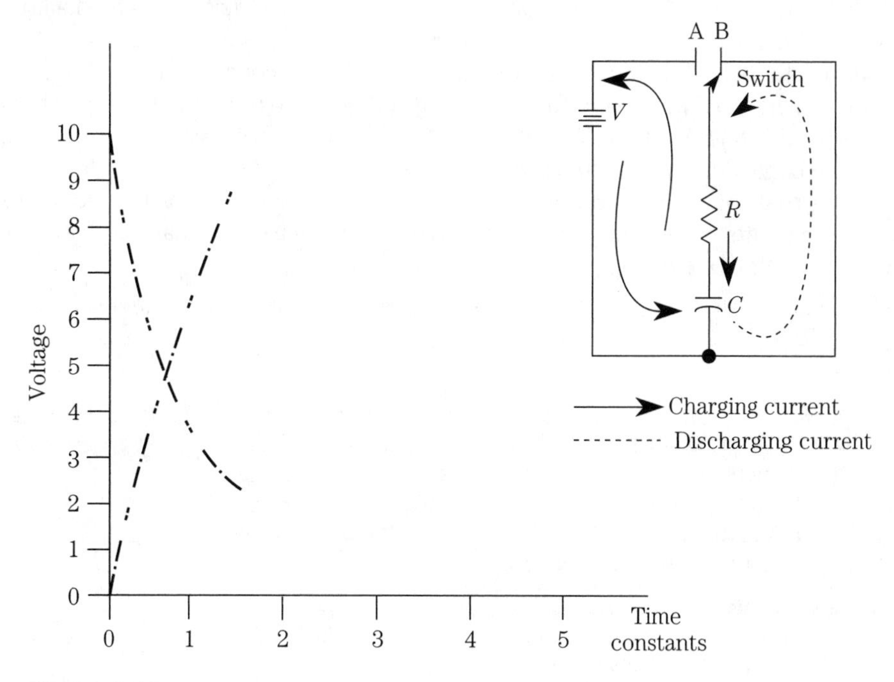

Figure 2-23

To make the coordinates on a plain piece of paper, start near the lower left corner of the paper. Make a dot to represent the zero value. Draw a horizontal and vertical line, as shown in the illustration. You will need five evenly spaced marks on the horizontal line. Make the first mark.

Using the edge of a second piece of paper, mark off the distance you used for the first mark. Then, repeat the distance for the remaining marks so that you have a total of five equally spaced marks.

On the vertical line you will need a total of 10 equally spaced marks. Make the first mark at a convenient distance. Use the same method as for the making horizontal marks. Set the remaining vertical marks so you have 10 equally spaced points.

You are now ready to draw a piecewise time constant curve. Figure 2-23 shows the circuit used to describe what is happening. When the switch is turned to position A, capacitor C begins to charge through resistor R. It has been shown that at the end of one time constant the voltage across the charging capacitor will be 63.2 V—that is, 63.2% of the applied dc voltage.

When the switch is in position B, the charged capacitor discharges through the resistor. It has also been shown that voltage across the discharging capacitor will drop to 36.8% of the original full-charge voltage in one time constant.

Piecewise charging curve

Our discussion starts at the moment the switch turned to position A. From that moment until the marking for the first time constant the capacitor charges to 63.2 V. Mark that value on the first time constant line. You will have to estimate the location about one-third of the distance between marks 6 and 7.

During the period between the first and second time constant, the voltage will increase another 63.2 percent of the remaining voltage. The remaining voltage at one time constant is 36.8 V, or, 36.8% of the applied voltage. During each time constant period, the capacitor charges to 63.2% of the remaining voltage.

Taking that into consideration, you have a simple step-by-step procedure for plotting the points on the time constant curve for the charging capacitor. Table 2-2 shows the result of each step.

Table 2-2

Step 1: Voltage at the previous step: __________ V.
Step 2: Subtract the voltage in step 1 from 100: ______________ V.
Step 3: Multiply the voltage in step 2 by 0.632: ______________ V.
Step 4: Add the voltage in step three to the voltage in step one ____________ V.
Step 5: Plot the value on the graph.

1st example

At the first time constant:
Step 1: Voltage of the previous step: 0 V.
Step 2: Subtract the voltage in step 1 from 100: 100 V.
Step 3: Multiply the voltage in step 2 by 0.632: 63.2 V.
Step 4: Add the voltage in step three to the voltage in step one: 63.2 V.
Step 5: Plot 63.2 V on the first time constant line.

2nd example

For the voltage at the second time constant line:
Step 1: Voltage of the previous step: 63.2 V.
Step 2: Subtract the voltage in step 1 from 100: 36.8 V.
Step 3: Multiply the voltage in step 2 by 0.632: 23.3 V.
Step 4: Add the voltage in step three to the voltage in step one: 86.4 V.
Step 5: Plot 86.4 V on the second time constant line.

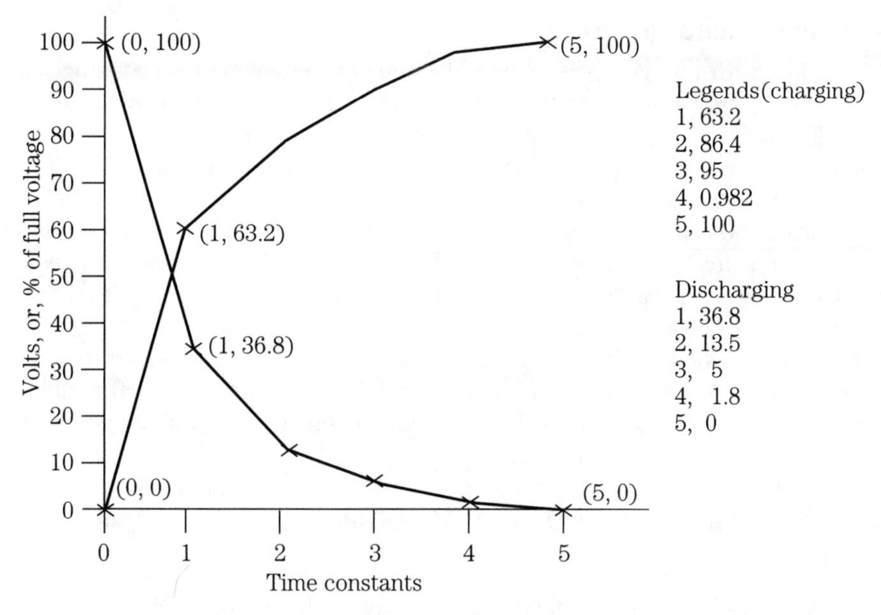

Figure 2-24

Continue with the remaining time constants. Then, sketch lines between the dots on the graph and you will have a piecewise graph of the time constant. The final result is shown in Fig. 2-24. If you have an artistic flair, you can curve the sketched lines between the points and get closer to the universal time constant curve.

Table 2-3

Charging capacitor	
To find the voltage at ↓	**Multiply applied voltage by** ↓
1st time constant	0.632
2nd time constant	0.864
3rd time constant	0.95
4th time constant	0.982
5th time constant	0.993
Discharging capacitor	
To find the voltage at ↓	**Multiply applied voltage by** ↓
1st time constant	0.368
2nd time constant	0.135
3rd time constant	0.05
4th time constant	0.018
5th time constant	0.007

Piecewise discharge curve

To make the discharge curve, you *could* simply take 36.8% of each voltage at the previous step. For example, on the discharge curve the voltage at the first line is 100 V. Take 36.8% of that to get the voltage at one time constant. That gives you a voltage of 36.8 V. (See Table 2-2.) For the second point you could take 36.8% of the voltage at that time constant value.

An easier and better way to do the discharge time constant curve is to just subtract the voltage for the charging time constant curve from the applied voltage. Mark that value on each vertical time constant line. Continue to do that until the time constant table is completed.

Now you can solve any time constant curve graphically. However, to solve problems on the FCC test, you do not need the time constant curve! All you really need is the table. In fact, go one step further: you don't even need the table if you remember how you got the values in the table!

The solutions to two time-constant problems in the GROL pool of questions are given here. All of the questions on an FCC GROL test must be taken from the GROL pool of questions.

1. How long does it take an initial charge of 20 Vdc to decrease to 7.36 V in a 0.01-μF capacitor when a 2-MΩ resistor is connected across it?

   Solution: Find voltage at one time constant.

   $$v = 0.368 \times 20 \text{ V} = 7.36 \text{ V}$$

   The question is really asking, "What is the time constant for the circuit?"

   $$T = RC = 2 \times 10^6 \times 0.01 \times 10^{-6} = 0.02 \text{ seconds (answer).}$$

2. How long does it take an initial charge of 20 Vdc to decrease to 2.71 V in a 0.01-μF capacitor when a 2-MΩ resistor is connected across it?

   Solution: This is the same circuit as in question number. 1. In the solution to that problem it was found that the voltage dropped to 7.26 V at one time constant. That is not low enough for this problem, so find the voltage at the next time constant.

   $$v = 0.368 \times 7.36 = 2.71 \text{ volts.}$$

   That is the voltage the question asks about. The question is really asking, "What is the value of two time constants for the circuit." One time constant was found to be 0.02 second. Therefore, two time constants is:

   $$2 \times 0.02 = 0.04 \text{ second (answer).}$$

Programmed review

Start with Block 1. Pick the answer that you believe is correct. Go to the next block and check your answer. All answers are in italics. There is only one choice for each block. There is some material covered in this section that was not covered in the chapter.

Block 1

What is the resistance of the circuit (R_{AB}) in Fig. 2-25 if all resistors have a resistance of 100 Ω?

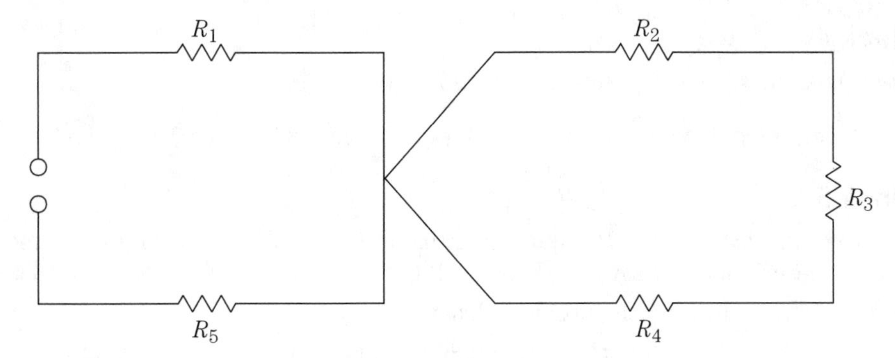

Figure 2-25

Block 2

The correct answer is 100 kΩ.

$$R_{AB} = R_1 + R_5 = 100 + 100 = 200\ \Omega$$

When you take any of the tests covered in this book or in the companion book, *The Complete TAB Book of Practice Tests for Communications Licenses*, you can expect to be asked to solve series, parallel, and series-parallel circuit problems. The problems might be for R, C, or L components in circuits.

In the circuits for Fig. 2-25 R_3, R_4, and R_5 are shorted together. Resistors R_1 and R_5 are added because they are in series.

Here is your next question: What is the capacitance of the component combination in Fig. 2-26?

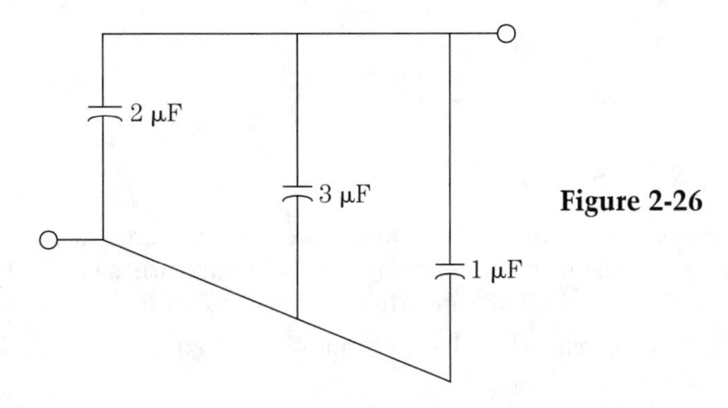

Figure 2-26

Block 3

The correct answer is 6 microfarads. When the capacitors are connected in parallel their capacitance values add.

Here is your next question: To increase the time constant of an RC circuit, increase

A. the resistance of R. C. choices A and B are both correct.
B. the capacitance of C. D. neither choice is correct.

Block 4

The correct answer is C. The equation is $T = RC$.

Here is your next question: How many degrees are in 2.86 radians?

Block 5

The correct answer is 163.9°. Remember that there are 57.3° in one radian. A more accurate description is to say that there are 180/π° in one radian.) $57.3 \times 2.86 = 163.9°$.

You might prefer to work the problem this way:

$$180/\pi°/\text{radian} \times 2.86 \text{ radians} = 163.9°$$

Note that the radians cancel. In electronics and other scientific applications, radian measure is preferred over degrees.

Here is your next question: Refer to Fig. 2-27. What is the value of applied voltage?

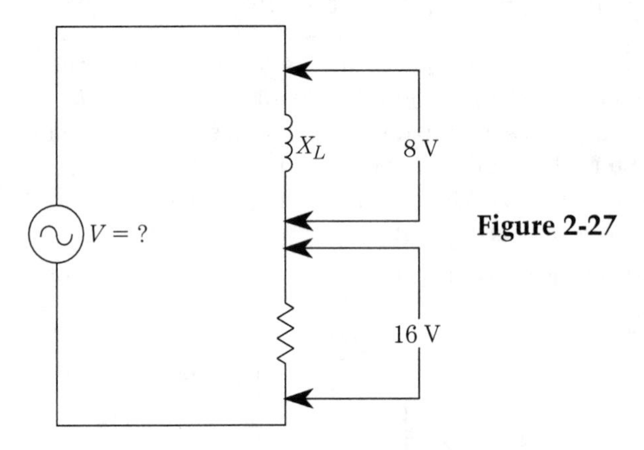

Figure 2-27

Block 6

The correct answer is 17.89 V. Remember that the current is the same in all parts of a series circuit. Resistance and reactance are in quadrature, so the voltages across R and X_C are in quadrature. The same thing is true for R and X_L in series.

The applied voltage is calculated by the same type of equation as used for impedance:

$$V = \sqrt{16^2 + 8^2} = 17.89 \text{ V (More about this in the next chapter.)}$$

Here is your next question: Which of the following materials has the lower reluctance?

A. iron. B. mica.

Block 7

The correct answer is A. Remember this mnemonic: *S*outhern *C*alifornia's *G*one *A*muck. The order of conductivity is: *S*ilver, *C*opper, *G*old, *A*luminum.

Here is your next question: The switch in the circuit of Fig. 2-28 is closed for 20 minutes. The energy is:

A. 269 watt-seconds. C. both choices are correct.
B. 269 joules. D. neither choice is correct.

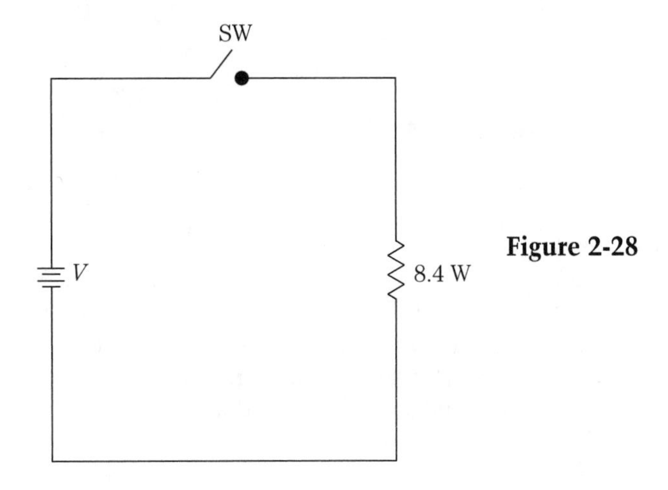

Figure 2-28

Block 8

The correct answer is D. Twenty minutes is 1200 seconds. The energy in watt-seconds equals watts multiplied by seconds. That gives 10,080 watt-seconds. (One joule = one watt-second.) So, choice D is correct.

Here is your next question: Two capacitors are connected in parallel. Their capacitance values are 5 μF and 2 μF. What is the capacitance of the parallel combination?

Block 9

The correct answer is: $5\ \mu F + 2\ \mu F = 7\ \mu F$

Here is your next question: An uncharged capacitor is connected to a dc source (V) as shown in Fig. 2-29. What is the value of V_x?

Block 10

The correct answer is 100 V. If the capacitor is uncharged there is no voltage across it. Therefore, V_x is equal to V.

Here is your next question: What is the term for energy that is stored in an electromagnetic or electrostatic field?

A. Potential energy. B. Amperes-joules. C. Joules-coulombs. D. Kinetic energy.

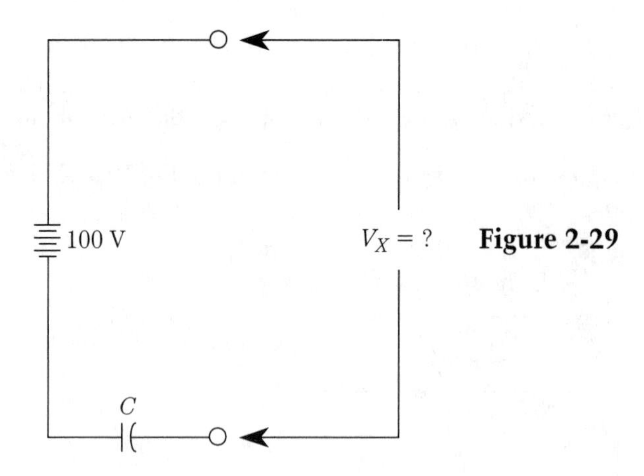

Figure 2-29

Block 11

The correct answer is A.

Here is your next question: How can ferrite beads be used to suppress ignition noise?

A. Install them in the resistive high-voltage cable every 2 years.
B. Install them between the starter solenoid and the starter motor.
C. Install them in the primary and secondary ignition leads.
D. Install them in the antenna lead to the radio.

Block 12

The correct answer is C.

Here is your next question: What is a magnetic field?

A. Current flow through space around a permanent magnet.
B. A force set up when current flows through a conductor.
C. The force between the plates of a charged capacitor.
D. The force that drives current through a resistor.

Block 13

The correct answer is B.

Here is your next question: In what direction is the magnetic field about a conductor when current is flowing?

A. In the same direction as the current.
B. In a direction opposite to the current flow.
C. In all directions; omnidirectional.
D. In a direction determined by the left-hand rule.

Block 14

The correct answer is D.

Here is your next question: What device is used to store electrical energy in an electrostatic field?

A. A battery. B. A transformer. C. A capacitor. D. An inductor.

Block 15

The correct answer is C.

Here is your next question: What is the term used to express the amount of electrical energy stored in an electrostatic field?

A. Coulombs. B. Joules. C. Watts. D. Volts.

Block 16

The correct answer is B.

Here is your next question: What factors determine the capacitance of a capacitor?

A. Area of the plates, voltage on the plates, and distance between the plates.
B. Area of the plates, distance between the plates, and the dielectric constant of the material between the plates.
C. Area of the plates, voltage on the plates, and the dielectric constant of the material between the plates.
D. Area of the plates, amount of charge on the plates, and the dielectric constant of the material between the plates.

Block 17

The correct answer is B.

Here is your next question: What is the dielectric constant for air?

A. Approximately 1. B. Approximately 2. C. Approximately 4. D. Approximately 0.

Block 18

The correct answer is A.

Here is your next question: What determines the strength of the magnetic field around a conductor?

A. The resistance divided by the current.
B. The ratio of the current to the resistance.
C. The diameter of the conductor.
D. The amount of current.

Block 19

The correct answer is D.

Here is your next question: What is the photoconductive effect?

A. The conversion of photon energy to electromotive energy.
B. The increased conductivity of an illuminated semiconductor junction.
C. The conversion of electromotive energy to photon energy.
D. The decreased conductivity of an illuminated semiconductor junction.

Block 20

The correct answer is B.

Here is your next question: What happens to photoconductive material when light shines on it?

A. The conductivity of the material increases.
B. The conductivity of the material decreases.
C. The conductivity of the material stays the same.
D. The conductivity of the material becomes temperature dependent.

Block 21

The correct answer is A.

Here is your next question: What happens to the resistance of a photoconductive material when light shines on it?

A. It increases.
B. It becomes temperature dependent.
C. It stays the same.
D. It decreases.

Block 22

The correct answer is D.

Here is your next question: What happens to the conductivity of a semiconductor junction when it is illuminated?

A. It stays the same.
B. It becomes temperature dependent.
C. It increases.
D. It decreases.

Block 23

The correct answer is C.

Here is your next question: What is the meaning of the term time constant of an RC circuit?

A. The time required to charge the capacitor in the circuit to 36.8% of the supply voltage.
B. The time required to charge the capacitor in the circuit to 36.8% of the supply current.
C. The time required to charge the capacitor in the circuit to 63.2% of the supply current.
D. The time required to charge the capacitor in the circuit to 63.2% of the supply voltage.

Block 24

The correct answer is D.

Here is your next question: What is the meaning of the term time constant of an RL circuit?

A. The time required for the current in the circuit to build up to 36.8% of the maximum value.
B. The time required for the voltage in the circuit to build up to 63.2% of the maximum value.
C. The time required for the current in the circuit to build up to 63.2% of the maximum value.
D. The time required for the voltage in the circuit to build up to 36.8% of the maximum value.

Block 25

The correct answer is C.

Here is your next question: What is the term for the time required for the capacitor in an RC circuit to be charged to 63.2% of the supply voltage?

A. An exponential rate of one.
B. One time constant.
C. One exponential period.
D. A time factor of one.

Block 26

The correct answer is B.

Here is your next question: What is the term for the time required for the current in an RL circuit to build up to 63.2% of the maximum value?

A. One time constant.
B. An exponential period of one.
C. A time factor of one.
D. One exponential rate.

Block 27

The correct answer is A.

Here is your next question: What is the term for the time it takes for a charged capacitor in an RC circuit to discharge to 36.8% of its initial value of stored charge?

A. One discharge period.
B. An exponential discharge rate of one.
C. A discharge factor of one.
D. One time constant.

Block 28

The correct answer is D.

Here is your next question: What is meant by back EMF?

A. A current equal to the applied EMF.
B. An opposing EMF equal to R time C (RC) percent of the applied EMF.
C. A current that opposes the applied EMF.
D. A voltage that opposes the applied EMF.

Block 29

The correct answer is D.

Here is your next question: After two time constants, the capacitor in an RC circuit is charged to what percentage of the supply voltage?

A. 36.8%. B. 63.2%. C. 86.5%. D. 95%.

Block 30

The correct answer is C.

Here is your next question: After two time constants, the capacitor in an RC circuit is discharged to what percentage of the starting voltage?

A. 86.5%. B. 63.2%. C. 36.8%. D. 13. 5%.

Block 31

The correct answer is D.

Here is your next question: What is the time constant of a circuit having a 100-μF capacitor in series with a 470-kΩ resistor?.

A. 4700 seconds. B. 470 seconds. C. 47 seconds. D. 0.47 second.

Block 32

The correct answer is C.

Here is your next question: What is the time constant of a circuit having a 220-μF capacitor in parallel with a 1-MΩ resistor?

A. 220 seconds. B. 22 seconds. C. 2.2 seconds. D. 0.22 second.

Block 33

The correct answer is A.

Here is your next question: What is the time constant of a circuit having two 100-μF capacitors and two 470-kΩ resistors all in series?

A. 470 seconds. B. 47 seconds. C. 4.7 seconds. D. 0.47 second.

Block 34

The correct answer is B.

Here is your next question: What is the time constant of a circuit having two 100-μF capacitors and two 470-kΩ resistors all in parallel?

A. 470 seconds. B. 47 seconds. C. 4.7 seconds. D. 0.47 second.

Block 35

The correct answer is B.

Here is your next question: What is the time constant of a circuit having two 220-μF capacitors and two 1-MΩ resistors all in series?

A. 55 seconds. B. 110 seconds. C. 220 seconds. D. 440 seconds.

Block 36

The correct answer is C.

Here is your next question: What is the time constant of a circuit having two 220-µF capacitors and two 1-MΩ resistors all in parallel?

A. 22 seconds. B. 44 seconds. C. 220 seconds. D. 440 seconds.

Block 37

The correct answer is C.

Here is your next question: What is the time constant of a circuit having one 100-µF capacitor, one 220-µF capacitor, one 470-kΩ resistor and one 1-MΩ resistor all in series?

A. 68.8 seconds. B. 101.1 seconds. C. 220.0 seconds. D. 470.0 seconds.

Block 38

The correct answer is B.

Here is your next question: What is the time constant of a circuit having a 470-µF capacitor and a 1-MΩ resistor in parallel?

A. 0.47 second. B. 47 seconds. C. 220 seconds. D. 470 seconds.

Block 39

The correct answer is D.

Here is your next question: What is the time constant of a circuit having a 470-µF capacitor in series with a 470-kΩ resistor?

A. 221 seconds. B. 221,000 seconds. C. 470 seconds. D. 470,000 seconds.

Block 40

The correct answer is A.

Here is your next question: What is the time constant of a circuit having a 220-µF capacitor in series with a 470-kΩ resistor?

A. 103 seconds. B. 220 seconds. C. 470 seconds. D. 470,000 seconds.

Block 41

The correct answer is A.

Here is your next question: How long does it take for an initial charge of 20 Vdc to decrease to 7.36 Vdc in a 0.01-µF capacitor when a 2-MΩ resistor is connected across it?

A. 12.64 seconds. B. 0.02 second. C. 1 second. D. 7.98 seconds.

Block 42

The correct answer is B.

Here is your next question: How long does it take for an initial charge of 20 Vdc to decrease to 2.71 Vdc in a 0.01-μF capacitor when a 2-MΩ resistor is connected across it?

A. 0.04 second. B. 0.02 second. C. 7.36 seconds. D. 12.64 seconds.

Block 43

The correct answer is A.

Here is your next question: How long does it take for an initial charge of 20 Vdc to decrease to 1 Vdc in a 0.01-μF capacitor when a 2-MΩ resistor is connected across it?

A. 0.01 second. B. 0.02 second. C. 0.04 second. D. 0.06 second.

Block 44

The correct answer is D.

Here is your next question: How long does it take for an initial charge of 20 Vdc to decrease to 0.37 Vdc in a 0.01-μF capacitor when a 2-MΩ resistor is connected across it?

A. 0.08 second. B. 0.6 second. C. 0.4 second. D. 0.2 second.

Block 45

The correct answer is A.

Here is your next question: How long does it take for an initial charge of 20 Vdc to decrease to 0.13 Vdc in a 0.01-μF capacitor when a 2-MΩ resistor is connected across it?

A. 0.06 second. B. 0.08 second. C. 0.1 second. D. 1.2 second.

Block 46

The correct answer is C.

Here is your next question: How long does it take for an initial charge of 800 Vdc to decrease to 294 Vdc in a 450-μF capacitor when a 1-MΩ resistor is connected across it?

A. 80 seconds. B. 294 seconds. C. 368 seconds. D. 450 seconds.

Block 47

The correct answer is D.

Here is your next question: How long does it take for an initial charge of 800 Vdc to decrease to 108 Vdc in a 450-μF capacitor when a 1-MΩ resistor is connected across it?

A. 225 seconds. B. 294 seconds. C. 450 seconds. D. 900 seconds.

Block 48

The correct answer is D.

Here is your next question: How long does it take for an initial charge of 800 Vdc to decrease to 39.9 Vdc in a 450-μF capacitor when a 1-MΩ resistor is connected across it?

A. 1,350 seconds. B. 900 seconds. C. 450 seconds. D. 225 seconds.

Block 49

The correct answer is A.

Here is your next question: How long does it take for an initial charge of 800 Vdc to decrease to 40.2 Vdc in a 450-μF capacitor when a 1-MΩ resistor is connected across it?

A. Approximately 225 seconds.
B. Approximately 450 seconds.
C. Approximately 900 seconds.
D. Approximately 1,350 seconds.

Block 50

The correct answer is D.

Here is your next question: How long does it take for an initial charge of 800 Vdc to decrease to 14.8 Vdc in a 450-μF capacitor when a 1-MΩ resistor is connected across it?

A. Approximately 900 seconds.
B. Approximately 1,350 seconds.
C. Approximately 1,804 seconds.
D. Approximately 2,000 seconds.

Block 51

The correct answer is C.

Here is your next question: When an emergency transmitter uses 325 watts and a receiver uses 50 watts, how many hours can a 12.6 V, 55 ampere-hour battery supply full power to both units?

A. 6 hours. B. 3 hours. C. 1.8 hours. D. 1.2 hours.

Block 52

The correct answer is C.

Here is your next question: A 12.6 V, 8 ampere-hour battery is supplying power to a receiver which uses 50 watts and a radar system that uses 300 watts. How long will the battery last?

A. 100.8 hours. B. 27.7 hours. C. 1 hour. D. 17 minutes or 0.3 hours.

Block 53

The correct answer is D. This type of question is based upon the relationship:

$$\text{Ampere} - \text{hours} = \text{ampere} \times \text{hours}$$

Therefore:

$$\text{Hours} = \text{ampere} - \text{hours} \div \text{amperes}$$

Remember, also, that $P = V \times 1$. Therefore:

$$I = \frac{P}{V} = \frac{P}{12.6}$$

The total current is:

$$I_T = \frac{50}{12.6} + \frac{300}{12.6} = 27.8 \text{ amperes}$$

So:

$$Hours = \frac{8 \text{ ampere hours}}{27.8 \text{ amperes}}$$

$$= 0.288 \text{ hours } (0.3 \text{ hours})$$

$$= 0.288 \times 60 = 17^{+} \text{ sec.}$$

Here is your next question: How long will a 12.6 V, 50 ampere-hour battery last if it supplies power to an emergency transmitter rated at 531 watts of plate input power and other emergency equipment with a combined power rating of 530 watts?

A. 6 hours. B. 4 hours. C. 1 hour. D. 35 minutes.

Block 54

The correct answer is D.

Here is your next question: A 6-V battery with 1.2 ohms internal resistance is connected across two 3 watt bulbs. What is the current flow?

A. 0.57 amp. B. 0.83 amp. C. 1.0 amps. D. 6.0 amps.

Block 55

The correct answer is B.

Quiz

You will probably find tests that require fill-in answers are more difficult than multiple-choice types, because the fill-in types require recall without the hints you get from the multiple choice types.

There is another important reason for reviewing with this type of quiz: lucky guesses will not mislead you into believing you know a subject that you probably should be reviewing.

To broaden the spectrum of subject matter covered, some of the answers are discussed.

1. According to a well-accepted model, a capacitor is a component that opposes any change in:

   A. voltage across its terminals. B. current through it.

2. When one-sixth of an ampere flows through 12 turns of wire the MMF is ________.

3. The wire lead to an amplifier passes through a small donut-shaped device. It is called a/an ______________ and it does the same job as a/an _________.

4. A variable resistor used to control current is called a ______________.

5. The magnetic north pole of the earth is located at the geographic

   A. North Pole. B. South Pole.

6. How many picofarads are there in one nanofarad?

7. What is the abbreviation for electric current?

8. Each cell in the combination shown in Fig. 2-30 has a voltage of 1.5 V. What is the terminal voltage?

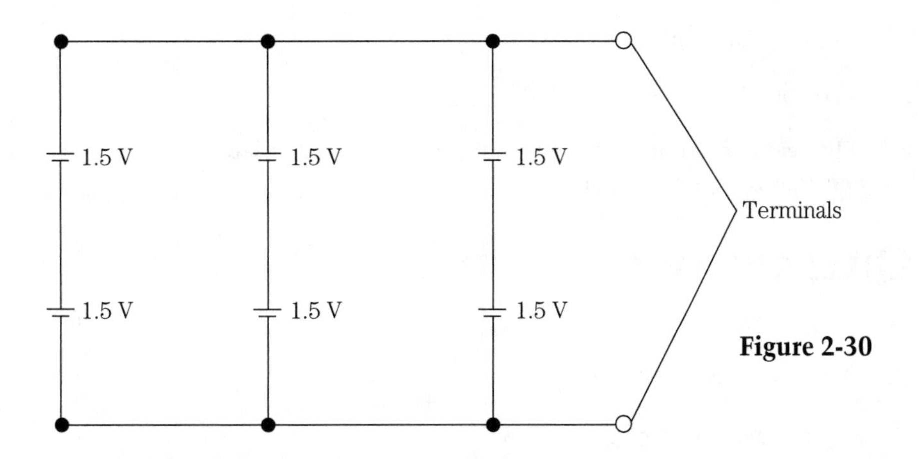

Figure 2-30

9. What is the capacitance of a 10-μF capacitor in series with a 5-μF capacitor?

10. What is the color code of a 10-Ω resistor if its tolerance is ±2%?

11. If a fuse is to operate properly its resistance value must be:

    A. zero ohms. B. very low. C. very high.

12. What are the three parameters that determine the capacitance of a capacitor?

13. How much resistance must be added in series with a 150 mH coil to get a time constant of 175 microseconds?

14. An inductor opposes any change in:

    A. voltage across it. B. current through it.

15. The inductance of a coil is 175 mH. What is its reactance in a 1000 hertz circuit?

16. If you charge a capacitor, then move the plates further apart, the voltage across the capacitor will:

    A. not change. B. increase. C. decrease.

17. The dielectric constant of a vacuum is ____________.

18. To increase the capacitance of a trimmer capacitor, adjust the screw so the plates are:

    A. closer together. B. farther apart.

19. Where is the charge stored in a capacitor?

20. Where is the energy stored in a capacitor?

21. Which type of material moves away from a magnetic field?

22. Convert 7 pF to μF.

23. A combination of cells is called a ______________.

24. What value of inductance is needed in series with a 1-K resistor to get a time constant of one millisecond?

Quiz answers

1. A
2. $\frac{1}{6} \times 12 = 2$ ampere-turns
3. Ferrite bead/inductor
4. Rheostat
5. B
6. 1000
7. 1 (for Intensity)
8. 3 V
9. $(10 \times 5)/(10 + 5) = 3.33$ microfarads
10. Brown, Black, Black, Red
11. B
12. Area of plates
    Distance between plates
    Type of dielectric
13. $T = L/R$

    $R = L/T = \dfrac{150 \times 10^{-3}}{175 \times 10^{-6}}$

    $=$ (about) 857 ohms

14. B
15. $X_L = 2\pi fL$
    $= 2 \times 3.14 \times 1000 \times 172 \times 10^{-3}$
    $X_L = 1080$ ohms (answer)
16. The correct answer is C. The capacitance varies inversely as the distance between the plates.
17. The correct answer is 1.0. It is the lowest dielectric constant.
18. A (See answer to number 16.)
19. The *charge* (accumulation of electrons) is stored in the plates. (See answer to number 20.)
20. The energy (capacity to do work) is stored in the dielectric.
21. Diamagnetic
22. You are looking for μF per 7 pF. In other words, you want 1 μF divided by 7 pF.

$$\frac{1\ \mu F}{7\ pF} = \frac{1 \times 10^{-6}}{7 \times 10^{-12}} = 0.143 \times 10^6$$

23. Battery
24. $T = \frac{L}{R}$ so $L = TR$

    $L = 1 \times 10^3 \times 1 \times 10^{-3}$
    $L = 1$ henry (answer)

3 CHAPTER

Passive circuits

Passive circuit components were reviewed in chapters 1 and 2. This chapter reviews some series and parallel circuits that use those components. As a technician, you are expected to know how to solve problems involving:

- series, parallel, or series-parallel resistance (R)
- series, parallel, or series-parallel inductance (L)
- series, parallel, or series-parallel capacitance (C)

Also, you are expected to know how to solve problems involving both dc and ac series and parallel RLC circuits. Calculation of impedance and resonant frequencies in ac circuits is especially important if you are planning to take a license or certification test.

Table 3-1 reviews some methods of evaluating combinations. This chapter deals with the more advanced methods. Remember, the questions in this book and in the companion book, *The Complete TAB Book of Practice Tests for Communications Licenses*, give extensive practice in working the required problems.

Complex numbers

The mathematics involved with use of RLC in ac circuits requires the use of *complex numbers.* A *complex number* is a number that is made by combining two parts: *real* and *imaginary*. It can be represented graphically two different ways. One way is to use an *argand diagram* and the other is to use *polar coordinates*. Both methods are illustrated in Fig. 3-1.

When the complex number is represented on argand diagram, the "real" part is plotted on the horizontal axis and the "imaginary" part is plotted on the vertical axis. The complex number represented by the graph is in the form $R \pm jX$ where X is inductive reactance ($+jX_L$), or capacitive reactance ($-jX_C$), or the difference between inductive and capacitive reactance ($jX_L - jX_C$). This is usually called the *rectangular form* of the complex number of complex impedance.

Table 3-1

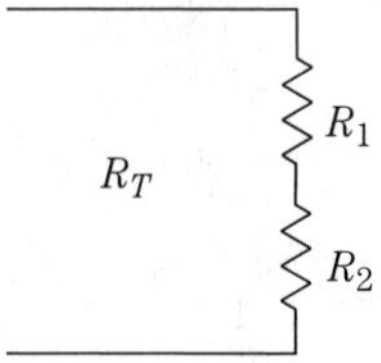

R_T (series R_1, R_2)	$R_T = R_1 + R_2$	X_L (L)	$X_L = 2\pi fL$
L_T (series L', L'')	$L_T = L' + L''$ (No coupling) Total $X_L = X_{L'} + X_{L''}$		
X_C (C)	$X_C = \dfrac{1}{2\pi fc}$		
C_{eq} (series C', C'')	$C_{eq} = \dfrac{1}{\dfrac{1}{C'} + \dfrac{1}{C''}}$ $X_C = X_{C'} + X_{C''}$		For two capacitors only $C_{eq} = \dfrac{C_1 \; C_2}{C_1 + C_2}$
Parallel R_1, R_2	$R_{eq} = \dfrac{R_1 R_2}{R_1 + R_2}$ Two resistors only $R_{eq} = \dfrac{1}{\dfrac{1}{R_1} + \dfrac{1}{R_2} + \dfrac{1}{R_3} \cdots + \dfrac{1}{R_n}}$ Any number of resistors		
Parallel L', L''	$L_{eq} = \dfrac{L'L''}{L' + L''}$ No coupling—two inductors only $L_{eq} = \dfrac{1}{\dfrac{1}{L_1} + \dfrac{1}{L_2} + \dfrac{1}{L_3} \cdots + \dfrac{1}{L_n}}$ Any number of inductors		
Parallel C_1, C_2	$C_T = C_1 + C_2$ $X_C = \dfrac{X'X''}{X' + X''}$ Two capacitors		

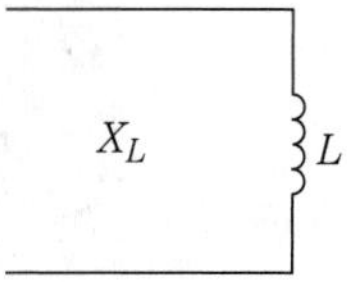

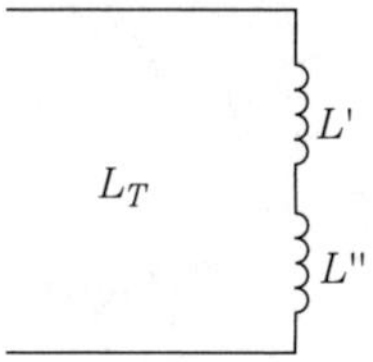

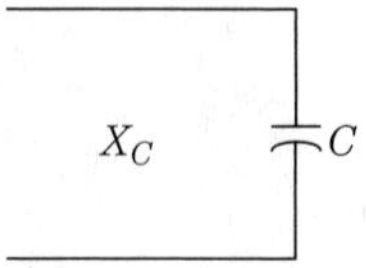

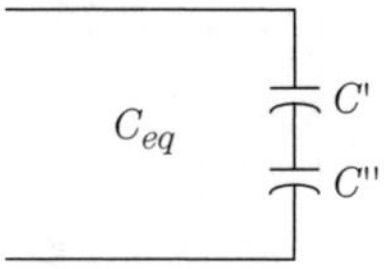

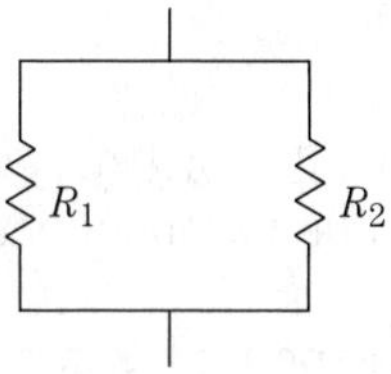

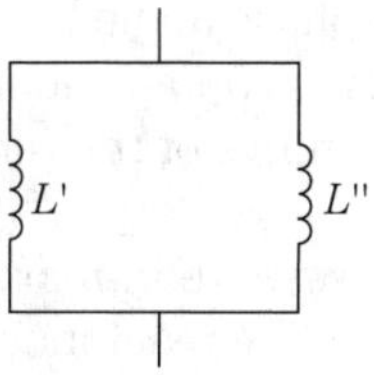

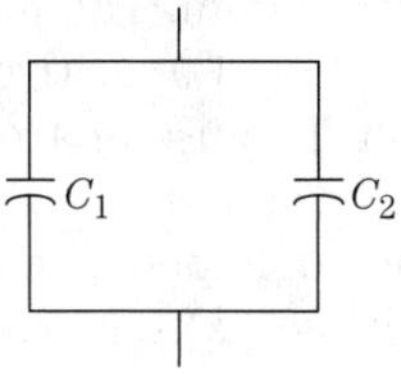

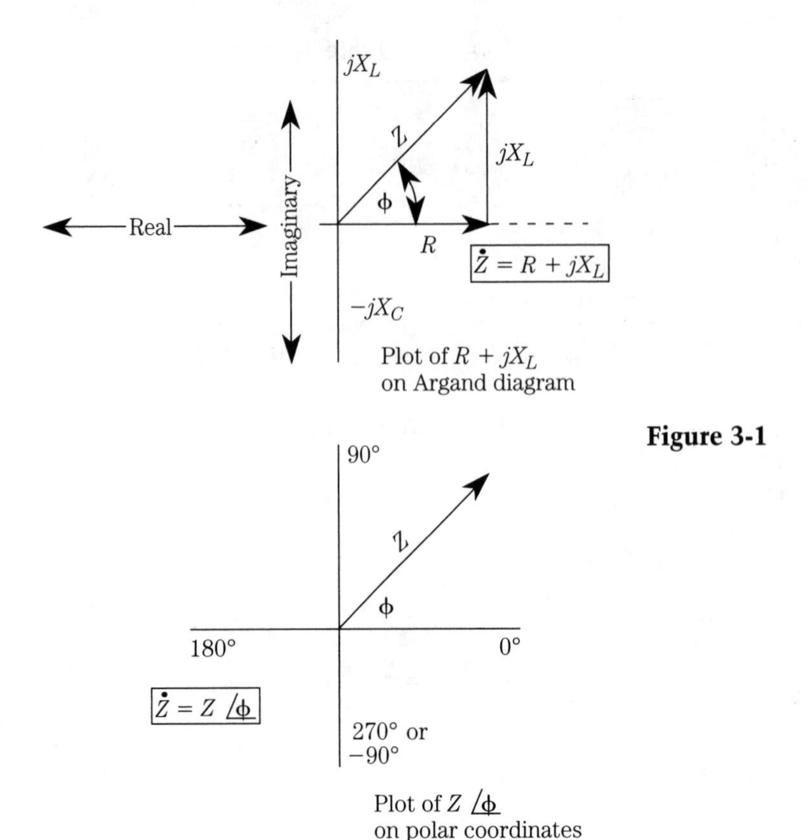

Figure 3-1

The lower value of X is subtracted from the higher value of X to get the resultant value. For example, if $X_L = 20\ \Omega$ and $X_C = 5\ \Omega$, the resultant is an inductive reactance of ($X_L = 15\ \Omega$); in that case X_C, is disregarded.

As shown in Fig. 3-1, complex impedance can also be represented on polar coordinates as $Z \angle \phi$. In that form, ϕ (phi) represents the phase angle between the voltage and current when the impedance is connected across an ac generator. In many cases, θ (theta) is used instead of ϕ. The graphical meanings of the terms are shown on the polar coordinates of Fig. 3-1.

Figure 3-1 shows how series RLC circuits are represented on argand diagrams and polar diagrams. Notice that the inductive and capacitive reactances are drawn at an angle of 90° with the resistance line on an argand diagram. The inductive reactance is drawn vertically (upward) in the positive direction. Capacitive reactance is drawn vertically (downward) in the negative direction. The use of arrows to represent R, X_L, and X_C is traditional.

A distinction must be made between the magnitude of impedance (Z) shown on the graphs and the complex representation of impedance ($\dot{z}$) enclosed in a box in Fig. 3-1. The magnitude of impedance gives no information about the phase angle between the voltage and current in an ac circuit. The rectangular and polar forms, represented by $\dot{z}$, give information about both the magnitude and phase angle of the impedance (Z). That is done directly with the polar form and indirectly with the rectangular form.

The meaning of *j*-operators

A short review of the mathematics of complex numbers will be helpful for working with R-L, R-C, and RLC in circuits with an applied ac voltage.

In mathematics books, complex numbers are worked with the symbol i, the symbol for *imaginaries*. It is a very unfortunate term because it leads one to believe there is no practical use for the subject.

In post-high school technical electronics studies, the i symbol is eliminated in favor of the symbol j. That is probably because i represents instantaneous current in an ac circuit. The study of *j-operators* is really the subject involved with here.

The letter j is not an abbreviation. It is an operator.

An *operator* in mathematics is a symbol that tells you what to do. For example:

$$77 + 6 = \qquad 228 \times 1.6 =$$

$$53 - 12 = \qquad 909 \div 33 =$$

These problems utilize the following operators: +, –, ×, and ÷. An operator tells you what to do with the next number. That is exactly what the j-operator does. It tells you to make a 90° left turn ($+j$) or a 90° right turn ($-j$) when you get to the next term.

Working with complex forms of impedance

You might be more used to calling the argand diagram an X-Y graph or an X-Y plot. Those expressions are usually used in technical work. However, *argand* is the correct mathematical name for the rectangular coordinates shown in Fig. 3-2.

Observe that the coordinates are divided into four quarters (called *quadrants*) that are numbered with Roman numerals. When working with these graphs in electricity and electronics most of the work is done in the first and fourth quadrants. When working a problem in ac, if you get an answer in the second or third quadrant, you will usually need to change it into the first or fourth quadrants. If you don't do that, the answer might be meaningless.

The arrow marked "+1" in the first diagram of Fig. 3-2 represents the magnitude (amount or value it has) and its angle (0°). It is a *phasor* because phasors represent both magnitude and phase angle. A phasor is not the same as a vector. A vector tells the magnitude and direction of a quantity. However, many of the techniques used for working with vectors are useful for dealing with phasors.

In the second illustration in Fig. 3-2, the arrow has been rotated 180° from its original position. That changes its magnitude for +1 to –1. It follows that the arrow in the 0° position has been multiplied by –1 to get it to rotate 180°.

An important question is this: what do you have to do to the arrow in the 0° position to get it to rotate 90°? It must be some "number" such that when you multiply the arrow at 90° by that same number, it will go another 90° to the 180° position.

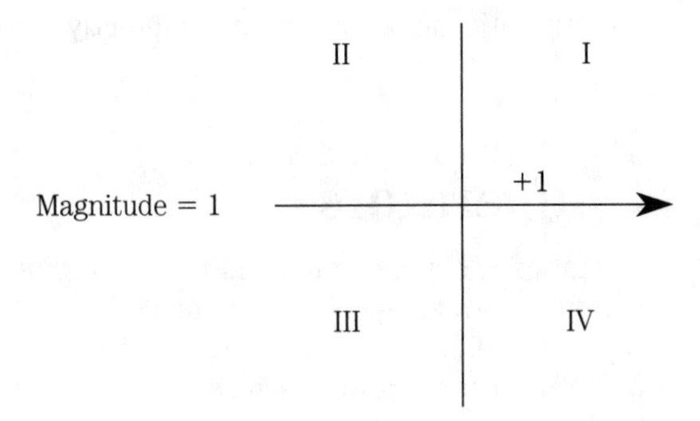

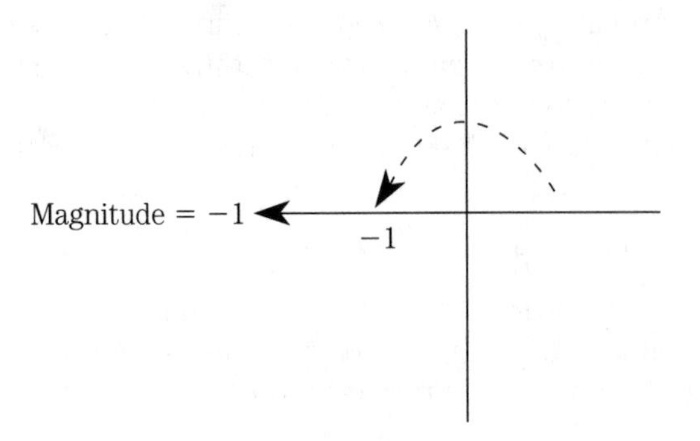

Figure 3-2

Putting it another way, what number when multiplied by itself equals –1? The only logical answer is $\sqrt{-1}$ because $\sqrt{-1} \times \sqrt{-1} = -1$. The symbol $\sqrt{-1}$ is called the j operator. On a rectangular diagram a positive j tells you to rotate the next number (which represents a phasor) an angle of 90° counterclockwise. A negative j tells you to rotate the next number (which is a phasor) clockwise 90°. In the United States, the standard direction of rotation is counterclockwise.

A value of $\sqrt{-1}$ is assigned to j even though it is an operator. The reason is that when phasors are manipulated with math operations, the value of j makes it possible to use basic mathematics. That will be demonstrated in the next section.

The following values of j are used for the horizontal and vertical lines on an argand diagram:

$$j = \sqrt{-1}$$
$$j^2 = j \times j = \sqrt{-1} \times \sqrt{-1} = -1$$
$$j^3 = j^2 \times j = -1 \times j = -j$$
$$j^4 = j^2 \times j^2 = -1 \times -1 = +1$$

Return again to Fig. 3-1. Note that resistance is represented by a phasor in the standard position. In the standard position, the phase angle is 0°. Inductive reactance is represented by a phasor in the $+j$ position and capacitive reactance is represented by a phasor in the $-j$ position.

Numerical examples of addition, subtraction, multiplication and division of complex numbers are given in Table 3-2.

Table 3-2

Addition and subtraction of complex numbers

(a) $\begin{array}{r} 3 + j6 \\ + 4 + j2 \\ \hline 7 + j8 \end{array}$ (b) $\begin{array}{r} 170 + j184 \\ + 35 - j\ 16 \\ \hline 205 + j168 \end{array}$

(c) $\begin{array}{r} 77 + j58 \\ - 17 + j\ 2 \\ \hline 60 + j56 \end{array}$ Change sign and add

(d) $(43 + j60) - (-j7 - 20) = 63 + j67$

Note: You cannot add or subtract numbers when they are in polar form. They must be converted to rectangular form first.

Example:

(e) $3\angle 30° + 8\angle -30° = (25 + j15) + (6.93 - j4)$
$= 31.93 + j11$

Multiplication in rectangular form

$3 + j7$
$5 - j1$ → Means obtained from
$-j3 - j^2 7$ → $[(-j1)(3 + j7)]$
$15 + j35$ → $[(5)(3 + j7)]$
$15 + j32 - j^2 7$ → $[(-j3 - j^2 7) + (15 + j32)]$
Substitute (–1) for j^2
$(-1)^2 = [-(-1)^2] = +1$
$15 + j32 + 7 = 22 + j32$

Multiplication in polar form

$3 + j7 = 7.62\ \angle 66.8°$
$5 - j = 5.1\ \angle -11.3°$
$(7.62\ \angle 66.8°)\ (5.1\ \angle -11.3) = \angle 55.5°$
Answer obtained by:
(66.8) + (–11.3) to get 55.5°

Division in complex form

$(3 + j7) \div (2 - j6)$

$\frac{3 + j7}{2 - j6} \times \frac{2 + j6}{2 + j6}$ = Multiply numerator and denominator by conjugate of denominator

$= \frac{-36 + j32}{4 + 36} = \frac{-36 + j32}{40}$
$= -0.9 + j0.8$
$= 1.2\ \angle -41.6°$

Division in polar form

Convert above rectangular values to polar form and work the same problem

$3 + j7 = 7.62\ \angle 66.8°$
$2 - j6 = 6.32\ \angle -71.6°$

$\frac{7.62\angle 66.8°}{6.32\ \angle -71.6} = \frac{7.62}{6.32}\ \angle 66.8-(-71.6) = 1.21\ \angle +138.4°$*

Table 3-2 continued.

When this answer is converted to rectangular form:

$1.21 \text{ Cos } 138.4° + j\, 1.21 \text{ Sin } 138.4°$

$= -0.9 + j\, 0.8 = 1.2 \angle -41.6°$

Same answer as for division in rectangular form:

$\tan(+138.4) = \tan - 41.6$

$-0.8878415 = -0.8878415$

$\tan^{-1} 0.8878415 = -41.6°$

Impedance triangles

The standard triangle shown in Fig. 3-3 is used to establish basic relationships in trigonometry. For our immediate purposes:

$$\text{Pythagorean Theorem: } a^2 + b^2 = C^2$$

This relationship holds true for all right triangles regardless of how they are lettered. For angle ϕ:

$$\text{Tangent } \phi = b/a$$

The tangent of angle ϕ equals the opposite side divided by the adjacent side.

It will be very useful to know that some triangles that represent combinations of R, L, and C in rectangular diagrams. These triangles are actually representations of impedances in rectangular form. A few examples will make this clear.

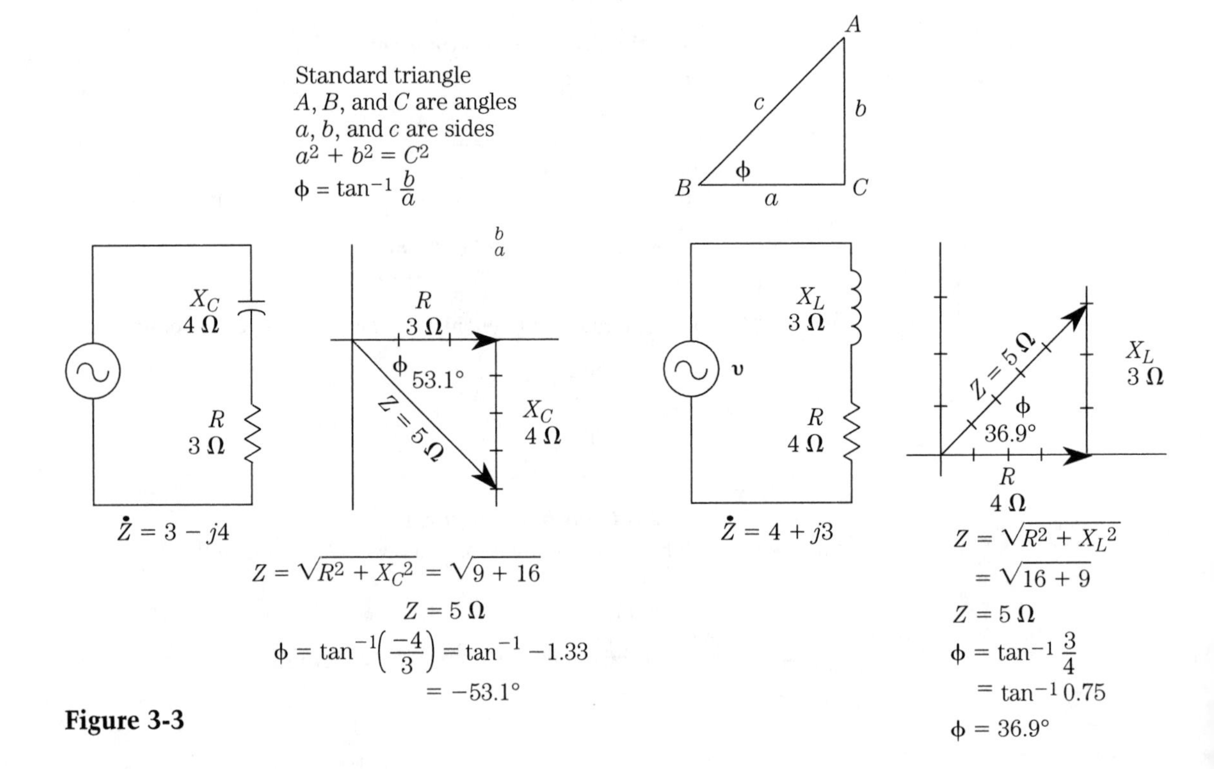

Figure 3-3

Example An inductive reactance is connected in series with a resistor across an ac source (Fig. 3-4A.) What is the impedance of the *RL* combination, and what is the phase angle between the voltage and current in the circuit?

The solution is based upon the impedance triangle shown in the illustration. The phasors are connected head-to-tail to get that triangle. The hypotenuse of the triangle gives the impedance, and the phase angle is the angle between R and Z. This type of problem can be solved graphically.

For the mathematical solution, the impedance is found by applying the Pythagorean equation. It applies to the impedance triangle for the RL circuit shown in Fig. 3-4A.

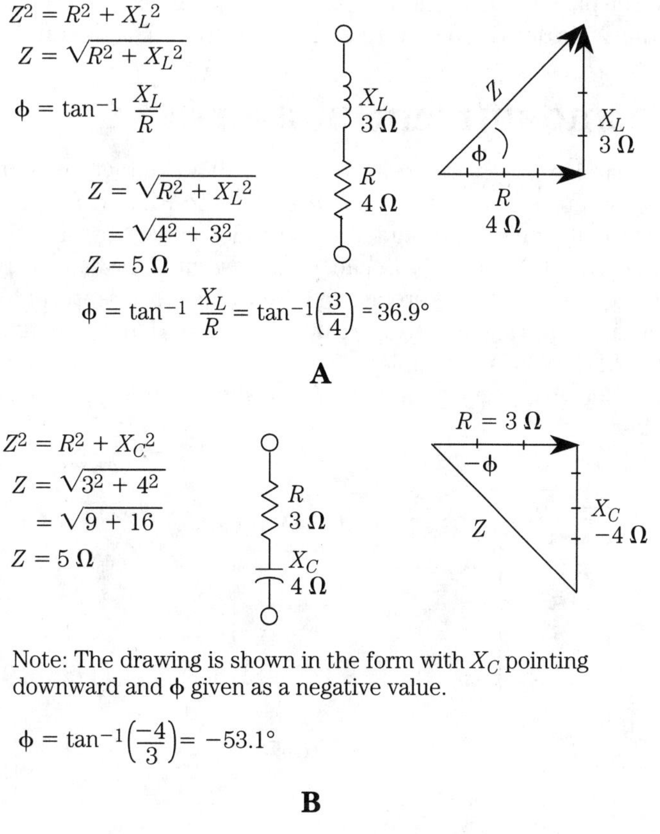

$$Z^2 = R^2 + X_L^2$$
$$Z = \sqrt{R^2 + X_L^2}$$
$$\phi = \tan^{-1}\frac{X_L}{R}$$

X_L 3 Ω
R 4 Ω
Z
ϕ
X_L 3 Ω
R 4 Ω

$$Z = \sqrt{R^2 + X_L^2}$$
$$= \sqrt{4^2 + 3^2}$$
$$Z = 5\ \Omega$$
$$\phi = \tan^{-1}\frac{X_L}{R} = \tan^{-1}\left(\frac{3}{4}\right) = 36.9°$$

A

Figure 3-4

$$Z^2 = R^2 + X_C^2$$
$$Z = \sqrt{3^2 + 4^2}$$
$$= \sqrt{9 + 16}$$
$$Z = 5\ \Omega$$

R 3 Ω
X_C 4 Ω
$R = 3\ \Omega$
$-\phi$
Z
X_C $-4\ \Omega$

Note: The drawing is shown in the form with X_C pointing downward and ϕ given as a negative value.

$$\phi = \tan^{-1}\left(\frac{-4}{3}\right) = -53.1°$$

B

By comparison to that equation the impedance in ohms is obtained as follows:

$$Z^2 = R^2 + X_L^2$$
$$Z^2 = 4^2 + 3^2$$
$$Z = \sqrt{16 + 9} = 5\ \Omega$$

Remember that this equation gives you Z, not $\dot{z}$. The solution using the equation is shown in the illustration.

The phase angle (ϕ) is calculated from the basic trigonometry relationship given for the standard triangle. It has been solved to get the *arc tangent*, which is sometimes written $\tan^{-1}$. On a calculator, it is usually obtained by punching INV TAN.

$$\phi = \tan^{-1}(X_L/R)$$

Read that equation as follows: "ϕ is the angle that has a tangent equal to X_L/R." The procedure on a calculator is to divide the value of X_L by the value of R and then punch the *INV*erse *TANgent* to get ϕ.

Example Find the impedance (Z) of the RC combination shown in Fig. 3-4B. Also, find the phase angle between the voltage and current in the circuit. This problem is solved mathematically in Fig. 3-4.

Note that the phasor representing X_L is drawn upward in the $+j$ position. The phasor representing X_C is drawn downward in the $-j$ position. That is standard procedure.

Voltage and current phasors

Do not get careless when answering questions about voltage and current in ac circuits. Voltage and current are in *quadrature* for both RL and RC circuits; in other words, they are at an angle of 90° and must be added as phasors in the same manner as X_L and X_C in RL and RC circuits. Because the current is the same through all parts of a series circuit, it is used for reference and the voltages are represented by phasors.

A type of question that has tripped up many experienced technicians is illustrated in Fig. 3-5. Here is an example:

Example The voltages across R and C are shown in Fig. 3-5. What is the value of applied voltage?

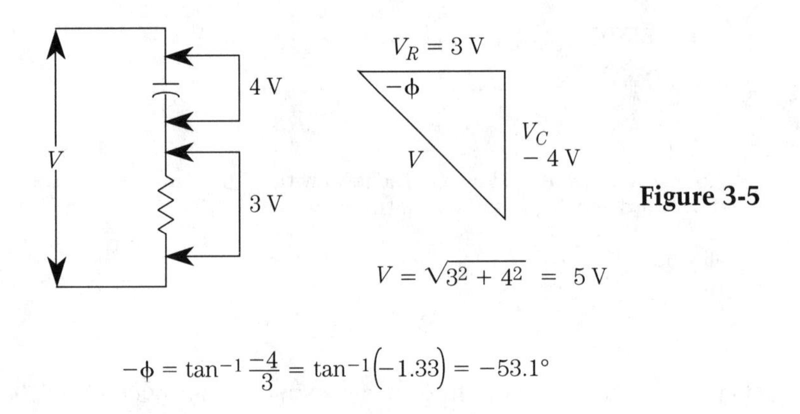

Figure 3-5

Kirchhoff's voltage law for dc circuits says that the algebraic sum of the voltages around any closed loop must be equal to the applied voltage. In an ac circuit, that law is modified to say the phasor sum of the voltages around any closed loop must be equal in magnitude and phase angle of the applied voltage.

The algebraic sum of the voltages in the circuit of Fig. 3-5 would be 7 V. However, that is not the applied voltage! The phasor sum is equal to 5 V. The voltages have to be added in quadrature! Here is the calculation:

$$V^2 = 3^2 + (-4^2)$$
$$V = \sqrt{3^2 + (-4^2)} = 5\text{ V}$$

The phase angle between the generator voltage and generator current is determined by using the ARC TAN the same way as for impedance problems above.

$$\phi = \text{Tan}^{-1}(-4/3) = -53.13°$$

The negative sign in the phase angle solution indicates that the solution is in the fourth quadrant. It also tells you that the voltage is lagging behind the current by an angle of 53.13°. That follows because current is the reference and it is in the same direction as the resistance. Therefore, a current phasor for this series circuit would be in the 0° position.

Example For the circuit in Fig. 3-6, write the following: Z and $\dot{z}$ in polar and rectangular forms. (Give numerical answers.) The solutions are given in Fig. 3-6.

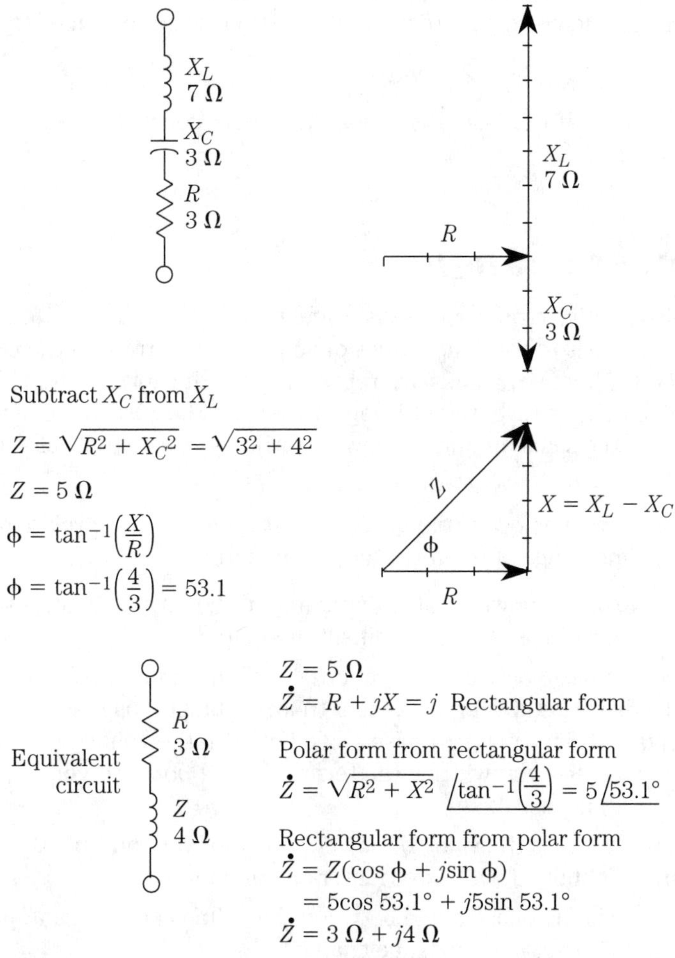

Figure 3-6

Polar to rectangular and rectangular to polar conversions

Figure 3-6 also shows two very important conversions that you should memorize. Conversion of polar to rectangular form and conversion from rectangular form can be used to greatly simplify the solution of some types of problems in complex numbers. The reason is that the polar form cannot be used for addition and subtraction of complex numbers. Both forms can be used for multiplication and division of complex numbers, however multiplication and division of polar forms is easier than for rectangular forms.

Most scientific calculators can be used for direct conversions without resorting to the equations in Fig. 3-6. However, you should know the equations for those conversions.

In the example just given, it was demonstrated that when X_L and X_C are both in a series circuit their values must be a subtracted rectangular form to get the resultant reactance. This leads to a more general equation for series impedance:

$$Z = \sqrt{R^2 + (X_L - X_C)^2}$$

It doesn't matter if $(X_L - X_C)$ is a negative value because it becomes positive when it is squared.

The power triangle

In a purely resistive circuit, power is calculated the same way as for dc circuits. The usual procedure is to use RMS values of voltage and current in the calculations. Remember, RMS (effective) values are represented with capital letters. That is also the way dc values are represented. It is logical because the RMS value of voltage and current produces the same amount of power (or heat) as an equal value of dc.

$$P = V \times I = V^2/R = I^2R$$

Figure 3-7 shows the power triangle that illustrates the power relationships in an RLC circuit. The meanings of the terms are given here:

- *True Power* is the power that is dissipated in the form of heat. If there is no resistance in a circuit, there is no heat dissipated.
- *VARS* stands for *R*eactive *V*olt-*A*mpere*S*. It is the amount of power that would be dissipated by a reactance if that reactance was resistance. Excessive VARS means the phase angle between the voltage is high. Instead of the term VARS, you will see the term *reactive power* if you take the GROL (FCC) license exam.
- *Apparent Power* is the value of power you get if you multiply the voltage and current obtained with a voltmeter and ammeter.
- *The angle* ϕ is the phase angle between the voltage and current. It is the same angle you get in impedance triangles.

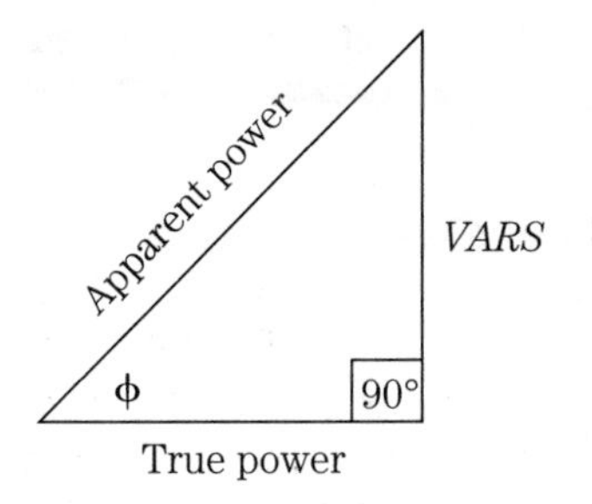

ϕ is the phase angle between voltage and current

Apparent power $= V \times I$

VARS $= V \times I \sin \phi$

True power $= V \times I \times \cos \phi$

Power factor $= \cos \phi$

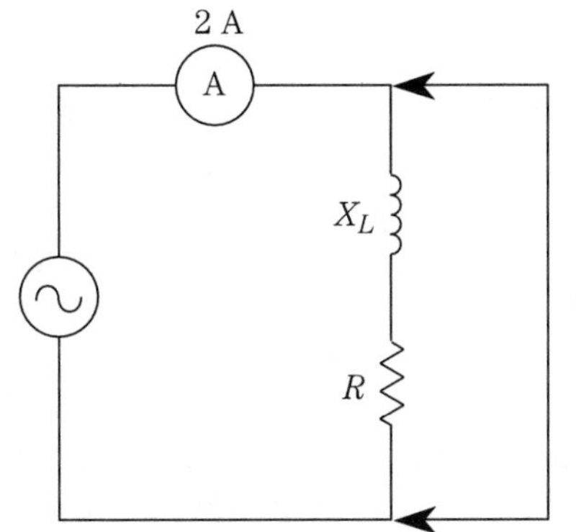

Figure 3-7

$P = V \times I$

$P = 50 \times 2 = 100$ W

The cosine of the phase angle is called the *power factor*. In a right triangle, the cosine of angle ϕ is equal to the adjacent side divided by the hypotenuse. From the power triangle it is obvious that the cosine of the phase angle is equal to the true power divided by the apparent power, sometimes expressed as a percent value.

Power Factor = True Power/Apparent Power

%Power Factor = True Power/Apparent Power × 100

The maximum value of power factor is 1.0, representing the best possible condition as far as the generator is concerned, which means the phase angle is 0°. ($\cos \phi = \cos 0° = 1.0$).

When there is inductive reactance or capacitive reactance in ac circuit the reactive component "borrows" power from the generator during one half cycle and then returns it during the next half cycle. The returned power adds to the power the generator is producing. The overall result is high current and high internal heat in the generator. That condition can cause generator destruction, explaining why power companies levy fines against industries that have a low power factor (that is, a power factor less than 1.0).

Remember this important point: when the circuit contains L or C, you must calculate the true power and apparent power by using the power triangle relationships.

Example A voltmeter measures 50 V across a series RL circuit. An ammeter measures 2 A in the same circuit. If the power factor is 0.80, what is the true power being dissipated in the circuit?

Solution The power factor is given as 0.80 (sometimes expressed as 80%.) From Fig. 3-7:

$$\textit{Apparent Power} = V \times I$$
$$= 50 \times 2 = 100 \text{ volt amperes}$$
$$\text{Power Factor } (P.F.) = \textit{True power/Apparent Power } (V \times I)$$

Multiply both sides of the equation by $(V \times I)$.

$$\textit{True Power} = (P.F.) \times (V \times I)$$
$$\textit{True Power} = 0.80 \times 100 = 80 \text{ W (answer)}$$

Resonance

Resonance is a special condition in RLC circuits in which the following conditions are met:

- The phase angle between the voltage and current is 0°.
- The impedance is either maximum or minimum depending upon whether it is a series or parallel circuit.

Resonance in a series circuit

Resonance in a series RLC circuit occurs when the inductive and capacitive reactances cancel. When that happens there is only resistance in that circuit. If the resistance is very low a series-resonant circuit can draw a destructively high current from a generator.

All of the following conditions are met when a series RLC circuit is in resonance:

- The reactances cancel so that there is only resistance in the circuit.
- The phase angle between the voltage and current is 0°.
- The power factor (which is the cosine of the phase angle) is equal to 1.0 or 100%.
- The voltage across the inductive reactance is equal to the voltage across the capacitive reactance.
- The voltage across the RLC circuit is minimum.
- The impedance of the RLC circuit is minimum.
- The current through the RLC circuit is maximum.

The frequency when the circuit is at resonance is called the *resonant frequency* and it is usually designated f_r.

The graph called the *response curve* is marked "X_L vs. X_C" in Fig. 3-8 is an example of a frequency domain display. This graph shows a plot of X_L and X_C on a frequency domain axis. The dotted line shows the frequency where $X_L = X_C$. Regardless of the values of L and C in a series circuit, there will always be a resonant frequency where that dotted line can be drawn on a similar graph.

The response curve for X_L and X_C and R in Fig. 3-8 shows that the voltage across the circuit and the impedance of the circuit are minimum at the resonant frequency (f_r). It also shows that the current through the circuit is maximum at resonance. You should memorize this graph!

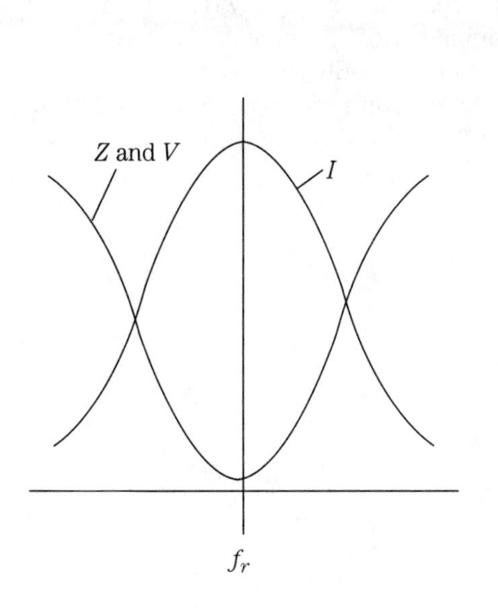

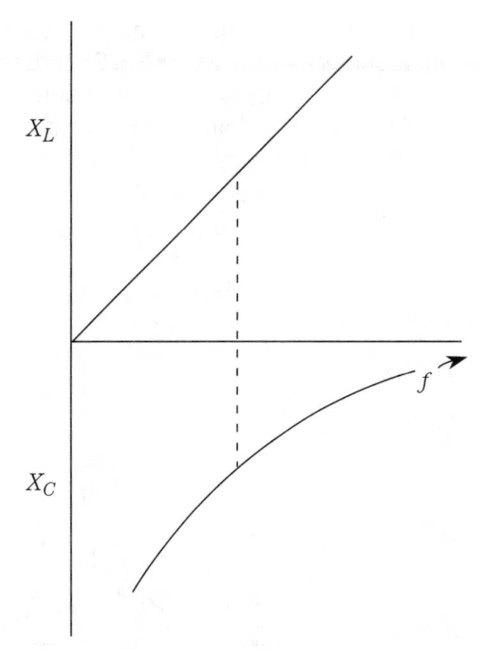

Figure 3-8

Response curve for series RLC circuit

Resonant frequency can be determined mathematically by setting $X_L = X_C$. The detailed solution is shown in Table 3-3. You are expected to be able to understand (not perform) that type of simple algebra problem in most of the certification and license tests. Also, given the values of L and C, you should be able to find the value of f_r using a calculator and the equation just derived.

Table 3-3

$$X_L = X_c$$

but $X_L = 2\ \pi f L$

and $X_c = \dfrac{1}{2\pi f c}$

By substitution:

$$2\pi f L = \frac{1}{2\pi f c}$$

Multiply both sides by f:

$$2\pi f^2 L = \frac{1}{2\pi c}$$

Divide both sides by $2\pi L$:

$$f^2 = \frac{1}{4\pi^2 Lc}$$

Take the square root of both sides L

$$f = \frac{1}{2\pi\sqrt{LC}}$$

There is no R in the equation for the series resonant frequency. That means you can discount the importance of R in determining series-resonance. However, resistance is a very important factor in determining how broadly the circuit tunes.

The frequency domain graph marked "Effect of R on series resonance" in Fig. 3-9 shows how high resistance and low resistance affects the shape of the response curve.

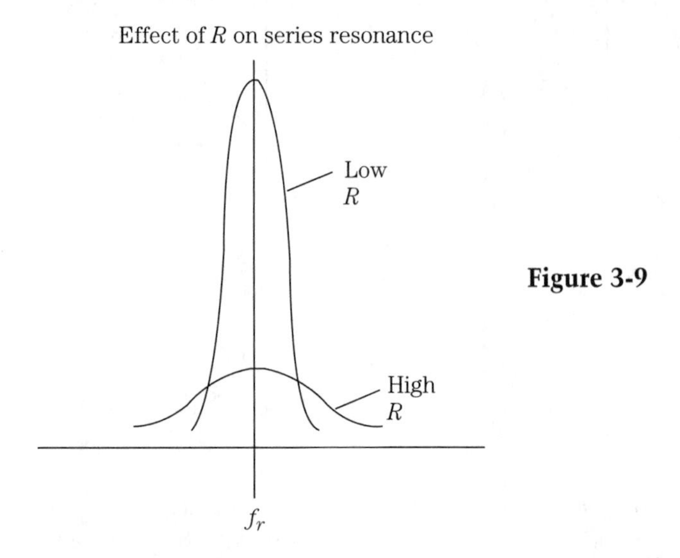

Figure 3-9

It shows that a high resistance causes the circuit to tune over a wider range of frequencies than a low resistance. Those relationships are very important!

Circuit Q

In the early days of radio, experimenters were aware of the effect of resistance on the shape of the frequency response curve. In those days, a sharp resonance curve (obtained with low resistance) was highly desirable.

In a series-tuned LC circuit, most of the resistance is concentrated in the coil. There are two ways to reduce the effect of that resistance: change the L/C ratio or use a larger diameter of wire for the coil.

The *quality factor (Q)* of a series-resonant LC combination was considered to be directly related to how sharply the circuit tuned. Sharp tuning is needed to separate the stations. A method of evaluating coils, called Q, was developed by the early radio enthusiasts. Mathematically:

$$Q = X_L/R = 2\pi f L/R, \text{ also}$$

$$Q = X_C/R \quad \text{This equation is not often used.}$$

The equations are based upon the assumption that R and X are in series. For R and X in parallel, the equations are:

$$Q = R/X_L, \text{ and}$$

$$Q = R/X_C$$

For these equations, it is convenient to think of the Q as a relationship between energy stored and energy lost. The energy lost is in the form of heat dissipated by the resistor.

At first, it might seem to be a simple matter of increasing L to get a higher X_L and therefore a higher Q for an inductor. However, increasing L by adding more turns to the coil leads to an eventual tradeoff because the additional length of wire increases the coil resistance.

Another way of defining Q is to relate it to the bandwidth of the tuned circuit. When you review the curves showing the relationship between low and high circuit resistance it is apparent that greater resistance makes it possible to pass a wider range of frequencies.

The range of frequencies that a tuned circuit will pass is called its *bandpass*. Bandpass (also called bandwidth or passband) is defined in two ways (Fig. 3-10).

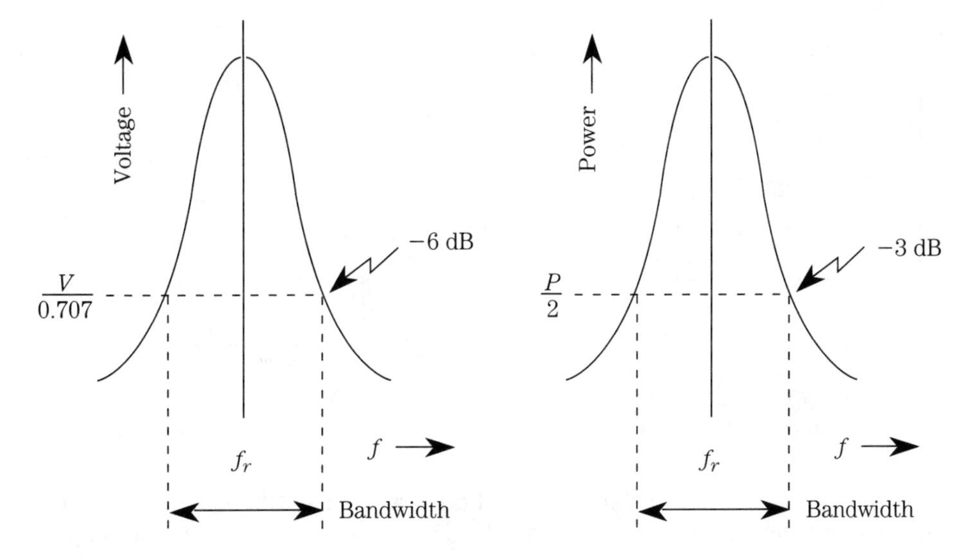

Figure 3-10

If the response curve shows the power vs. frequency, then by definition the bandwidth is the range of frequencies between the half-power points on the power response curve, 3 dB down from the maximum power point.

When the response curve shows the voltage vs. frequency, then, by definition, the bandwidth is the range of frequencies between the points on the voltage response curve that are 6 dB down from the maximum voltage points. The graph shows that the –6-dB points are located where the voltage is: $0.707 \times V_{max}$.

Based upon the information just given, the relationship between Q and bandwidth is:

$$Q = f_r / Bandwidth$$

As mentioned, early experimenters were very much interested in high Q in order to get good selectivity in tuned circuits. Selectivity is a measure of how well a receiver is able to select one station and reject all others.

There are cases in today's technology where a high Q and resulting sharply tuned circuits are definitely not desirable. As an example, there are tuned circuits in television receivers that must pass a 6-MHz range of frequencies. A high-Q series-tuned circuit is unable to do that.

Parallel resonance

Parallel resonance is also known as *antiresonance*. Figure 3-11 is used to explain the conditions for parallel resonance. In the special case where there is only L and C in the parallel circuit, the equation for f_r is the same as for series resonance.

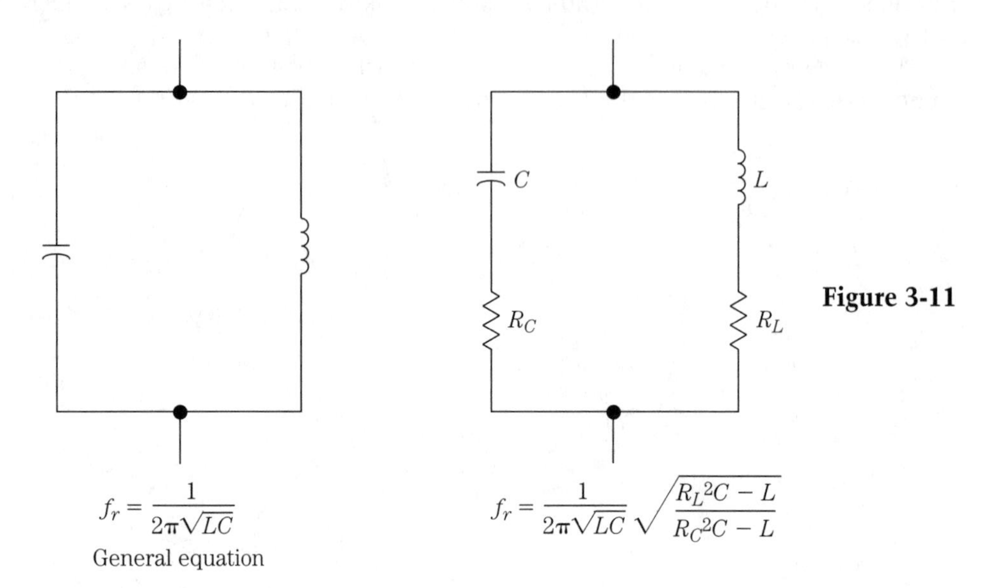

Figure 3-11

$$f_r = \frac{1}{2\pi\sqrt{LC}}$$

General equation

$$f_r = \frac{1}{2\pi\sqrt{LC}}\sqrt{\frac{R_L{}^2C - L}{R_C{}^2C - L}}$$

When there is resistance in either or both branches of the parallel resonant circuit that simple equation is not applicable. The general equation that does apply is given in Fig. 3-11.

If there is resistance in only one of the branches, you can let the resistance of the other branch equal zero. For example, suppose there is resistance (R_L) in the inductive branch, but the resistance in the capacitive branch is so low that it can be disregarded. In that case, the general equation can be used, but let $R_C = 0$ ohms.

Important note—Working problems in the GROL exam requires only a knowledge of the simpler special-case equation.

The general resonance equation is not always covered in technical literature. That sometimes causes confusion in understanding certain parallel LCR circuits. A parallel resonant *trap* was once used in a television receiver. A trap is a resonant circuit used to eliminate an undesired frequency or range of frequencies.

The trap in the TV receiver was tuned by a variable resistor. Using only the simpler (special-case) equation, it appears that the resonant frequency is independent of resistance. Based upon that information, the trap circuit was not clearly understood by those who had never seen the general equation for parallel resonance. However, the general-case equation makes it clear that the resonant frequency of a parallel resonant circuit can be tuned by a variable resistor.

There are some important considerations for the general equation shown in Fig. 3-11:

- If the equation under the radical yields a negative number, or if the denominator equals zero, it means that there is no resonant frequency.
- If the equation under the radical equals zero (because the numerator is equal to zero), the value of the resonant frequency is technically 0 Hz. In the real world, that is the same as saying there is no resonant frequency.

Compare that information with the case of series resonance. For a series resonant circuit, there is always a resonant frequency.

Effect of *R* on parallel circuit bandpass

It has been shown that increasing the series resistance in a resonant series RLC circuit decreases the circuit Q. That, in turn, causes the circuit to tune over a wider range of frequencies.

In the parallel circuit, the resistance is in parallel with L and C as shown in Fig. 3-12. The effect of R on the resonant curve is also shown in that illustration. Note that lowering the resistance increases the bandwidth. Think of it this way. If the resistance was lowered to that of a straight piece of wire there would be a short circuit across the tuned circuit, but the bandwidth of the wire used for the short circuit would be very wide.

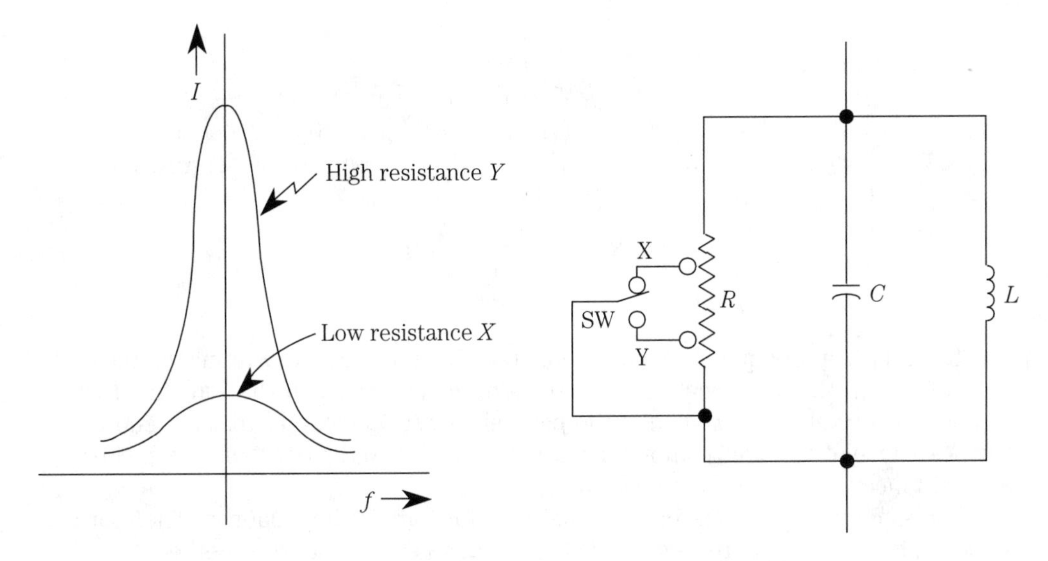

Figure 3-12

The overall result is that the bandwidth can be controlled by a variable resistor. In amateur radio jargon, the resistor across the parallel tuned circuit has been called a "swamping resistor."

Impedance of parallel circuits

The impedance of parallel RLC circuits can be calculated by using the same procedures as for calculating parallel resistor circuits provided that you are always sure to use the rectangular or polar versions of the branch impedances!

Recall that the equivalent resistance of two resistors in parallel can be calculated by the method called the *product over the sum.*

$$R_{eq} = R_1R_2/(R_1 + R_2)$$

This method is used in Table 3-3. Here is an equation for calculating parallel impedance:

$$\dot{Z}_{eq} = (\dot{Z}_1)(\dot{Z}_2)/(\dot{Z}_1 + \dot{Z}_2)$$

The equation is for two parallel impedances. To calculate the impedance of a circuit with more than two branches proceed by working on two branches at a time.

Remember that you can use the polar form for multiplication and division, but you must use the rectangular form for addition and subtraction. That was demonstrated in Table 3-1.

There is another method of calculating parallel impedance, which follows the procedure for calculating any number of parallel resistances. Called the *reciprocal method*; it involves taking "the reciprocal of the sum of the reciprocals."

Refer to Fig. 3-13. The resistance of the parallel circuit is determined by first taking the sum of the reciprocals. The reciprocal key on a calculator is usually marked X^{-1} or $1/x$. That key can be used in this step.

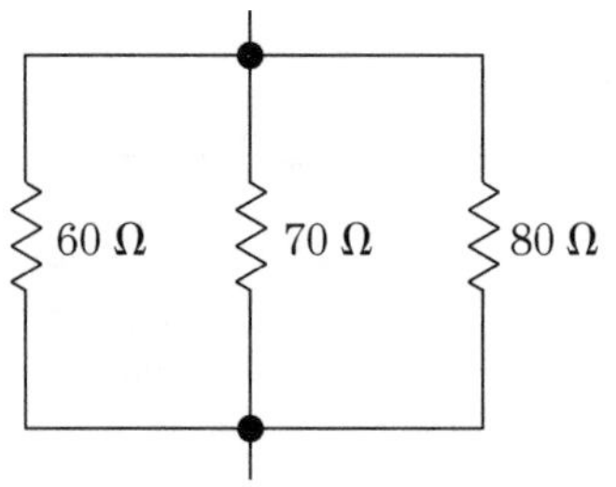

60 Ω 70 Ω 80 Ω

$$\frac{1}{\text{Req}} = \frac{1}{60} + \frac{1}{70} + \frac{1}{80}$$

$$= 0.0167 + 0.0143 + 0.0125$$

$$\frac{1}{\text{Req}} = 0.156$$

$$\text{Req} = \frac{1}{0.156} = 23\ \Omega$$

Figure 3-13

To work the problem in the illustration, take the reciprocal of R_1 plus the reciprocal of R_2 plus the reciprocal of R_3. Then, take the reciprocal of the sum, and that will be the equivalent resistance of the parallel circuit. Remember that the equivalent resistance of a parallel combination is always less than the resistance of the lowest resistance value in the combination.

The same reciprocal method can be used for parallel impedances. There are some important terms to review before looking at an example of this method.

- The reciprocal of impedance ($\dot{Z}$) is called *admittance* ($\dot{Y}$).
- The reciprocal of reactance (X) is called *susceptance* (B).
- The reciprocal of resistance (R) is called *conductance* (G).

Consider the circuit of Fig. 3-14. A series RL circuit is in parallel with a resistor. The problem is to find the impedance of the parallel circuit.

Given:

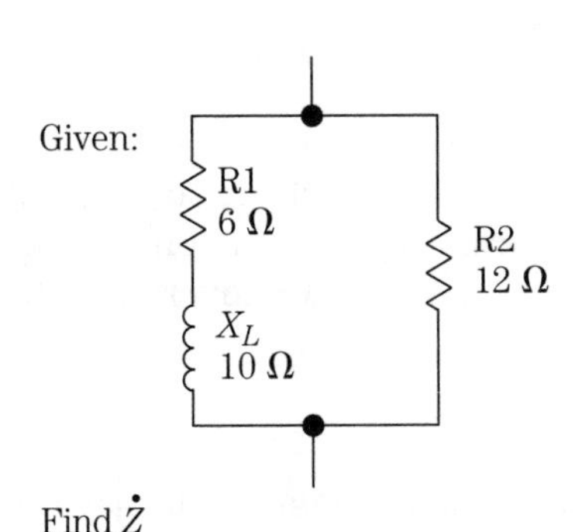

Find $\dot{Z}$

By the product over the sum method

$$\dot{Y} = \frac{1}{\dot{Z}_1} + \frac{1}{\dot{Z}_2} = \frac{1}{6 + j10} + \frac{1}{12 + j0}$$

$$= \frac{1}{11.6\underline{/59^\circ}} + \frac{1}{12\,\underline{/0^\circ}}$$

Take reciprocals:

$$\dot{Y} = 0.0857\underline{/-59^\circ} + 0.0833 + j0$$

Change to rectangular:

$$\dot{Y} = 0.0441 - j0.0734 + 0.0833$$

$$\dot{Y} = 0.1274 - j0.0734$$

The reciprocal of $\dot{Y}$ is $\dot{Z}$.

$$\dot{Z} = \frac{1}{\dot{Y}} = \frac{1}{0.1274 - j0.0734} = \frac{1}{0.147\underline{/-30^\circ}}$$

$$\dot{Z} = 6.8\underline{/30^\circ}$$

By the product over the sum method

$$\dot{Z}_{eq} = \frac{(6 + j10)(12 + j0)}{6 + j10 + 12 + j0} = \frac{72 + j20}{18 + j10}$$

$$= \frac{139.9\underline{/54^\circ}}{20.59\underline{/29^\circ}}$$

$$\dot{Z}_{eq} = 6.8\underline{/30^\circ}$$

Note: in order to get exact values for answers, it would be necessary to carry out each step many decimal places.

Figure 3-14

The first of the two methods used in Fig. 3-14 uses the reciprocal of the sum of the reciprocals. As shown in the illustration, the admittance of the series RL circuit is written first:

$$1/(R_1 + jX_L) = 1/(6 + j10)$$

The same procedure is used for the second impedance:

$$R_2 = 12 + j0$$

The impedances are converted to polar form and then the reciprocal of the impedance for each branch is determined. That gives the admittance for each branch.

In the next step, the branch admittances are converted to rectangular form so they can be added to get the total admittance ($\dot{y}$). The reciprocal of that admittance is the impedance of the parallel branch.

The second method of solving parallel impedance problems uses the product over the solution. The solution by the admittance method is checked by using the product over the sum method.

When solving problems with that many steps, it is the usual practice to round-answers off at various steps. If solutions to each step could be carried out an infinite number of decimal places, the answers would be identical. In these solutions, we rounded off at the answer.

Programmed review

Start with Block No. 1. Pick the answer you believe is correct. Go to the next block and check your answer. All answers are in italics. There is only one choice for each block. There is some material in this section that was not covered in the chapter.

Block 1

Using a graph that shows power vs. frequency on a frequency domain display the bandwidth is the range of frequency between:

A. minimum power points. C. 10-dB points.
B. half power points. D. points that are 70.7% of maximum.

Block 2

The correct answer is B.

Here is your next question: Using a graph that shows voltage vs. frequency on a frequency domain display the bandwidth is the range of frequencies between:

A. minimum power points. C. 10-dB points.
B. half power points. D. points that are 70.7% of maximum.

Note: If you can't tell from the illustration or drawing whether it is a voltage or power curve, assume it is a power curve.

Block 3

The correct answer is D.

Here is your next question: What is the resonant frequency of the circuit in Fig. 3-15?

A. 7.17 MHz. B. 12.2 MHz. C. 100 kHz. D. 33.3 kHz.

Block 4

The correct answer is C.

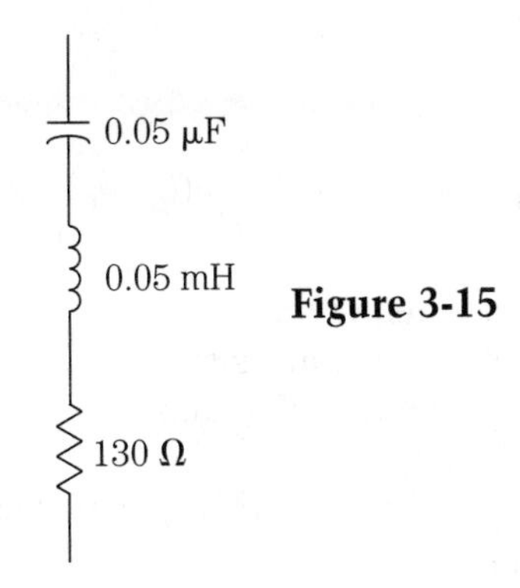

Figure 3-15

Here is your next question: Determine the inductive reactance of a 75-μH coil when the frequency is 1100 kHz.

A. 518 Ω. B. 663 Ω. C. 707 Ω. D. 141 Ω.

Block 5

The correct answer is A.

Here is your next question: What is the capacitive reactance of a 0.022-μF capacitor when the frequency is 2000 kHz?

A. 2.884 Ω. B. 3.346 Ω. C. 3.617 Ω. D. 3.884 Ω.

Block 6

The correct answer is C.

Here is your next question: What is the term for an out-of-phase, nonproductive power associated with inductors and capacitors?

A. Effective power. B. True power. C. Peak envelope power. D. Reactive power.

Block 7

The correct answer is D.

Here is your next question: What is the term for energy that is stored in an electromagnetic or electrostatic field?

A. Potential energy. B. Amperes-joules. C. Joules-coulombs. D. Kinetic energy.

Block 8

The correct answer is A.

Here is your next question: What is responsible for the phenomenon when voltages across reactances in series can often be larger than the voltages applied to them?

A. Capacitance. B. Resonance. C. Conductance. D. Resistance.

Block 9

The correct answer is B.

Here is your next question: Under what conditions does resonance occur in an electrical circuit?

A. When the power factor is at a minimum.
B. When inductive and capacitive reactances are equal.
C. When the square root of the sum of the capacitive and inductive reactances is equal to the resonant frequency.
D. When the square root of the product of the capacitive and inductive reactances is equal to the resonant frequency.

Block 10

The correct answer is B. (See Fig. 3-8.)

Here is your next question: What is the characteristic of the current flow in a series RLC circuit at resonance?

A. It is at a minimum. B. It is at a maximum. C. It is dc. D. It is zero.

Block 11

The correct answer is B.

Here is your next question: Why would the rate at which electrical energy is used in a circuit be less than the product of the magnitudes of the ac voltage and current?

A. Because there is a phase angle that is greater than zero between the current and voltage.
B. Because there are only resistances in the circuit.
C. Because there are no reactances in the circuit.
D. Because there is a phase angle that is equal to zero between the current and voltage.

Block 12

The correct answer is A.

Here is your next question: What does the power factor equal in an RL circuit having a 60° phase angle between the voltage and the current?

A. 1.414. B. 0.866. C. 0.5. D. 1.73.

Block 13

The correct answer is C. (P.F. = Cos 60 degrees)

Here is your next question: What does the power factor equal in an RL circuit having a 30° phase angle between the voltage and the current?

A. 1.73. B. 0.5. C. 0.866. D. 0.577.

Block 14

The correct answer is C.

Here is your next question: How many watts are being consumed in a circuit having a power factor of 0.6 when the input is 200 Vac and 5 A is being drawn?

A. 100. B. 200. C. 400. D. 600.

Block 15

The correct answer is D.

Here is your next question: What is the impedance of a network comprised of a 0.1-μH inductor in series with a 30-Ω resistor, at 5 MHz? (Specify your answer in rectangular coordinates.)

A. $30 - j3$. B. $30 + j3$. C. $3 + j30$. D. $3 - j30$.

Block 16

The correct answer is B.

Here is your next question: What is the impedance of a network comprised of a 10-μH inductor in series with a 40-Ω resistor, at 500 MHz? (Specify your answer in rectangular coordinates.)

A. $40 + j31400$. B. $40 - j31400$. C. $31400 + j40$. D. $31400 - j40$.

Block 17

The correct answer is A.

Here is your next question: What is the impedance of a network comprised of a 100-Ω reactance inductor in series with a 100-Ω resistor? (Specify your answer in polar coordinates.)

A. $121\ \Omega\angle 35°$. B. $141\ \Omega\angle 45°$. C. $161\ \Omega\angle 55°$. D. $181\ \Omega\angle 65°$.

Block 18

The correct answer is B.

Here is your next question: What is the impedance of a network comprised of a 400-Ω reactance capacitor in series with a 300-Ω resistor? (Specify your answer in polar coordinates.)

A. $240\ \Omega\angle 36.9°$. B. $240\ \Omega\angle -36.9°$. C. $500\ \Omega\angle 53.1°$. D. $500\ \Omega\angle -53.1°$.

Block 19

The correct answer is D.

Here is your next question: What is the impedance of a network comprised of a 400-Ω reactance inductor in parallel with a 300-Ω resistor? (Specify your answer in polar coordinates.)

A. 240 Ω/36.9°. B. 240 Ω/–36.9°. C. 500 Ω/53.1°. D. 500 Ω/–53.1°.

Block 20

The correct answer is A.

Here is your next question: What is the impedance of a network comprised of a 10-μH inductor in series with a 600-Ω resistor, at 10 kHz? (Specify your answer in rectangular coordinates.)

A. 628 + *j*600. B. 628 – *j*600. C. 600 + *j*0.628. D. 600 – *j*0.628.

Block 21

The correct answer is C.

Here is your next question: What is the impedance of a network comprised of a 0.01-μF capacitor in parallel with a 300-Ω resistor, at 50 kHz? (Specify your answer in rectangular coordinates.)

A. 150 – *j*159. B. 150 + *j*159. C. 159 + *j*150. D. 159 – *j*150.

Block 22

The correct answer is D.

Here is your next question: What is the impedance of a network comprised of a 0.1-μF capacitor in series with a 40-Ω resistor, at 50 kHz? (Specify your answer in rectangular coordinates.)

A. 40 + *j*32. B. 40 – *j*32. C. 32 – *j*40. D. 32 + *j*40.

Block 23

The correct answer is B.

Here is your next question: What is the impedance of a network comprised of a 100-Ω reactance capacitor in series with a 100-Ω resistor? (Specify your answer in polar coordinates.)

A. 121 Ω∠–25°. B. 141∠–45°. C. 161 Ω∠–65°. D. 191 Ω∠–85°.

Block 24

The correct answer is B.

Here is your next question: What is the impedance of a network comprised of a 300-Ω reactance inductor in series with a 400-Ω resistor? (Specify your answer in polar coordinates.)

A. 400 Ω∠27°. B. 500 Ω∠37°. C. 600 Ω∠47°. D. 700 Ω∠57°.

Block 25

The correct answer is B.

Here is your next question: What is the impedance of a network comprised of a 300-Ω reactance capacitor in series with a 400-Ω resistor? (Specify your answer in polar coordinates.)

A. 200 Ω$\angle -10°$. B. 300 Ω$\angle -17°$. C. 400 Ω$\angle -27°$. D. 500 Ω$\angle -37°$.

Block 26

The correct answer is D.

Quiz

As a technician, you should be able to answer all of the questions in this quiz without referring to the chapter material or any other reference. If there are questions you cannot answer, or questions where you had to guess answers, you should study the related subject material.

1. Write the equations for converting polar-to-rectangular and rectangular-to-polar impedance. Convert to circuits with both inductive reactance and capacitive reactance.
2. Draw a power triangle and label all of the parts.
3. What term is used in place of VARS?
4. Write the equation for calculating true power when you know the apparent power and the phase angle.
5. Write the equation for determining power factor.
6. When a circuit is connected across an ac generator what is the best value of power factor (as far as efficiency is concerned)?
7. Is resistance normally graphed on the real or imaginary axis?
8. What is the maximum possible phase angle between V and I in an ac circuit?
9. The power factor in a certain circuit is 0.85 and the true power is 100 watts. What is the power factor in percent?
10. What is the reciprocal of reactance called?
11. What is the reciprocal of admittance called?
12. In a certain RLC circuit, the phase angle is 0°. What is the condition of the circuit called?

13. If the time for one cycle is 500 μs, what is the frequency?

14. What is the value of TAN^{-1} (1.4) in degrees?

15. If the sine-wave voltage in an electronic circuit is 25 kHz, how long does it take the voltage to go from $t = 0$ to the next positive value?

16. An oscilloscope shows the peak-to-peak voltage across a resistor to be 130 V. What is the RMS (effective) value of the voltage?

17. An oscilloscope display shows that it takes 5 ms for 10 complete cycles of sinewave voltage. What is the frequency of the sinewave voltage?

18. A resistor is connected in series with a switch and an inductor across a battery. Write the equation for the time constant.

19. What is the 5th harmonic of a pure sine-wave voltage that has a frequency of 430 Hz?

20. Write z = $7\angle 1800°$ ohms in rectangular form.

Quiz answers

1. For inductive reactance in series with resistance:

$$Z \angle\phi = Z \cos\phi + jZ \sin\phi \quad \text{(polar to rectangular)}$$
$$R + jX_L = \sqrt{R^2 + (X_L)^2}\angle\phi \quad \text{(rectangular to polar)}$$

For capacitive reactance in series with resistance:

$$Z - \phi = Z \cos\phi - jZ \sin\phi \quad \text{(polar to rectangular)}$$
$$R - jX_C = \sqrt{R^2 + (-X_C)^2}\angle-\phi \quad \text{(rectangular to polar)}$$

2. See Fig. 3-16.

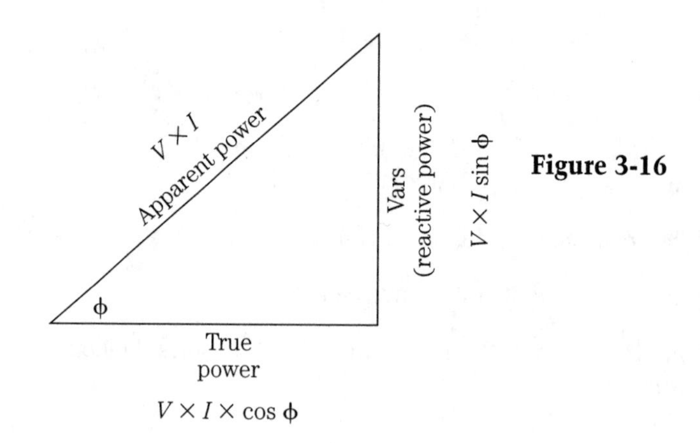

Figure 3-16

3. Reactive power

4. You only need to remember one equation:

$$\cos \angle\phi = \textit{true power/apparent power}$$

The other equations can be derived. For example:

$$\textit{true power} = \textit{apparent power} \times \cos \phi$$

$$(\textit{apparent power} = V \times I)$$

$$\textit{true power} = V \times I \cos \phi$$

5. *Power factor* = cos ϕ (where ϕ is the phase angle)

6. *P.F.* = 1.0, or 100%

7. On the real axis

8. 90°

9. 85% (true power not required)

10. Susceptance (B)

11. Impedance (Z)

12. Resonance

13. $T = 1/f = 1/(5\mu s) = 2000$ Hz

14. 54.46°

15. $T = 1/f = 1/(25 \text{ kHz}) = 0.00004$ sec

That is the time for one cycle. You are solving for the time of one-fourth cycle:

$$T = 0.00004/4 = 0.00001 \text{ s} = 10 \ \mu s$$

16. $V_{max} = V_{p\text{-}p} \div 2 = 65V$

$$V \text{ (rms value)} = 0.707 \times 65 = 49.96 \text{ V}$$

17. If 10 cycles takes 5 ms, then two cycles will occur in one ms. If two cycles occur in one ms, then one cycle will occur in half of that time ($T = 0.5$ ms)

$$T = 1/(0.5 \text{ ms}) = 2000 \text{ Hz}$$

18. $T = L/R$

19. Pure sine waves have no harmonics

20. With this type of problem it is useful to know how many complete cycles occur and what decimal part of a cycle remains:

1800 ÷ 360 = exactly 5 complete cycles. That means the angle is 0°.

$$7 \angle 1800° = 7 \angle 0° = 7 \cos 0° + j7 \sin 0° = 7 + j0$$

4

CHAPTER

Signals, transmission lines, and antennas

This chapter is devoted to the definitions of terms found in the GROL pool of questions. These terms are also used in other tests involving the transmission of signals.

These are the exact same questions that you will get on the GROL test. However, you should concentrate on the subject matter rather than try to memorize questions and answers. If there are terms not familiar to you, flag the related questions. Then, conduct further study on those subjects.

In some cases, the definition of a term is the answer to the question. The same questions are given in the companion Q&A book. However, in that book the questions and choices are scrambled to give you a better way to test your understanding of the subject matter.

Programmed review

Start with Block number 1. Pick the answer that you believe is correct. Go to the next block and check your answer. Be sure you understand the terms and the definitions.

Here is your first question:

Block 1

What is emission A3C?

A. Facsimile. B. RTTY. C. ATV. D. Slow scan TV.

Block 2

The correct answer is A.

Here is your next question: What type of emission is produced when an amplitude modulated transmitter is modulated by a facsimile signal?

A. A3F. B. A3C. C. F3F. D. F3C.

Block 3

The correct answer is B.

Here is your next question: What is facsimile?

A. The transmission of tone-modulated telegraphy.
B. The transmission of a pattern of printed characters designed to form a picture.
C. The transmission of printed pictures by electrical means.
D. The transmission of moving pictures by electrical means.

Block 4

The correct answer is C.

Here is your next question: What is emission F3C?

A. Voice transmission. B. Slow Scan TV. C. RTTY. D. Facsimile.

Block 5

The correct answer is D.

Here is your next question: What type of emission is produced when a frequency modulated transmitter is modulated by a facsimile signal?

A. F3C. B. A3C. C. F3F. D. A3F.

Block 6

The correct answer is A.

Here is your next question: What is emission A3F?

A. RTTY. B. Television. C. SSB. D. Modulated CW.

Block 7

The correct answer is B.

Here is your next question: What type of emission is produced when an amplitude modulated transmitter is modulated by a television signal?

A. F3F. B. A3F. C. A3C. D. F3C.

Block 8

The correct answer is B.

Here is your next question: What is emission F3F?

A. Modulated CW. B. Facsimile. C. RTTY. D. Television.

Block 9

The correct answer is D.

Here is your next question: What type of emission is produced when a frequency modulated transmitter is modulated by a television signal?

A. A3F. B. A3C. C. F3F. D. F3C.

Block 10

The correct answer is C.

Here is your next question: How can an FM-phone signal be produced?

A. By modulating the supply voltage to a class-B amplifier.
B. By modulating the supply voltage to a class-C amplifier.
C. By using a reactance modulator on an oscillator.
D. By using a balanced modulator on an oscillator.

Block 11

The correct answer is C.

Here is your next question: How can a double-sideband phone signal be produced?

A. By using a reactance modulator on an oscillator.
B. By varying the voltage to the varactor in an oscillator circuit.
C. By using a phase detector, oscillator, and filter in a feedback loop.
D. By modulating the plate supply voltage to a class-C amplifier.

Block 12

The correct answer is D.

Here is your next question: How can a single-sideband phone signal be produced?

A. By producing a double-sideband signal with a balanced modulator and then removing the unwanted sideband by filtering.
B. By producing a double-sideband signal with a balanced modulator and then removing the unwanted sideband by heterodyning.
C. By producing a double-sideband signal with a balanced modulator and then removing the unwanted sideband by mixing.
D. By producing a double-sideband signal with a balanced modulator and then removing the unwanted sideband by neutralization.

Block 13

The correct answer is A.

Here is your next question: What is meant by the term *deviation ratio?*

A. The ratio of the audio modulating frequency to the center carrier frequency.
B. The ratio of the maximum carrier frequency deviation to the highest audio modulating frequency.

C. The ratio of the carrier center frequency to the audio modulating frequency.
D. The ratio of the highest audio modulating frequency to the average audio modulating frequency.

Block 14

The correct answer is B.

Here is your next question: In an FM-phone signal, what is the term for the maximum deviation from the carrier frequency divided by the maximum audio modulating frequency?

A. Deviation index. B. Modulation index. C. Deviation ratio. D. Modulation ratio.

Block 15

The correct answer is C.

Here is your next question: What is the deviation ratio for an FM-phone signal having a maximum frequency swing of plus or minus 5 kHz and accepting a maximum modulation rate of 3 kHz?

A. 60. B. 0.16. C. 0.6. D. 1.66.

Block 16

The correct answer is D.

Here is your next question: What is the deviation ratio of an FM-phone signal having a maximum frequency swing of plus or minus 7.5 kHz and accepting a maximum modulation rate of 3.5 kHz?

A. 2.14. B. 0.214. C. 0.47. D. 47.

Block 17

The correct answer is A.

Here is your next question: What is meant by the term modulation index?

A. The processor index.
B. The ratio between the deviation of a frequency-modulated signal and the modulating frequency.
C. The FM signal-to-noise ratio.
D. The ratio of the maximum carrier frequency deviation to the highest audio modulating frequency.

Block 18

The correct answer is B.

Here is your next question: In an FM-phone signal, what is the term for the ratio between the deviation of the frequency-modulated signal and the modulating frequency?

A. FM compressibility. C. Percentage of modulation.
B. Quieting index. D. Modulation index.

Block 19

The correct answer is D.

Here is your next question: How does the modulation index of a phase-modulated emission vary with the modulated frequency?

A. The modulation index increases as the RF carrier frequency (the modulated frequency) increases.
B. The modulation index decreases as the RF carrier frequency (the modulated frequency) increases.
C. The modulation index varies with the square root of the RF carrier frequency (the modulated frequency).
D. The modulation index does not depend on the RF carrier frequency (the modulated frequency).

Block 20

The correct answer is D.

Here is your next question: In an FM-phone signal having a maximum frequency deviation of 3.0 kHz either side of the carrier frequency, what is the modulation index when the modulating frequency is 1.0 kHz?

A. 3. B. 0.3. C. 3000. D. 1000.

Block 21

The correct answer is A.

Here is your next question: What is the modulation index of an FM-phone transmitter producing an instantaneous carrier deviation of 6 kHz when modulated with a 2-kHz modulating frequency?

A. 6000. B. 3. C. 2000. D. ⅓.

Block 22

The correct answer is B.

Here is your next question: What are electromagnetic waves?

A. Alternating currents in the core of an electromagnet.
B. A wave consisting of two electric fields at right angles to each other.
C. A wave consisting of an electric field and a magnetic field at right angles to each other.
D. A wave consisting of two magnetic fields at right angles to each other.

Block 23

The correct answer is C.

Here is your next question: What is a wave front?

A. A voltage pulse in a conductor.
B. A current pulse in a conductor.
C. A voltage pulse across a resistor.
D. A fixed point in an electromagnetic wave.

Block 24

The correct answer is D.

Here is your next question: At what speed do electromagnetic waves travel in free space?

A. Approximately 300 million meters per second.
B. Approximately 468 million meters per second.
C. Approximately 186,300 feet per second.
D. Approximately 300 million miles per second.

Block 25

The correct answer is A.

Here is your next question: What are the two interrelated fields considered to make up an electromagnetic wave?

A. An electric field and a current field.
B. An electric field and a magnetic field.
C. An electric field and a voltage field.
D. A voltage field and a current field.

Block 26

The correct answer is B.

Here is your next question: Why do electromagnetic waves not penetrate a good conductor to any great extent?

A. The electromagnetic field induces currents in the insulator.
B. The oxide on the conductor surface acts as a shield.
C. Because of eddy currents.
D. The resistivity of the conductor dissipates the field.

Block 27

The correct answer is C.

Here is your next question: What is meant by referring to electromagnetic waves as horizontally polarized?

A. The electric field is parallel to the earth.
B. The magnetic field is parallel to the earth.
C. Both the electric and magnetic fields are horizontal.
D. Both the electric and magnetic fields are vertical.

Block 28

The correct answer is A.

Here is your next question: What is meant by referring to electromagnetic waves as having circular polarization?

A. The electric field is bent into a circular shape.
B. The electric field rotates.
C. The electromagnetic wave continues to circle the earth.
D. The electromagnetic wave has been generated by a quad antenna.

Block 29

The correct answer is B.

Here is your next question: When the electric field is perpendicular to the surface of the earth, what is the polarization of the electromagnetic wave?

A. Circular. B. Horizontal. C. Vertical. D. Elliptical.

Block 30

The correct answer is C.

Here is your next question: When the magnetic field is parallel to the surface of the earth, what is the polarization of the electromagnetic wave?

A. Circular. B. Horizontal. C. Elliptical. D. Vertical.

Block 31

The correct answer is D.

Here is your next question: When the magnetic field is perpendicular to the surface of the earth, what is the polarization of the electromagnetic field?

A. Horizontal. B. Circular. C. Elliptical. D. Vertical.

Block 32

The correct answer is A.

Here is your next question: When the electric field is parallel to the surface of the earth, what is the polarization of the electromagnetic wave?

A. Vertical. B. Horizontal. C. Circular. D. Elliptical.

Block 33

The correct answer is B.

Here is your next question: What is the RMS voltage of a 165 V peak pure sine wave?

A. 223 Vac. B. 104.9 Vac. C. 58.3 Vac. D. 117 Vac.

Block 34

The correct answer is D.

Here is your next question: What is the average voltage of a 165 V peak pure sine wave?

A. 233 Vac. B. 104.9 Vac. C. 58.3 Vac. D. 117 Vac.

Block 35

The correct answer is B.

Here is your next question: For many types of voices, what is the ratio of PEP to average power during a modulation peak in a single-sideband phone signal?

A. Approximately 1.0 to 1. C. Approximately 2.5 to 1.
B. Approximately 25 to 1. D. Approximately 100 to 1.

Block 36

The correct answer is C.

Here is your next question: In a single-sideband phone signal, what determines the PEP-to-average power ratio?

A. The frequency of the modulating signal. C. The speech characteristics.
B. The degree of carrier suppression. D. The amplifier power.

Block 37

The correct answer is C.

Here is your next question: What is the approximate dc input power to a class-B RF power amplifier stage in an FM-phone transmitter when the PEP output power is 1500 W?

A. Approximately 900 W. C. Approximately 2500 W.
B. Approximately 1765 W. D. Approximately 3000 W.

Block 38

The correct answer is C.

Here is your next question: What is the approximate dc input power to a class-C RF amplifier stage is an RTTY transmitter when the PEP output power is 1000 W?

A. Approximately 850 W. C. Approximately 1667 W.
B. Approximately 1250 W. D. Approximately 2000 W.

Block 39

The correct answer is B.

Here is your next question: What is the approximate dc input power to a class-AB RF power amplifier stage in an unmodulated carrier transmitter when the PEP output power is 500 W?

A. Approximately 250 W. C. Approximately 800 W.
B. Approximately 600 W. D. Approximately 1000 W.

Block 40

The correct answer is D.

Here is your next question: Where is the noise generated which primarily determines the signal-to-noise ratio in a VHF (150 MHz) marine band receiver?

A. In the receiver front end. C. In the atmosphere.
B. Man-made noise. D. In the ionosphere.

Block 41

The correct answer is A.

Here is your next question: In a pulse-width modulation system, what parameter does the modulating signal vary?

A. Pulse duration. B. Pulse frequency. C. Pulse amplitude. D. Pulse intensity.

Block 42

The correct answer is A.

Here is your next question: What is the type of modulation in which the modulating signal varies the duration of the transmitted pulse?

A. Amplitude modulation. C. Pulse-width modulation.
B. Frequency modulation. D. Pulse-height modulation.

Block 43

The correct answer is C.

Here is your next question: In a pulse-position modulation system, what parameter does the modulating signal vary?

A. The number of pulses per second.
B. Both the frequency and amplitude of the pulses.
C. The duration of the pulses.
D. The time at which each pulse occurs.

Block 44

The correct answer is D.

Here is your next question: Why is the transmitter peak power in a pulse-modulation system much greater than its average power?

A. The signal duty cycle is less than 100%.
B. The signal reaches peak amplitude only when voice modulated.

C. The signal reaches peak amplitude only when voltage spikes are generated within the modulator.
D. The signal reaches peak amplitude only when the pulses are also amplitude modulated.

Block 45

The correct answer is A.

Here is your next question: What is one way that voice is transmitted in a pulse-width modulation system?

A. A standard pulse is varied in amplitude by an amount depending on the voice waveform at that instant.
B. The position of a standard pulse is varied by an amount depending on the voice waveform at that instant.
C. A standard pulse is varied in duration by an amount depending on the voice waveform at that instant.
D. The number of standard pulses per second varies depending on the voice waveform at that instant.

Block 46

The correct answer is C.

Here is your next question: The International Organization for Standardization has developed a seven-level reference model for a packet-radio communications structure. What level is responsible for the actual transmission of data and handshaking signals?

A. The physical layer. C. The communications layer.
B. The transport layer. D. The synchronization layer.

Block 47

The correct answer is A.

Here is your next question: The International Organization for Standardization has developed a seven-level reference model for a packet-radio communications structure. What level arranges the bits into frames and controls data flow?

A. The transport layer. C. The communications layer.
B. The link layer. D. The synchronization layer.

Block 48

The correct answer is B.

Here is your next question: What is one advantage of using the ASCII code, with its larger character set, instead of the Baudot code?

A. ASCII includes built-in error-correction features.
B. ASCII characters contain fewer information bits than Baudot characters.

C. It is possible to transmit upper- and lowercase text.
D. The larger character set allows store-and-forward control characters to be added to a message.

Block 49

The correct answer is C.

Here is your next question: How many more voice transmissions can be packed into a given frequency band for amplitude-compandored single-sideband systems over conventional FM-phone systems?

A. 2. B. 4. C. 8. D. 16.

Block 50

The correct answer is B.

Here is your next question: What term describes a wide-bandwidth communications stem in which the RF carrier varies according to some predetermined sequence?

A. Amplitude compandored single sideband.
B. SITOR.
C. Time-domain frequency modulation.
D. Spread spectrum communication.

Block 51

The correct answer is D.

Here is your next question: What is the term used to describe a spread spectrum communications system where the center frequency of a conventional carrier is altered many times per second in accordance with a pseudorandom list of channels?

A. Frequency hopping.
B. Direct sequence.
C. Time-domain frequency modulation.
D. Frequency compandored spread spectrum.

Block 52

The correct answer is A.

Here is your next question: What term is used to describe a spread spectrum communications system in which a very fast binary bit stream is used to shift the phase of an RF carrier?

A. Frequency hopping.
B. Direct sequence.
C. Binary phase-shift keying.
D. Phase compandored spread spectrum.

Block 53

The correct answer is B.

Here is your next question: What is the term for the amplitude of the maximum positive excursion of a signal as viewed on an oscilloscope?

A. Peak-to-peak voltage.
B. Inverse peak negative voltage.
C. RMS voltage.
D. Peak positive voltage.

Block 54

The correct answer is D.

Here is your next question: What is the term for the amplitude of the maximum negative excursion of a signal as viewed on an oscilloscope?

A. Peak-to-peak voltage. C. RMS voltage.
B. Inverse peak positive voltage. D. Peak negative voltage.

Block 55

The correct answer is D.

Here is your next question: What is the easiest voltage amplitude dimension to measure by viewing a pulse sine wave signal on an oscilloscope?

A. Peak-to-peak voltage. B. RMS voltage. C. Average voltage. D. dc voltage.

Block 56

The correct answer is A.

Here is your next question: What is the relationship between the peak-to-peak voltage and the peak voltage amplitude in a symmetrical waveform?

A. 1:1. B. 2:1. C. 3:1. D. 4:1.

Block 57

The correct answer is B.

Here is your next question: What is an L-network?

A. A network consisting entirely of four inductors.
B. A network consisting of an inductor and a capacitor.
C. A network used to generate a leading phase angle.
D. A network used to generate a lagging phase angle.

Block 58

The correct answer is B.

Here is your next question: What is a pi-network?

A. A network consisting entirely of four inductors or four capacitors.
B. A power incidence network.
C. An antenna matching network that is isolated from ground.
D. A network consisting of one inductor and two capacitors or two inductors and one capacitor.

Block 59

The correct answer is D.

Here is your next question: What is a pi-L-network?

A. A phase inverter load network.
B. A network consisting of two inductors and two capacitors.
C. A network with only three discrete parts.
D. A matching network in which all components are isolated from ground.

Block 60

The correct answer is B.

Here is your next question: Does the L-, pi-, or pi-L-network provide the greatest harmonic suppression?

A. L-network. B. Pi-network. C. Inverse L-network. D. Pi-L-network.

Block 61

The correct answer is D.

Here is your next question: What are the three most commonly used networks to accomplish a match between an amplifying device and a transmission line?

A. M-network, pi-network, and T-network.
B. T-network, M-network, and Q-network.
C. L-network, pi-network, and pi-L-network.
D. L-network, M-network, and C-network.

Block 62

The correct answer is C.

Here is your next question: How are networks able to transform one impedance to another?

A. Resistances in the networks substitute for resistances in the load.
B. The matching network introduces negative resistance to cancel the resistive part of an impedance.
C. The matching network introduces transconductance to cancel the reactive part of an impedance.
D. The matching network can cancel the reactive part of an impedance and change the value of the resistive part of an impedance.

Block 63

The correct answer is D.

Here is your next question: Which type of network offers the greater transformation ratio?

A. L-network. B. Pi-network. C. Constant-K. D. M-derived.

Block 64

The correct answer is B.

Here is your next question: Why is the L-network of limited utility in impedance matching?

A. It matches a small impedance range.
B. It has limited power handling capabilities.
C. It is thermally unstable.
D. It is prone to self resonance.

Block 65

The correct answer is A.

Here is your next question: What is an advantage of using a pi-L-network instead of a pi-network for impedance matching between the final amplifier of a vacuum tube transmitter and a multiband antenna?

A. Greater transformation range.
B. Higher efficiency.
C. Lower losses.
D. Greater harmonic suppression.

Block 66

The correct answer is D.

Here is your next question: Which type of network provides the greatest harmonic suppression?

A. L-network. B. Pi-network. C. Pi-L-network. D. Inverse-pi network.

Block 67

The correct answer is C.

Here is your next question: What is meant by the term modulation?

A. The squelching of a signal until a critical signal-to-noise ratio is reached.
B. Carrier rejection through phase nulling.
C. A linear amplification mode.
D. A mixing process whereby information is imposed upon a carrier.

Block 68

The correct answer is D.

Here is your next question: How is a G3E FM-phone emission produced?

A. With a balanced modulator on the audio amplifier.
B. With a reactance modulator on the oscillator.
C. With a reactance modulator on the final amplifier.
D. With a balanced modulator on the oscillator.

Block 69

The correct answer is B.

Here is your next question: What is a reactance modulator?

A. A circuit that acts as a variable resistance or capacitance to produce FM signals.
B. A circuit that acts as a variable resistance or capacitance to produce AM signals.
C. A circuit that acts as a variable inductance or capacitance to produce FM signals.
D. A circuit that acts as a variable inductance or capacitance to produce AM signals.

Block 70

The correct answer is C.

Here is your next question: What is a balanced modulator?

A. An FM modulator that produces a balanced deviation.
B. A modulator that produces a double-sideband, suppressed carrier signal.
C. A modulator that produces a single-sideband, suppressed carrier signal.
D. A modulator that produces a full carrier signal.

Block 71

The correct answer is B.

Here is your next question: How can a single-sideband phone signal be generated?

A. By driving a product detector with a DSB signal.
B. By using a reactance modulator followed by a mixer.
C. By using a loop modulator followed by a mixer.
D. By using a balanced modulator followed by a filter.

Block 72

The correct answer is D.

Here is your next question: How can a double-sideband phone signal be generated?

A. By feeding a phase-modulated signal into a low-pass filter.
B. By using a balanced modulator followed by a filter.
C. By detuning a Hartley oscillator.
D. By modulating the plate voltage of a class-C amplifier.

Block 73

The correct answer is D.

Here is your next question: What is blanking in a video signal?

A. Synchronization of the horizontal and vertical sync-pulses.
B. Turning off the scanning beam while it is traveling from right to left and from bottom to top.
C. Turning off the scanning beam at the conclusion of a transmission.
D. Transmitting a black-and-white test pattern.

Block 74

The correct answer is B.

Here is your next question: What is the standard video voltage level between the sync tip and the whitest white at TV camera outputs and modulator inputs?

A. 1 V peak-to-peak. B. 120 IEEE units. C. 12 Vdc. D. 5 V RMS.

Block 75

The correct answer is A.

Here is your next question: What is the standard video level, in percent PEV, for black?

A. 0%. B. 12.5%. C. 70%. D. 100%.

Block 76

The correct answer is C.

Here is your next question: What is the standard video level, in percent PEV, for white?

A. 0%. B. 12.5%. C. 70%. D. 100%.

Block 77

The correct answer is A.

Here is your next question: What is the standard video level, in percent PEV, for blanking?

A. 0%. B. 12.5%. C. 75%. D. 100%.

Block 78

The correct answer is C.

Here is your next question: What is meant by the term antenna gain?

A. The numerical ratio relating the radiated signal strength of an antenna to that of another antenna.
B. The ratio of the signal in the forward direction to the signal in the back direction.
C. The ratio of the amount of power produced by the antenna compared to the output power of the transmitter.
D. The final amplifier gain minus the transmission line losses (including any phasing lines present).

Block 79

The correct answer is A.

Here is your next question: What is the term for a numerical ratio that relates the performance of one antenna to that of another real or theoretical antenna?

A. Effective radiated power. C. Conversion gain.
B. Antenna gain. D. Peak effective power.

Block 80

The correct answer is B.

Here is your next question: What is meant by the term antenna bandwidth?

A. Antenna length divided by the number of elements.
B. The frequency range over which an antenna can be expected to perform well.
C. The angle between the half-power radiation points.
D. The angle formed between two imaginary lines drawn through the ends of the elements.

Block 81

The correct answer is B.

Here is your next question: What is the wavelength of a shorted stub used to absorb even harmonics?

A. ½ wavelength. B. ⅓ wavelength. C. ¼ wavelength. D. ⅛ wavelength.

Block 82

The correct answer is C.

Here is your next question: What is a trap antenna?

A. An antenna for rejecting interfering signals.
B. A highly sensitive antenna with maximum gain in all directions.
C. An antenna capable of being used on more than one band because of the presence of parallel LC networks.
D. An antenna with a large capture area.

Block 83

The correct answer is C.

Here is your next question: What is an advantage of using a trap antenna?

A. It has high directivity in the high-frequency bands.
B. It has high gain.
C. It minimizes harmonic radiation.
D. It can be used for multiband operation.

Block 84

The correct answer is D.

Here is your next question: What is a disadvantage of using a trap antenna?

A. It will radiate harmonics.
B. It can only be used for single-band operation.
C. It is too sharply directional at lower frequencies.
D. It must be neutralized.

Block 85

The correct answer is A.

Here is your next question: What is the principle of a trap antenna?

A. Beamwidth may be controlled by nonlinear impedances.
B. The traps form a high impedance to isolate parts of the antenna.
C. The effective radiated power can be increased if the space around the antenna "sees" a high impedance.
D. The traps increase the antenna gain.

Block 86

The correct answer is B.

Here is your next question: How does the length of the reflector element of a parasitic element beam antenna compare with that of the driven element?

A. It is about 5% longer. C. It is twice as long.
B. It is about 5% shorter. D. It is one-half as long.

Block 87

The correct answer is A.

Here is your next question: How does the length of the director element of a parasitic element beam antenna compare with that of the driven element?

A. It is about 5% longer. C. It is one-half as long.
B. It is about 5% shorter. D. It is twice as long.

Block 88

The correct answer is B.

Here is your next question: What is meant by the term radiation resistance for an antenna?

A. Losses in the antenna elements and feed line.
B. The specific impedance of the antenna.
C. An equivalent resistance that would dissipate the same amount of power as that radiated from an antenna.
D. The resistance in the trap coils to received signals.

Block 89

The correct answer is C.

Here is your next question: What are the factors that determine the radiation resistance of an antenna?

A. Transmission line length and height of antenna.
B. The location of the antenna with respect to nearby objects and the length/diameter ratio of the conductors.
C. It is a constant for all antennas since it is a physical constant.
D. Sunspot activity and the time of day.

Block 90

The correct answer is B.

Here is your next question: What is a driven element of an antenna?

A. Always the rearmost element.
B. Always the forward most element.
C. The element fed by the transmission line.
D. The element connected to the rotator.

Block 91

The correct answer is C.

Here is your next question: What is the usual electrical length of a driven element in an HF beam antenna?

A. ¼ wavelength. B. ½ wavelength. C. ¾ wavelength. D. 1 wavelength.

Block 92

The correct answer is B.

Here is your next question: What is the term for an antenna element which is supplied power from a transmitter through a transmission line?

A. Driven element. B. Director element. C. Reflector element. D. Parasitic element.

Block 93

The correct answer is A.

Here is your next question: How is antenna "efficiency" computed?

A. *Efficiency* = (*radiation resistance/transmission resistance*) × 100%.
B. *Efficiency* = (*radiation resistance/total resistance*) × 100%.
C. *Efficiency* = (*total resistance/radiation resistance*) × 100%.
D. *Efficiency* = (*effective radiated power/transmitter output*) × 100%.

Block 94

The correct answer is B.

Here is your next question: What is the term for the ratio of the radiation resistance of an antenna to the total resistance of the system?

A. Effective radiated power. C. Antenna efficiency.
B. Radiation conversion loss. D. Beamwidth.

Block 95

The correct answer is C.

Here is your next question: What is included in the total resistance of an antenna system?

A. Radiation resistance plus space impedance.
B. Radiation resistance plus transmission resistance.
C. Transmission line resistance plus radiation resistance.
D. Radiation resistance plus ohmic resistance.

Block 96

The correct answer is D.

Here is your next question: How can the antenna efficiency of an HF grounded vertical antenna be made comparable to that of a half-wave antenna?

A. By installing a good ground radial system. C. By shortening the vertical.
B. By isolating the coax shield from ground. D. By lengthening the vertical.

Block 97

The correct answer is A.

Here is your next question: Why does a half-wave antenna efficiency of an HF grounded vertical antenna needs to be made comparable to that of a half-wave antenna?

A. Because it is non-resonant.
B. Because the conductor resistance is low compared to the radiation resistance.
C. Because earth-induced currents add to its radiated power.
D. Because it has less corona from the element ends than other types of antennas.

Block 98

The correct answer is B.

Here is your next question: What is a folded dipole antenna?

A. A dipole that is one-quarter wavelength long.
B. A ground plane antenna.
C. A dipole whose ends are connected by another one-half wavelength piece of wire.
D. A fictional antenna used in theoretical discussions to replace the radiation resistance.

Block 99

The correct answer is C.

Here is your next question: How does the bandwidth of a folded dipole antenna compare with that of a simple dipole antenna?

A. It is 0.707 times the simple dipole bandwidth.
B. It is essentially the same.
C. It is less than 50% that of a simple dipole.
D. It is greater.

Block 100

The correct answer is D.

Here is your next question: What is the input terminal impedance at the center of a folded dipole antenna?

A. 300 Ω. B. 72 Ω. C. 50 Ω. D. 450 Ω.

Block 101

The correct answer is A.

Here is your next question: What is the meaning of the term velocity factor of a transmission line?

A. The ratio of the characteristic impedance of the line to the terminating impedance.
B. The index of shielding for coaxial cable.
C. The velocity of the wave on the transmission line multiplied by the velocity of light in a vacuum.
D. The velocity of the wave on the transmission line divided by the velocity of light in a vacuum.

Block 102

The correct answer is D.

Here is your next question: What is the term for the ratio of actual velocity at which a signal travels through a line to the speed of light in a vacuum?

A. Velocity factor.
B. Characteristic impedance.
C. Surge impedance.
D. Standing wave ratio

Block 103

The correct answer is A.

Here is your next question: What is the velocity factor for non-foam dielectric 50 or 75 ohms flexible coaxial cable such as RG 8, 11, 58 and 59?

A. 2.70. B. 0.66. C. 0.30. D. 0.10.

Block 104

The correct answer is B.

Here is your next question: What determines the velocity factor in a transmission line?

A. The termination impedance.
B. The line length.
C. Dielectrics in the line.
D. The center conductor resistivity.

Block 105

The correct answer is C.

Here is your next question: Why is the physical length of a coaxial cable transmission line shorter than its electrical length?

A. Skin effect is less pronounced in the coaxial cable.
B. RF energy moves slower along the coaxial cable.
C. The surge impedance is higher in the parallel feed line.
D. The characteristic impedance is higher in the parallel feed line.

Block 106

The correct answer is B.

Here is your next question: What would be the physical length of a typical coaxial transmission line which is electrically one-quarter wavelength long at 14.1 MHz?

A. 20 meters. B. 3.51 meters. C. 2.33 meters. D. 0.25 meters.

Block 107

The correct answer is B.

Here is your next question: What would be the physical length of a typical coaxial transmission line which is electrically one-quarter wavelength long at 7.2 MHz?

A. 10.5 meters. B. 6.88 meters. C. 24 meters. D. 50 meters.

Block 108

The correct answer is B.

Here is your next question: What is the physical length of a parallel antenna feed line which is electrically one-half wavelength long at 14.10 MHz? (Assume a velocity factor of 0.82.)

A. 15 meters. B. 24.3 meters. C. 8.7 meters. D. 70.8 meters.

Block 109

The correct answer is C.

Here is your next question: What is the physical length of a twin lead transmission feed line at 36.5 MHz? (Assume a velocity factor of 0.80.)

A. Electrical length times 0.8.
B. Electrical length divided by 0.8.
C. 80 meters.
D. 160 meters.

Block 110

The correct answer is A.

Here is your next question: In a half-wave antenna, where are the current nodes?

A. At the ends.
B. At the center.
C. Three-quarters of the way from the feed point toward the end.
D. One-half of the way from the feed point toward the end.

Block 111

The correct answer is A.

Here is your next question: In a half-wave antenna, where are the voltage nodes?

A. At the ends.
B. At the feed point.
C. Three-quarters of the way from the feed point toward the end.
D. One-half of the way from the feed point toward the end.

Block 112

The correct answer is B.

Here is your next question: At the ends of a half-wave antenna, what values of current and voltage exist compared to the remainder of the antenna?

A. Equal voltage and current.
B. Minimum voltage and maximum current.
C. Maximum voltage and minimum current.
D. Minimum voltage and minimum current.

Block 113

The correct answer is C.

Here is your next question: At the center of a half-wave antenna, what values of voltage and current exist compared to the remainder of the antenna?

A. Equal voltage and current.
B. Maximum voltage and minimum current.
C. Minimum voltage and minimum current.
D. Minimum voltage and maximum current.

Block 114

The correct answer is D.

Here is your next question: What happens to the base feed point of a fixed length mobile antenna as the frequency of operation is lowered?

A. The resistance decreases and the capacitive reactance decreases.
B. The resistance decreases and the capacitive reactance increases.

C. The resistance increases and the capacitive reactance decreases.
D. The resistance increases and the capacitive reactance increases.

Block 115

The correct answer is B.

Here is your next question: Why should an HF mobile antenna loading coil have a high ratio of reactance to resistance?

A. To swamp out harmonics. C. To minimize losses.
B. To maximize losses. D. To minimize the *Q*.

Block 116

The correct answer is C.

Here is your next question: Why is a loading coil often used with an HF mobile antenna?

A. To improve reception. C. To lower the *Q*.
B. To lower the losses. D. To tune out the capacitive reactance.

Block 117

The correct answer is D.

Here is your next question: For a shortened vertical antenna, where should a loading coil be placed to minimize losses and produce the most effective performance?

A. Near the center of the vertical radiator.
B. As low as possible on the vertical radiator.
C. As close to the transmitter as possible.
D. At a voltage node.

Block 118

The correct answer is A.

Here is your next question: What happens to the bandwidth of an antenna as it is shortened through the use of loading coils?

A. It is increased. B. It is decreased. C. No change occurs. D. It becomes flat.

Block 119

The correct answer is B.

Here is your next question: Why are self-resonant antennas popular in many applications?

A. They are very broad banded.
B. They have high gain in all azimuthal directions.
C. They are the most efficient radiators.
D. They require no calculations.

Block 120

The correct answer is C.

Here is your next question: What is the advantage of using top loading in a shortened HF vertical antenna?

A. Lower Q.
B. Greater structural strength.
C. Higher losses.
D. Improved radiation efficiency.

Block 121

The correct answer is D.

Here is your next question: What is an isotropic radiator?

A. A hypothetical, omnidirectional antenna.
B. In the northern hemisphere, an antenna whose directive pattern is constant in southern directions.
C. An antenna high enough in the air that its directive pattern is substantially unaffected by the ground beneath it.
D. An antenna whose directive pattern is substantially unaffected by the spacing of the elements.

Block 122

The correct answer is A.

Here is your next question: When is it useful to refer to an isotropic radiator?

A. When comparing the gains of directional antennas.
B. When testing a transmission line for standing wave ratio.
C. When (in the northern hemisphere) directing the transmission in a southerly direction.
D. When using a dummy load to tune a transmitter.

Block 123

The correct answer is A.

Here is your next question: What theoretical reference antenna provides a comparison for antenna measurements?

A. Quarter-wave vertical. B. Yagi-Uda array. C. Bobtail curtain. D. Isotropic radiator.

Block 124

The correct answer is D.

Here is your next question: What purpose does an isotropic radiator serve?

A. It is used to compare signal strengths (at a distant point) of different transmitters.
B. It is used as a reference for antenna gain measurements.
C. It is used as a dummy load for tuning transmitters.
D. It is used to measure the standing-wave-ratio on a transmission line.

Block 125

The correct answer is B.

Here is your next question: How much gain does a ½-wavelength dipole have over an isotropic radiator?

A. About 1.5 dB. B. About 2.1 dB. C. About 3.0 dB. D. About 6.0 dB.

Block 126

The correct answer is B.

Here is your next question: How much gain does an antenna have over a ½-wavelength dipole when it has 6 dB gain over an isotropic radiator?

A. About 3.9 dB. B. About 6.0 dB. C. About 8.1 dB. D. About 10.0 dB.

Block 127

The correct answer is A.

Here is your next question: How much gain does an antenna have over a ½-wavelength dipole when it has 12 dB gain over an isotropic radiator?

A. About 6.1 dB. B. About 9.9 dB. C. About 12.0 dB. D. About 14.1 dB.

Block 128

The correct answer is B.

Here is your next question: What is the antenna pattern for an isotropic radiator?

A. A figure-8. B. A unidirectional cardioid. C. A parabola. D. A sphere.

Block 129

The correct answer is D.

Here is your next question: What type of directivity pattern does an isotropic radiator have?

A. A figure-8. B. A unidirectional cardioid. C. A parabola. D. A sphere.

Block 130

The correct answer is D.

Here is your next question: What factors determine the receiving antenna gain required at a station in earth operation?

A. Height, transmitter power, and antennas of satellite.
B. Length of transmission line and impedance match between receiver and transmission line.
C. Preamplifier location on transmission line and presence or absence of RF amplifier stages.
D. Height of earth antenna and satellite orbit.

Block 131

The correct answer is A.

Here is your next question: What factors determine the EIRP required by a station in earth operation?

A. Satellite antennas and height, satellite receiver sensitivity.
B. Path loss, earth antenna gain, signal-to-noise ratio.
C. Satellite transmitter power and orientation of ground receiving antenna.
D. Elevation of satellite above horizon, signal-to-noise ratio, satellite transmitter power.

Block 132

The correct answer is B.

Here is your next question: How does the gain of a parabolic dish type antenna change when the operating frequency is doubled?

A. Gain does not change. C. Gain increases 6 dB.
B. Gain is multiplied by 0.707. D. Gain increases 3 dB.

Block 133

The correct answer is C.

Here is your next question: What happens to the beamwidth of an antenna as the gain is increased?

A. The beamwidth increases geometrically as the gain is increased.
B. The beamwidth increases arithmetically as the gain is increased.
C. The beamwidth is essentially unaffected by the gain of the antenna.
D. The beamwidth decreases as the gain is increased.

Block 134

The correct answer is D.

Here is your next question: What is the beamwidth of a symmetrical pattern antenna with a gain of 20 dB as compared to an isotropic radiator?

A. 10.1°. B. 20.3°. C. 45.0°. D. 60.9°.

Block 135

The correct answer is B.

Here is your next question: What is the beamwidth of a symmetrical pattern antenna with a gain of 30 dB as compared to an isotropic radiator?

A. 3.2°. B. 6.4°. C. 37°. D. 60.4°.

Block 136

The correct answer is B.

Here is your next question: What is the beamwidth of a symmetrical pattern antenna with a gain of 15 dB as compared to an isotropic radiator?

A. 72°. B. 52°. C. 36°. D. 3.61°.

Block 137

The correct answer is C.

Here is your next question: What is the beamwidth of a symmetrical pattern antenna with a gain of 12 dB as compared to an isotropic radiator?

A. 34.8°. B. 45.0°. C. 58.0°. D. 51.0°.

Block 138

The correct answer is D.

Here is your next question: How is circular polarization produced using linearly-polarized antennas?

A. Stack two Yagis, fed 90° out of phase, to form an array with the respective elements in parallel planes.
B. Stack two Yagis, fed in phase, to form an array with the respective elements in parallel planes.
C. Arrange two Yagis perpendicular to each other, with the driven elements in the same plane, fed 90° out of phase.
D. Arrange two Yagis perpendicular to each other, with the driven elements in the same plane, fed in phase.

Block 139

The correct answer is C.

Here is your next question: Why does an antenna system for earth operation (for communications through a satellite) need to have rotators for both azimuth and elevation control?

A. In order to point the antenna above the horizon to avoid terrestrial interference.
B. Satellite antennas require two rotators because they are so large and heavy.
C. In order to track the satellite as it orbits the earth.
D. The elevation rotator points the antenna at the satellite and the azimuth rotator changes the antenna polarization.

Block 140

The correct answer is C.

Here is your next question: What term describes a method used to match a high-impedance transmission line to a lower impedance antenna by connecting the line to the driven element in two places, spaced a fraction of a wavelength on each side of the driven element center?

A. The gamma matching system. C. The omega matching system.
B. The delta matching system. D. The stub matching system.

Block 141

The correct answer is B.

Here is your next question: What term describes an unbalanced feed system in which the driven element is fed both at the center of that element and a fraction of a wavelength to one side of center.

A. The gamma matching system. C. The omega matching system.
B. The delta matching system. D. The stub matching system.

Block 142

The correct answer is A.

Here is your next question: What term describes a method of antenna impedance matching that uses a short section of transmission line connected to the antenna feed line near the antenna and perpendicular to the feed line?

A. The gamma matching system. C. The omega matching system.
B. The delta matching system. D. The stub matching system.

Block 143

The correct answer is D.

Here is your next question: What kind of impedance does a ⅛-wavelength transmission line present to a generator when the line is shorted at the far end?

A. A capacitive reactance.
B. The same as the characteristic impedance of the line.
C. An inductive reactance.
D. The same as the input impedance to the final generator stage.

Block 144

The correct answer is C.

Here is your next question: What kind of impedance does a ⅛-wavelength transmission line present to a generator when the line is open at the far end?

A. The same as the characteristic impedance of the line.
B. An inductive reactance.
C. A capacitive reactance.
D. The same as the input impedance of the final generator stage.

Block 145

The correct answer is C.

Here is your next question: What kind of impedance does a ¼-wavelength transmission line present to a generator when the line is shorted at the far end?

A. A very high impedance.
B. A very low impedance.
C. The same as the characteristic impedance of the transmission line.
D. The same as the generator output impedance.

Block 146

The correct answer is A.

Here is your next question: What kind of impedance does a ¼-wavelength transmission line present to a generator when the line is open at the far end?

A. A very high impedance.
B. A very low impedance.
C. The same as the characteristic impedance of the line.
D. The same as the input impedance to the final generator stage.

Block 147

The correct answer is B.

Here is your next question: What kind of impedance does a ⅜-wavelength transmission line present to a generator when the line is shorted at the far end?

A. The same as the characteristic impedance of the line.
B. An inductive reactance.
C. A capacitive reactance.
D. The same as the input impedance to the final generator stage.

Block 148

The correct answer is C.

Here is your next question: What kind of impedance does a ⅜-wavelength transmission line present to a generator when the line is open at the far end?

A. A capacitive reactance.
B. The same as the characteristic impedance of the line.
C. An inductive reactance.
D. The same as the input impedance to the final generator stage.

Block 149

The correct answer is C.

Here is your next question: What kind of impedance does a ½-wavelength transmission line present to a generator when the line is shorted at the far end?

A. A very high impedance.
B. A very low impedance.
C. The same as the characteristic impedance of the line.
D. The same as the output impedance of the generator.

Block 150

The correct answer is B.

Here is your next question: What kind of impedance does a ½-wavelength transmission line represent to a generator when the line is open at the far end?

A. A very high impedance. C. The same as the characteristic impedance of the line.
B. A very low impedance. D. The same as the output impedance of the generator.

Block 151

The correct answer is A.

Here is your next question: What is the term used for an equivalent resistance which would dissipate the same amount of energy as that radiated from an antenna?

A. Space resistance. C. Transmission line loss.
B. Loss resistance. D. Radiation resistance.

Block 152

The correct answer is D.

Here is your next question: Why is the value of the radiation resistance of an antenna important?

A. Knowing the radiation resistance makes it possible to match impedances for maximum power transfer.
B. Knowing the radiation resistance makes it possible to measure the near-field radiation density from a transmitting antenna.
C. The value of the radiation resistance represents the front-to-side ratio of the antenna.
D. The value of the radiation resistance represents the front-to-back ratio of the antenna.

Block 153

The correct answer is A.

Here is your next question: Adding parasitic elements to an antenna will:

A. decrease its directional characteristics. C. increase its directional characteristics.
B. decrease its sensitivity. D. increase its sensitivity.

Block 154

The correct answer is C.

Here is your next question: What ferrite rod device prevents the formation of reflected waves on a waveguide transmission line?

A. Reflector. B. Isolator. C. Wave-trap. D. SWR refractor.

Block 155

The correct answer is B.

Here is your next question: Frequencies most affected by knife-edge refraction are:

A. low and medium frequencies.
B. high frequencies.
C. very high and ultra high frequencies.
D. 100 kHz to 3.0 MHz.

Block 156

The correct answer is C.

Here is your next question: When measuring I and V along a ½-wavelength Hertz antenna, where would you find the points where I and V are maximum and minimum?

A. V and I are high at the ends.
B. V and I are high in the middle.
C. V and I are uniform throughout the antenna.
D. V is maximum at both ends, I is maximum in the middle.

Block 157

The correct answer is D.

Here is your next question: To increase the resonant frequency of a ¼-wavelength antenna:

A. add a capacitor. B. lower capacitor value. C. cut antenna. D. add an inductor.

Block 158

The correct answer is A.

Here is your next question: Why are concentric transmission lines sometimes filled with nitrogen?

A. Reduce resistance at high frequencies.
B. Prevent water damage underground.
C. Keep moisture out and prevent oxidation.
D. Reduce microwave line losses.

Block 159

The correct answer is C.

Here is your next question: A vertical ¼-wave antenna receives signals:

A. in the microwave band.
B. in one vertical direction.
C. in one horizontal direction.
D. equally from all horizontal directions.

Block 160

The correct answer is D.

Here is your next question: Which of the following represents the best standing wave ratio (SWR)?

A. 1:1. B. 1:1.5. C. 1:3. D. 1:4.

Block 161

The correct answer is A.

Here is your next question: What is the purpose of stacking elements on an antenna?

A. Sharper directional pattern.
B. Increased gain.
C. Improved bandpass.
D. All of the above.

Block 162

The correct answer is D.

Here is your next question: In a quarter-wave *Marconi antenna*, where is the voltage and current concentrated?

A. At the ends.
B. Voltage at the ends, current in the middle.
C. Current at the ends, voltage in the middle.
D. Evenly throughout.

Block 163

The correct answer is B.

Here is your next question: On a half-wave Hertz antenna:

A. voltage is maximum at both ends and current is maximum at the center of the antenna.
B. current is maximum at both ends and voltage in the center.
C. voltage and current are uniform throughout the antenna.
D. voltage and current are high at the ends.

Block 164

The correct answer is A.

Here is your next question: What type of antenna is designed for minimum radiation?

A. Dummy antenna.
B. Quarter-wave antenna.
C. Half-wave antenna.
D. Directional antenna.

Block 165

The correct answer is A.

Here is your next question: What is the outcome when you stack antennas at various angles?

A. A more omnidirectional reception.
B. A more unidirectional reception.
C. An overall reception signal increase.
D. Both A and C.

Block 166

The correct answer is D.

Here is your next question: Adding parasitic elements to a quarter-wavelength antenna will:

A. reduce its directional characteristics. C. increase its sensitivity.
B. increase its directional characteristics. D. reduce its effectiveness.

Block 167

The correct answer is B.

Here is your next question: Ignoring line losses, voltage at a point on a transmission line without standing waves is:

A. equal to the product of the line current and impedance.
B. equal to the product of the line current and power factor.
C. equal to the product of the line current and the surge impedance.
D. zero at both ends.

Block 168

The correct answer is C.

Here is your next question: Stacking antenna elements:

A. will suppress odd harmonics. C. increases sensitivity to weak signals.
B. decrease signal to noise ratio. D. increases selectivity.

Block 169

The correct answer is C.

Here is your next question: What allows microwaves to pass in only one direction?

A. RF emitter. B. Ferrite isolator. C. Capacitor. D. Varactor-triac.

Block 170

The correct answer is B.

Here is your next question: What would be added to make a receiving antenna more directional?

A. Inductor. B. Capacitor. C. Parasitic elements. D. Height.

Block 171

The correct answer is C.

Here is your next question: Nitrogen is placed in transmission lines to:

A. improve the "skin-effect" of microwaves.
B. reduce arcing in the line.

C. reduce the standing wave ratio of the line.
D. prevent moisture from entering the line.

Block 172

The correct answer is D.

Here is your next question: Neglecting line losses, the voltage at any point along a transmission line, having no standing waves, will be equal to the:

A. transmitter output.
B. product of the line voltage and the surge impedance of the line.
C. product of the line current and the surge impedance of the line.
D. product of the resistance and surge impedance of the line.

Block 173

The correct answer is C.

Here is your next question: Adding a capacitor in series with a Marconi antenna:

A. increases the antenna circuit resonant frequency.
B. decreases the antenna circuit resonant frequency.
C. blocks the transmission of signals from the antenna.
D. increases the power handling capacity of the antenna.

Block 174

The correct answer is A.

Here is your next question: An antenna is carrying an unmodulated signal; when 100% modulation is impressed, the antenna current:

A. goes up 50%. B. goes down one half. C. stays the same. D. goes up 22.5%.

Block 175

The correct answer is D.

Here is your next question: An excited ½-wavelength antenna produces:

A. residual fields.
B. an electromagnetic field only.
C. both electromagnetic and electrostatic fields.
D. an electro-flux field sometimes.

Block 176

The correct answer is C.

Here is your next question: An antenna which intercepts signals equally from all horizontal directions is:

A. parabolic. B. vertical loop. C. horizontal Marconi. D. vertical ¼ wave.

Block 177

The correct answer is D.

Here is your next question: Loop antenna:

A. is bidirectional. C. is more often used as a receiving antenna.
B. is usually vertical. D. any of the above.

Block 178

The correct answer is D.

Here is your next question: Referred to the fundamental frequency, a shorted stub line attached to the transmission line to absorb even harmonics could have a wavelength of:

A. 1.41 wavelength. B. ½ wavelength. C. ¼ wavelength. D. ⅙ wavelength.

Block 179

The correct answer is C.

Here is your next question: Nitrogen gas in concentric RF transmission lines is used to:

A. keep moisture out. B. prevent oxidation. C. act as insulator. D. both A and B.

Block 180

The correct answer is D.

Here is your next question: "Stacking" elements on an antenna:

A. makes for better reception. C. decreases antenna current.
B. makes for poorer reception. D. decreases directivity.

Block 181

The correct answer is A.

Here is your next question: The following figure shows the radiated: (Please refer to Fig. 4-1.)

A. voltage along a straightened out balanced loop.
B. voltage along a ¼-wave Hertz antenna.
C. voltage along a ½-wave Hertz antenna.
D. current along a ½-wave Hertz antenna.

Block 182

The correct answer is D.

Here is your next question: The parasitic elements on a receiving antenna:

A. increases its directivity. C. have no effect on its impedance.
B. decreases its directivity. D. make it more nearly omnidirectional.

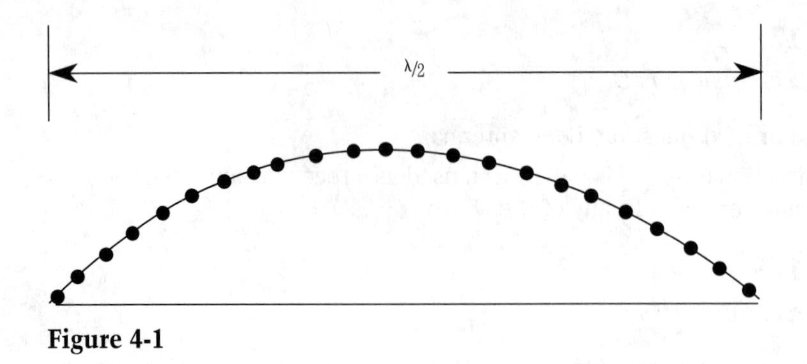

Figure 4-1

Block 183

The correct answer is A.

Here is your next question: Antenna voltage is:

A. inversely proportional to the square of its length.
B. proportional to its effective height.
C. measured voltage times length in feet.
D. always proportional to field strength.

Block 184

The correct answer is B.

Here is your next question: The resonant frequency of a Hertz antenna can be lowered by:

A. lowering the frequency of the transmitter.
B. placing a capacitor in series with the antenna.
C. placing a resistor in series with the antenna.
D. placing an inductance in series with the antenna.

Block 185

The correct answer is D.

Here is your next question: Parasitic elements are useful in a receiving antenna for:

A. increasing directivity. C. increasing sensitivity.
B. increasing selectivity. D. both A and C.

Block 186

The correct answer is C.

Here is your next question: In regards to shipboard satellite dish antenna systems, azimuth is:

A. vertical aiming of the antenna. C. 0 to 90°.
B. horizontal aiming of the antenna. D. north to east.

Block 187

The correct answer is B.

Here is your next question: What is the effect of adding a capacitor in series to an antenna?

A. Resonant frequency will decrease.
B. Resonant frequency will increase.
C. Resonant frequency will remain same.
D. Electrical length will be longer.

Block 188

The correct answer is B.

Here is your next question: If a transmission line has a power loss of 6 dB per 100 feet, what is the power at the feed point to the antenna at the end of a 200-foot transmission line fed by a 100-W transmitter?

A. 70 W. B. 50 W. C. 25 W. D. 6 W.

Block 189

The correct answer is D.

Here is your next question: Waveguides are:

A. used exclusively in high-frequency power supplies.
B. ceramic couplers attached to antenna terminals.
C. high-pass filters used at low radio frequencies.
D. hollow metal conductors used to carry high-frequency current.

Block 190

The correct answer is D.

Here is your next question: Which of the following represents the best SWR?

A. 1:1. B. 1:2. C. 1:15. D. 2:1.

Block 191

The correct answer is A.

Here is your next question: A 520-kHz signal is fed to a ½-wave Hertz antenna. The fifth harmonic will be:

A. 2.65 MHz. B. 2650 kHz. C. 2600 kHz. D. 104 kHz.

Block 192

The correct answer is C.

Here is your next question: When a capacitor is connected in series with a Marconi antenna:

A. an inductor of equal value must be added.
B. no change occurs to antenna.

C. antenna open circuit stops transmission.
D. antenna resonant frequency increases.

Block 193

The correct answer is D.

Here is your next question: How do you increase the electrical length of an antenna?

A. Add an inductor in parallel. C. Add a capacitor in series.
B. Add an inductor in series. D. Add a resistor in series.

Block 194

The correct answer is B.

Here is your next question: A coaxial cable has 7 dB of reflected power when the input is 5 W. What is the output of the transmission line?

A. 5 W. B. 2.5 W. C. 1.25 W. D. 1 W.

Block 195

The correct answer is D.

Here is your next question: What is the 7th harmonic of 450 kHz when fed through a ¼-wavelength vertical antenna?

A. 3150 Hz. B. 3150 MHz. C. 787.5 kHz. D. 3.15 MHz.

Block 196

The correct answer is D.

Here is your next question: What is the 5th harmonic of a 450 kHz transmitter carrier fed to a ¼-wave antenna?

A. 562.5 MHz. B. 1125 kHz. C. 2250 MHz. D. 2.25 MHz.

Block 197

The correct answer is D.

Here is your next question: Waveguide construction:

A. should not use silver plating. C. should have short vertical runs.
B. should not use copper. D. should not have long horizontal runs.

Block 198

The correct answer is D.

Here is your next question: This drawing indicates: (Please refer to Fig. 4-2.)

A. current along a ½-wave Marconi antenna.
B. current along a ¼-wave Hertz antenna.

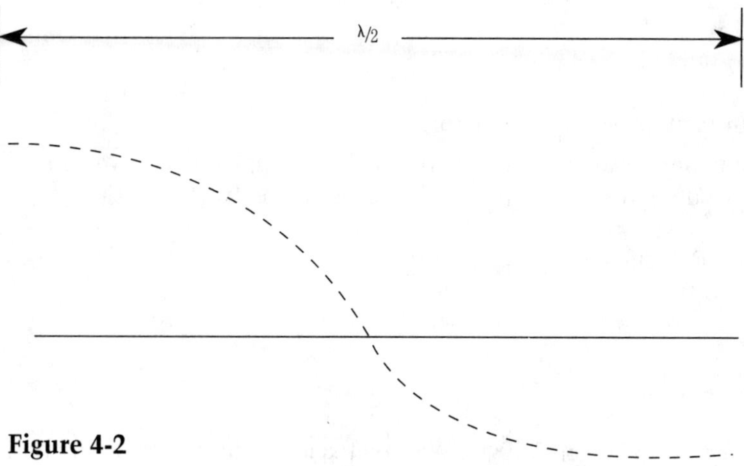

Figure 4-2

C. voltage along a ½-wave Hertz antenna.
D. voltage along a ¼-wave Hertz antenna.

Block 199

The correct answer is C.

Here is your next question: To lengthen an antenna electrically, add a:

A. coil. B. resistor. C. battery. D. conduit.

Block 200

The correct answer is A.

Here is your next question: How do you electrically decrease the length of an antenna?

A. Add an inductor in series. C. Add an inductor in parallel.
B. Add a capacitor in series. D. Add a resistor in series.

Block 201

The correct answer is B.

Here is your next question: If the length of an antenna is changed from 1.5 feet to 1.6 feet, its resonant frequency will:

A. decrease. B. increase. C. be 6.7% higher. D. be 6% lower.

Block 202

The correct answer is A.

Here is your next question: To couple energy into and out of a waveguide:

A. use wide copper sheeting. C. use capacitive coupling.
B. use an LC circuit. D. use a thin piece of wire as an antenna.

Block 203

The correct answer is D.

Here is your next question: An isolator:

A. acts as a buffer between a microwave oscillator coupled to a waveguide.
B. acts as a buffer to protect a microwave oscillator from variations in line load changes.
C. shields UHF circuits from RF transfer.
D. both A and B.

Block 204

The correct answer is D.

Here is your next question: A high SWR creates losses in transmission lines. A high standing wave ratio might be caused by:

A. improper turns ratio between primary and secondary in the plate tank transformer.
B. screen grid current flow.
C. an antenna electrically too long for its frequency.
D. an impedance mismatch.

Block 205

The correct answer is D.

Here is your next question: A properly installed shunt-fed, ¼-wave Marconi antenna:

A. has zero resistance to ground.
B. has high resistance to ground.
C. should be cut to ½ wave.
D. should not be shunt-fed.

Block 206

The correct answer is A.

Here is your next question: When a capacitor is connected in series with a Marconi antenna:

A. an inductor of equal value must be added.
B. no change occurs to antenna.
C. antenna open circuit stops transmission.
D. antenna resonant frequency increases.

Block 207

The correct answer is D.

Here is your next question: When excited by RF, a half-wave antenna will radiate:

A. a space wave.
B. a ground wave.
C. electromagnetic fields.
D. both electromagnetic and electrostatic fields.

Block 208

The correct answer is D.

Here is your next question: Waveguides are:

A. used exclusively in high frequency power supplies.
B. ceramic couplers attached to antenna terminals.
C. high-pass filters used at low radio frequencies.
D. hollow metal conductors used to carry high-frequency current.

Block 209

The correct answer is D.

Here is your next question: A 260-kHz signal is fed to a $\frac{1}{2}$-wave Hertz antenna. The fifth harmonic will be:

A. 2.65 MHz. B. 2650 kHz. C. 2600 kHz. D. 1300 kHz.

Block 210

The correct answer is D.

Here is your next question: The voltage produced in a receiving antenna is:

A. out of phase with the current if connected properly.
B. out of phase with the current if cut to $\frac{1}{8}$ wavelength.
C. variable depending on the station's SWR.
D. always proportional to the received field strength.

Block 211

The correct answer is D.

Here is your next question: A properly connected transmission line:

A. is grounded at the transmitter end.
B. is cut to a harmonic of the carrier frequency.
C. is cut to an even harmonic of the carrier frequency.
D. has an SWR as near as 1:1 as possible.

Block 212

The correct answer is D.

Here is your next question: Conductance takes place in a waveguide:

A. by interelectron delay.
B. through electrostatic field reluctance.
C. in the same manner as a transmission line.
D. through electromagnetic and electrostatic fields in the walls of the waveguide.

Block 213

The correct answer is D.

Here is your next question: In regards to shunt-fed ¼-wavelength Marconi antenna:

A. dc resistance of the antenna to ground is zero.
B. RF resistance from antenna feed point to ground is zero.
C. harmonic radiation is zero under all conditions.
D. it must be grounded at both feed and far ends.

Block 214

The correct answer is A.

Here is your next question: If a ¾-wavelength transmission line is shortened at one end, impedance at the open will be:

A. zero. B. infinite. C. decreased. D. increased.

Block 215

The correct answer is B.

Here is your next question: A dummy antenna is a:

A. nondirectional receiver antenna.
B. wide bandwidth directional receiver antenna.
C. transmitter test antenna designed for minimum radiation.
D. transmitter nondirectional narrow-band antenna.

Block 216

The correct answer is C.

5
CHAPTER

Electronic components and circuits

The usual method of putting electronic equipment into categories is to go from the simplest to the most complicated. An accepted order is:

- components
- circuits
- systems

Components

Components are individual parts like resistors, capacitors, and diodes. In this category, it is the practice to separate components into *active* and *passive* types. Active types generate a voltage and passive types do not.

Another way of classifying components is by whether they are *linear* or *nonlinear*. Linear components act according to Ohm's Law. If you double the voltage across a linear component, the current through it will also double. A carbon composition resistor is a linear component as long as it is operated within the manufacturer's specifications.

Doubling the voltage across a diode will not cause the current through it to double, so a diode is a nonlinear component. (Doubling the voltage across a semiconductor diode will likely destroy it if the diode is in the conducting state.)

Components are also classified as *bilateral* or *unilateral*. A bilateral component can conduct equally well in two directions. Carbon composition resistors are bilateral components. By way of contrast, a diode can conduct in only one direction, so it is unilateral.

Another distinction made between components is by the number of leads. Diodes are *two-terminal components*, and transistors are *three-terminal components*.

Circuits

Circuits are combinations of components that accomplish a specific purpose. Oscillators are circuits that change dc to ac. An audio amplifier circuit amplifies frequencies in the range of 20 Hz to 100 kHz.

Circuits can be sub-classified as being *analog* or *digital*. When the input to an analog (linear) circuit undergoes a continuous change, the output will follow the continuous change. There are a few exceptions to this and they will be discussed when amplifiers are considered.

A digital circuit has two outputs that can be called *on* or *off*, or *1* or *0*.

Systems

Systems are combinations of components and circuits that often accomplish a number of purposes. A computer is an example of a system; a radio receiver is another example.

Gray areas

There are some gray areas between the previous categories. In the right circumstance, a carbon composition resistor can be an active component. In other words, it can generate a voltage. For example, at any temperature above –273.3°C, a resistor generates a noise voltage that is very important in communications. The noise voltage increases as the temperature increases. (*Note:* –273.3°C is called *absolute zero*, or zero degrees Kelvin.)

Diodes, like resistors, are usually considered to be passive components. However, all diodes generate noise voltage at temperatures above absolute zero.

Noise voltage

To understand what causes the noise voltage in a resistor, consider the fact that at any temperature above 0°K there are electrons that escape from their atoms. For a very short time those electrons move in random directions inside the resistor. That motion of free electrons is called *intrinsic current*.

At any one instant of time, more electrons are moving toward one resistor lead and away from the other. At that instant, a low voltage is across the resistor (Fig. 5-1).

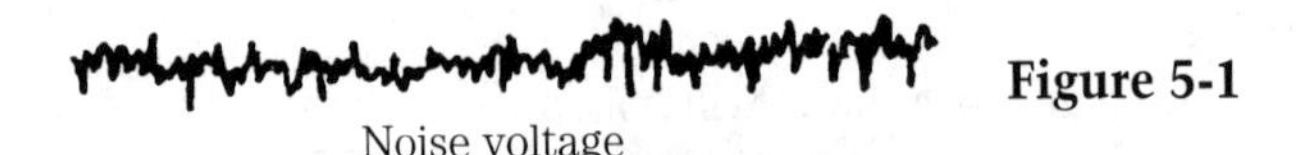

Noise voltage

Figure 5-1

At the next instant, more electrons are moving toward the opposite terminal and the polarity of the voltage across the resistor is reversed.

Over a short period of time, the random motion of electrons produces an ac voltage across the resistor. That ac voltage is called the *noise voltage* and it has a very wide range of amplitudes and frequencies.

Because of its random frequency distribution, the noise voltage is sometimes referred to as *white noise*. That is supposed to be a comparison to the color white, which is sometimes thought to be a combination of all color frequencies.

If you connect a resistor across the antenna input terminals of a receiver, you introduce a noise voltage into that receiver. In fact, noise is always introduced into a receiver by resistors in the RF amplifiers of the receivers. That noise is heard as a hissing sound when you turn up the volume control of a receiver. If it is a television receiver, it will also increase the "snow," or white spots on the screen. (The receiver should be tuned off station to hear and see the noise.)

Note: Communications receivers often employ a *squelch* circuit to eliminate the annoying receiver noise. However, television receivers do not utilize squelch circuits and the receiver noise is easily seen and heard. Again, turn up the volume control with the television receiver tuned to a channel not in use. The snow on the screen is also due to noise voltages.

To summarize, resistors are two-terminal bilateral components that can be either active or passive. In some applications the noise voltage of the resistor is disregarded.

In some audio systems, the term *component* is used to mean a smaller system that is part of a larger system. Therefore, a tape player and a compact disc player are examples of components when that way of categorizing is used.

Diodes

Despite the fact that semiconductor diodes are strongly preferred over tube types in newer designs, there are still a few tube types around. It is not a good practice to know only the semiconductor types.

Vacuum-tube rectifier diodes

The vacuum tube diode is easy to understand and its essential features are shown in Fig. 5-2. Either a filament or cathode sleeve surrounding the filament is heated to the point where electrons escape from its surface. The emitted electrons form a *space charge*, sometimes referred to as a *space cloud*, around the cathode.

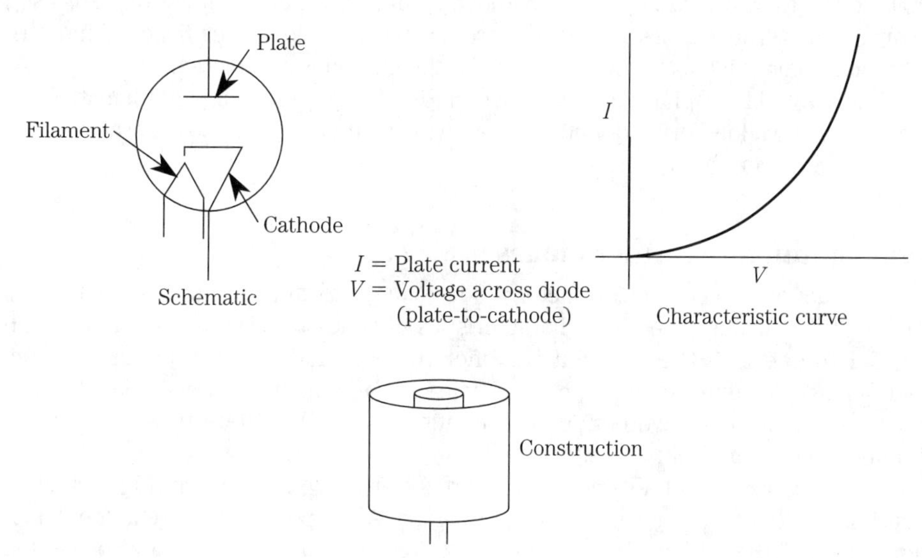

Figure 5-2

As shown in the illustration, the filament (or cathode) and plate of the diode are concentric. When the plate is made positive with respect to the cathode it attracts electrons from the space charge. When the plate is made sufficiently positive to deplete the space charge, the tube is said to be entering saturation. Complete loss of the space charge can destroy the tube.

An important feature of the vacuum-tube diode is that electron current can flow from the filament (or cathode) to the positive plate. However, when the plate is made negative with respect to the cathode (or filament), electron current cannot flow from the plate to the cathode during normal operation of the tube. See Fig. 5-3.

This feature makes it possible to rectify an ac voltage. This is also illustrated in Fig. 5-3. With an ac input that has positive and negative half cycles applied to the plate, only the positive half cycles can produce electron flow through the tube.

Rectifier circuits are covered in greater detail in the chapter that includes a discussion on power supplies.

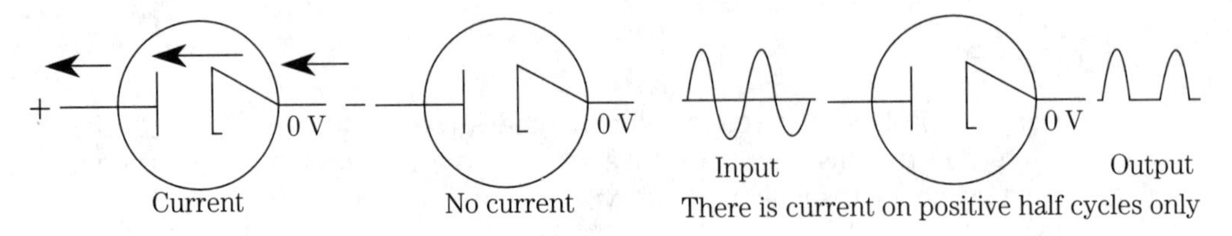

Figure 5-3

Phanatrons

Phanatrons are gas-filled diodes. Mercury vapor is used as a gas for high-current diodes. Inert gases, such as neon and krypton, are other examples of gases used in phanatrons. Those gases are called inert because they do not readily combine with other materials, such as the metals used for the cathodes and plates.

The symbol for a phanatron is shown in Fig. 7-2. An advantage of phanatrons is that they have a low forward voltage drop, usually about 15 volts. Also, these tubes can conduct a very high current.

Semiconductor rectifier diodes

As a general rule, semiconductor diodes can perform the same rectification as vacuum tube diodes. The main characteristics of solid-state rectifiers are shown in Fig. 5-4. At one time, the solid-state rectifier diodes included such types as selenium rectifiers, germanium rectifiers, copper oxide rectifiers, and others. Today, the silicon types are strongly favored for most rectifier circuits, but there are still some germanium types being sold.

The construction of a semiconductor rectifier diode is shown in Fig. 5-4. Note that it is made with P-type and N-type materials. An important feature is the depletion region at the junction, which is formed during the manufacturing process and plays an important part in the operation of semiconductor diodes.

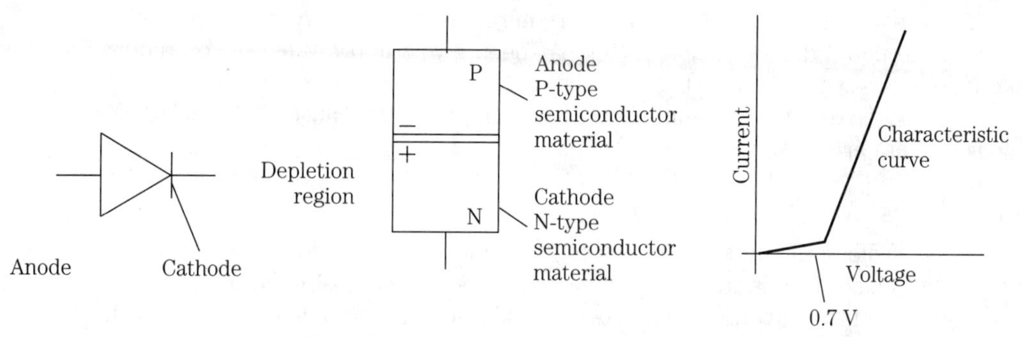

Figure 5-4

The P- and N-type materials are made of a semi-conductor material that has been doped with certain impurities. As a result of that doping, there are negative charge carriers available in the N-type material, and positive charge carriers available in the P-type material.

The available charge carriers in N-type material do not make it negatively charged. Think of it this way: copper is a material that has negative-charge carriers (electrons) available. However, copper is not a negative-charged material. In the same way, the positive-charge carriers (holes) in P-type material do not make it a positively charged material.

When the diode is first manufactured there is a motion of charge carriers that results in a voltage across the junction between the N- and P-type materials. The junction itself is called the *depletion region* because it has very few charge carriers.

For some applications, the depletion region is *considered to be* an insulating region. In reality, it is a semiconducting region. That means it is a better conductor than insulating materials, but it is not as good for conducting electricity as a conductor.

Compare the characteristic curves of the vacuum-tube type with solid-state (silicon) rectifier in Figs. 5-2 and 5-4. Observe that current in the tube type starts to flow almost as soon as the plate is made positive. By contrast, the silicon type requires a forward voltage of about 0.7 V before current flows through the diode. That is true for all devices made with a silicon PN junction.

The reason a forward voltage is required before conduction is that the charge carriers must be given an impetus to cross the depletion region. There is an internal voltage across the depletion region, as shown in its construction (Fig. 5-4), which is a reverse voltage in the opposite direction as the external voltage required for current flow through the diode.

No current can flow until the external voltage is sufficiently high to overcome the junction potential, explaining the need for the 0.7-V junction potential needed to get the current started.

For a germanium PN junction, the required forward voltage is about 0.2 V. Gallium Arsenide (GaAs) types require a forward voltage of about 1.5 V.

The important parameters of junction rectifier diodes are:

- The maximum allowable forward current.
- The maximum allowable reverse voltage across the diode. (This is also called the *peak inverse voltage* or *PIV*.)

The breakover voltage is the voltage required at the junction to obtain forward current through a diode. Although this has been given as 0.7 V for silicon diodes, there are cases where that voltage is a slightly different value. Actually, values from 0.7 V to 1.1 V are often encountered. That is important because it can affect the diode circuit performance.

Load lines

Because diodes are not linear devices, you cannot calculate their current flow by using Ohm's Law! That is true for both vacuum tube and solid-state types.

If you need to know how much current will flow through a diode in a given circuit it is possible to use a special diode equation. However, it is usually just as quick and sufficiently accurate to plot a load line for the same purpose. The procedure is shown in Fig. 5-5. In this case, the diode characteristic curve is assumed to be available from the manufacturer.

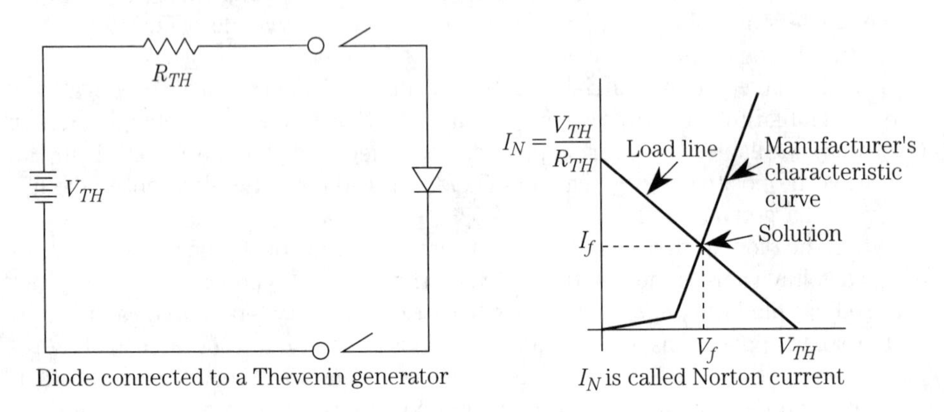

Figure 5-5

Thevenin's Theorem says that *any two-terminal network comprised of linear, bilateral circuit elements and one or more sources of voltage can be replaced with a single voltage source having no internal resistance and a single equivalent series resistor.*

The source is called a *constant-voltage* or *Thevenin voltage* source (V_{TH}). The resistance is called the *Thevenin equivalent resistance* (R_{TH}). The Thevenin equivalent circuit is shown in Fig. 5-5. It is important to understand that all complicated two-terminal active circuits can be replaced with this circuit.

While on the subject of equivalent generators, you should be familiar with the Norton generator, also illustrated in Fig. 5-5.

The Norton generator is made with a constant-current source I_N. There is a resistor (R_N) connected across its terminals. As with the Thevenin generator, the Norton generator can be used to replace any active two-terminal network for the purpose of making calculations.

The load line solution gives information about the diode when placed in a Thevenin equivalent circuit. Whatever it does in the equivalent circuit is the same thing that it will do in the original circuit.

The load line gives important information about the diode behavior in the original circuit. It tells the amount of current through the diode and the forward voltage across that diode when it is in operation.

As shown in Fig. 5-5, to draw the load line the first step is to locate the open-circuit voltage on the voltage line. The open-circuit voltage is the terminal voltage with nothing connected across the terminals, so it is equal to the Thevenin voltage (V_{TH}).

The second step is to locate the short-circuit current on the current line. The short-circuit is the current that flows when the terminals are shorted together. By Ohm's Law it is equal to V_{TH}/R_{TH}. (It is known as the *Norton Current* (I_N).

The load line is drawn between those points.

The point where the load line crosses the characteristic curve is called the *solution*. When that point is projected to the left it locates the actual diode current (I_f) on the current axis. When the point is projected downward the forward voltage across the conducting diode (V_f) is located on the voltage axis.

Piecewise analysis

The next logical question is, what happens if you don't have the diode characteristic curve? There is a simple technique, called *piecewise analysis*, that allows you to draw the characteristic curve from the typical operating characteristic supplied by the manufacturer. Figure 5-6 shows how it works.

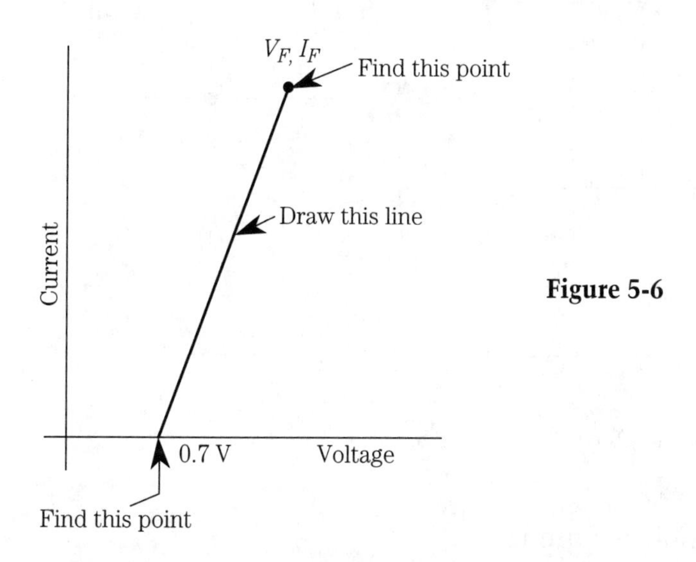

Figure 5-6

The forward breakover point is determined by the type of diode and the type of material it is constructed with. As stated before, it is usually considered to be about 0.7 volts for a silicon diode. The first step is to locate that point on the voltage line.

The second step is to locate the typical operating point on the graph. That point is supplied by the manufacturer and is a graphed point at V_F, I_F.

Once the piecewise characteristic curve is drawn, the load line is drawn as is shown before.

You will probably never be asked to draw a load line for a test question (it messes up the exam papers)! However, the procedure just given explains how diodes can be selected and what the typical operating conditions are needed for. Load lines are drawn for other devices as well as for diodes. You might be asked to interpret a load line for a specific device.

Series and parallel diode connections

Diodes can be connected in parallel to get a higher forward current rating and they may be connected in series to get a higher voltage rating (Fig. 5-7). Note that starting resistors are needed to assure that all of the parallel diodes are conducting.

As shown in Fig. 5-7, when diodes are connected in series it might be necessary to add parallel resistors to assure that the reverse voltage across the series diodes is evenly distributed. The parallel capacitors assure that unequal junction capacitance values cannot result in an excess voltage across a capacitor, and it can also serve to bypass high-frequency destructive transient voltages around the rectifiers.

In addition to rectifier applications, there are other circuits that use junction diodes. Also, there are specially-designed junction diodes, such as the varactor diode and the zener diode.

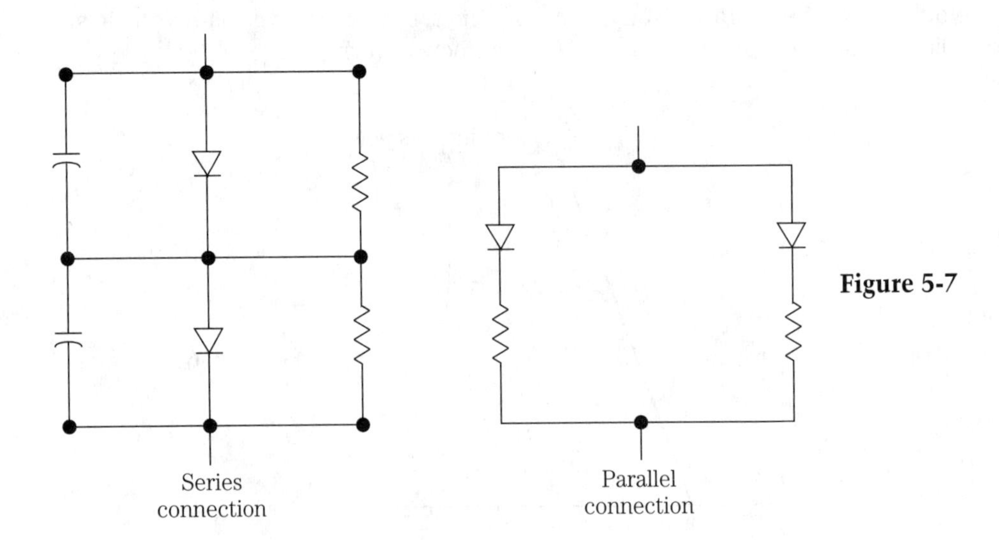

Figure 5-7

Reverse-biased diodes

A diode is reverse-biased when a positive voltage is applied to its cathode and a negative voltage is applied to its anode. Two types of diodes must be operated with reverse bias if they are to perform their required applications. They are *varactor diodes* and *zener diodes*.

For most diodes, reverse bias (+ on cathode, – on anode) means no current and, therefore, no operation. However, for the diodes that will be discussed in this section reverse bias is required.

Varactor is a name derived from *V*oltage v*AR*iable cap*ACiTOR*. As the name implies, it behaves like a capacitor in a circuit. The action depends upon the depletion region between the N-type and P-type materials.

As shown in Fig. 5-8, the depletion region can be made larger by using reverse bias on the diode. The reason is that the positive voltage on the N-type material attracts the negative charge carriers in N-type material away from the junction. At the same time, the negative voltage on the anode voltage attracts the positive charge carriers (holes) in the P-type material away from the junction. The overall result is that a reverse bias on a junction diode increases the size of the depletion region on the junction.

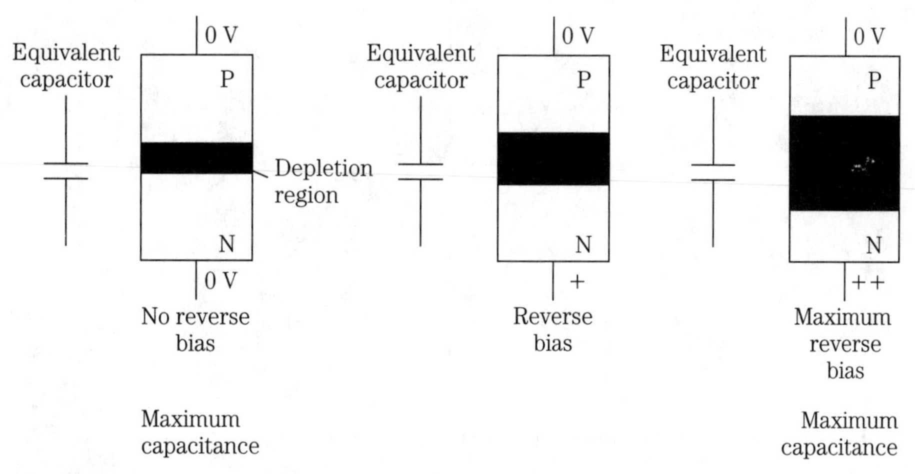

Figure 5-8

If you view the varactor diode as being two conducting regions (N-type and P-type materials) separated by an insulating (depletion) region, then the diode is a form of capacitor. It is because the distance between the capacitor plates can be controlled by the amount of reverse bias. That, in turn, means that the capacitance of the varactor diode can be controlled by the amount of reverse bias voltage across it.

When the plates of a capacitor are moved further apart, its capacitance is decreased. Saying it another way, the capacitance of a varactor diode decreases by increasing the amount of reverse bias.

When the reverse bias is decreased, the plates of the capacitor move closer together increasing the capacitance. The relationship between reverse bias and capacitance is illustrated in Fig. 5-8.

The reverse bias is not sufficient to break down the depletion region of the diode. If the depletion region was forced into conduction, that is, subjected to a voltage high enough to break down the depletion region, there could be a destructive reverse current, destroying the diode.

Varactor diodes are used for tuning receivers and for tuning inductors. They are also used for making the parametric amplifiers that are discussed later.

The zener diode

Although the depletion region of a junction diode is considered to be an insulator in the discussion of the varactor diode, it is actually a semiconductor. If a sufficiently high reverse bias is connected across a semiconductor diode, it will conduct in the reverse direction. When that happens, it is no longer acting like a traditional diode. The characteristic curve in Fig. 5-9 shows that the reverse current increases very rapidly once the breakdown voltage is reached. That breakdown voltage is called the *zener breakdown voltage*.

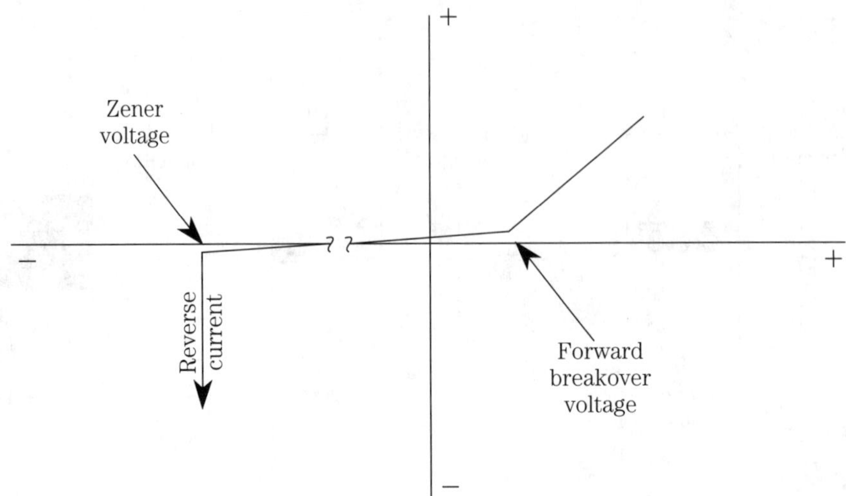

Figure 5-9 This device is operated in the 3rd quadrant

The reverse bias breakdown voltage (that is, the zener voltage) occurs with all semiconductor junction diodes. However, when it occurs in a germanium diode, the diode will be destroyed. Reverse bias will not destroy silicon or gallium arsenide diodes *provided they are connected into a circuit that limits the amount of reverse current!*

Zener diodes are specially constructed to allow them to operate continuously with a reverse current. Always remember that the reverse current must be limited, since excessive reverse current can destroy the diode. For that reason, there is always a series (current limiting) resistor used with operating zener diodes.

The reverse voltage required to produce a zener breakdown can be controlled by the amount of doping used for the PN regions. That means the zener diode can be tailor-made for a specific zener voltage value.

The characteristic curve for a zener diode in Fig. 5-9 shows that the voltage is very nearly constant when the reverse current is flowing. That permits zener diodes to be used for obtaining a fixed amount of voltage for a range of current values. They are called *voltage regulators* (or *voltage references*) in those applications.

Neon lamps can be used for the same purposes as zener diodes at higher voltages, that is, voltages above 60 V. As with zener diodes, neon lamps must be operated with a series-limiting resistor.

Figure 5-10 shows a circuit using the zener diode as a voltage regulator.

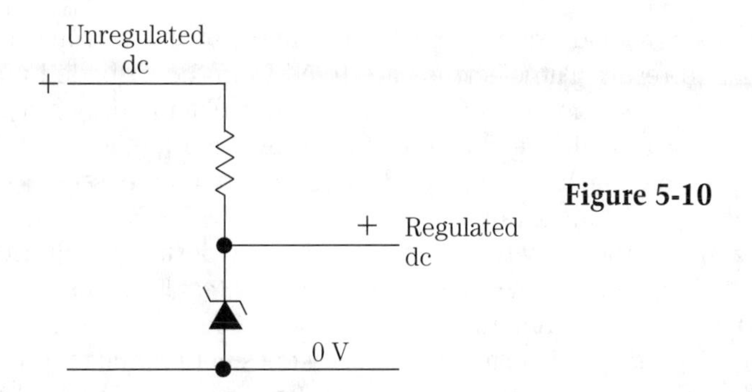

Figure 5-10

Noise diodes

Consider the case of a forward-biased junction-diode. When negative charge carriers cross the junction, they arrive at the P-type anode lead at random times. That means there are very low-amplitude variations in the current output from the diode. Those low-amplitude variations are called *noise*, which has nothing to do with sound that is noise. It has the same waveform as created by a resistor operated above absolute zero temperature.

There is always a noise voltage produced by current flowing through a semiconductor material. That was explained in chapter 1 with reference to resistor noise.

The combination of junction noise and semiconductor noise in a junction diode is usually thought of as being undesirable. However, there are *noise diodes* made for the purpose of utilizing and enhancing the generated noise when conducting a forward current. The noise produced by these diodes is used for circuit testing and for high-frequency measurements.

Production of x-rays and visible light by electron energy

In this discussion of x-rays and light obtained by electron energy, the term *kinetic energy* is used. Kinetic energy is energy by virtue of motion, and *potential energy* is energy by virtue of position. So, an electron moving at a high speed has kinetic energy. Conversion of that energy to other energy forms is one of the basic techniques involved in obtaining x-rays and visible light from moving electrons.

The amount of energy an electron has is not constant and can be increased or decreased by a number of different methods. For example, heating a material can increase the energy level of its electrons, a high-intensity electric field can increase the electron's energy, and a sudden decrease in speed can be used to decrease the energy of an electron.

When an electron loses energy it must get rid of its extra energy by some means. (Remember that energy cannot be created or destroyed, but energy can be changed from one form to another.)

An x-ray tube is a good example of how the energy of an electron can be used to produce a desired result. The electrons are freed from the cathode by heating the cathode material. Those electrons are focused and accelerated by a high-intensity electric field to a very high speed as they move toward a metal target. When they strike the target, they must give up the kinetic energy they possess as a result of their high speed by emitting x-rays.

In the example of the x-ray tube, the x-rays are the desired result. However, it is important to understand that any device that accelerates electrons to a high speed can be a producer of x-rays, and they can be undesirable.

In some devices, the rapid change in velocity does not produce x-rays. Instead, it produces light in the visible spectrum. The neon lamp shown in Fig. 5-11 is an example. When a high dc voltage is connected across the lamp, some free electrons in the neon gas are accelerated toward the more positive electrode. Those electrons strike gas molecules and free more electrons, which strike even more electrons. This process, called *avalanching*, continues until the electrons arrive at the positive electrode.

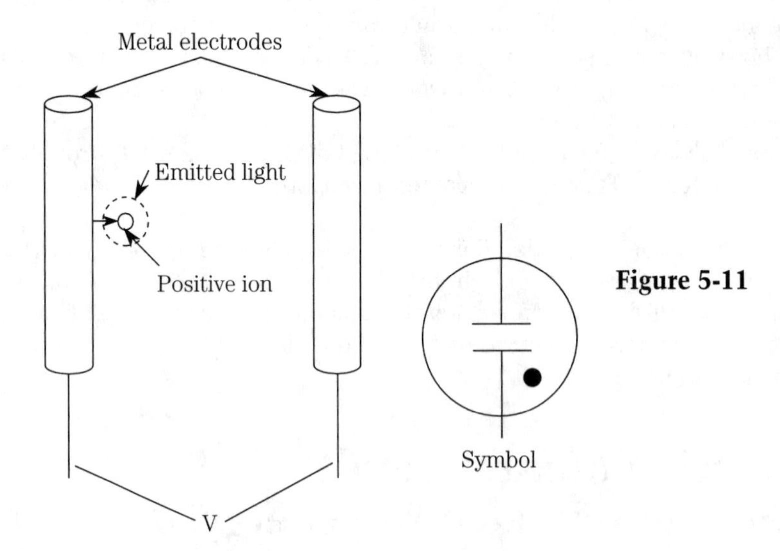

Figure 5-11

The atoms of gas that have lost an electron are called *positive ions*. (At one time the term "pions" was a more popular term.) The positive ions move toward the negative electrode. When they are very near that electrode, they pull electrons off its surface. This is called *field emission* of electrons.

The emitted electrons accelerate toward the positive ions. When they combine with those ions, they must give up their kinetic energy, which is released in the form of visible light. By mixing various gases with the neon, it is possible to obtain various colors of light.

LEDs and LADs

Figure 5-12 shows the cross section of a light-emitting diode (LED). It is forward biased so that electrons are crossing the junction. In order to get across that junc-

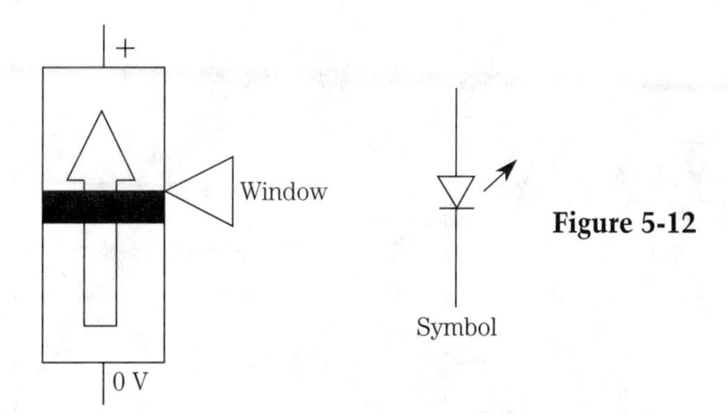

Figure 5-12

tion, they must be given extra energy, which comes from the positive voltage on the anode. When the electrons reach the P-type material used for the anode, the electrons release that extra energy in the form of light.

For silicon and germanium, the amount of light produced by the decelerating electrons is not significant. However, when the diode is made with gallium arsenide (GaAs), the electron must be given a greater impetus to cross the depletion region by using a higher forward voltage (1.5 V). The electrons thus have a higher energy and more light is produced when they give that energy up.

That is one reason why LEDs are made with gallium arsenide. Gallium aluminum oxide (GaAIAs) is also used.

As shown in Fig. 5-12, a very small window allows the light to pass to the outside world. That is how the LED produces visible light. Different doping materials of the semiconductor material produces the range of light colors available. Colored lenses can also be used for that purpose.

By proper design of the diode, it can be made to emit infrared light. Precise design and construction of the junction makes possible the laser diodes.

Light-activated diodes (LADs) are also called *photodiodes*. They are forward biased, but they do not conduct a forward current in the absence of light. When their relatively large depletion region is exposed to light, they conduct in the forward direction, but not in the reverse direction, so they are unilateral devices.

There are two ways to make photodiodes. They are compared in Fig. 5-13. In the first method, the diode has a very thin P-type material for its anode. When activated, the light shines through the P-type anode. Photons of light strike the molecules in the depletion region, releasing electrons that are attracted by the anode. Without the light there is not a sufficient number of charge carriers in the depletion region to support current flow.

The second photodiode device shown in Fig. 5-13, is called a *PIN diode*. The letters stand for *P*ositive-*I*ntrinsic-*N*egative. The term *intrinsic* is used to explain the fact that the middle material is a semiconductor material that has not been doped. *Remember:* N-type material is doped with donor impurities and P-type material has been doped with acceptors.

A small window exposes the depletion region of the PIN diode to the outside world. This type of diode is especially sensitive to infrared energy. For example, PIN

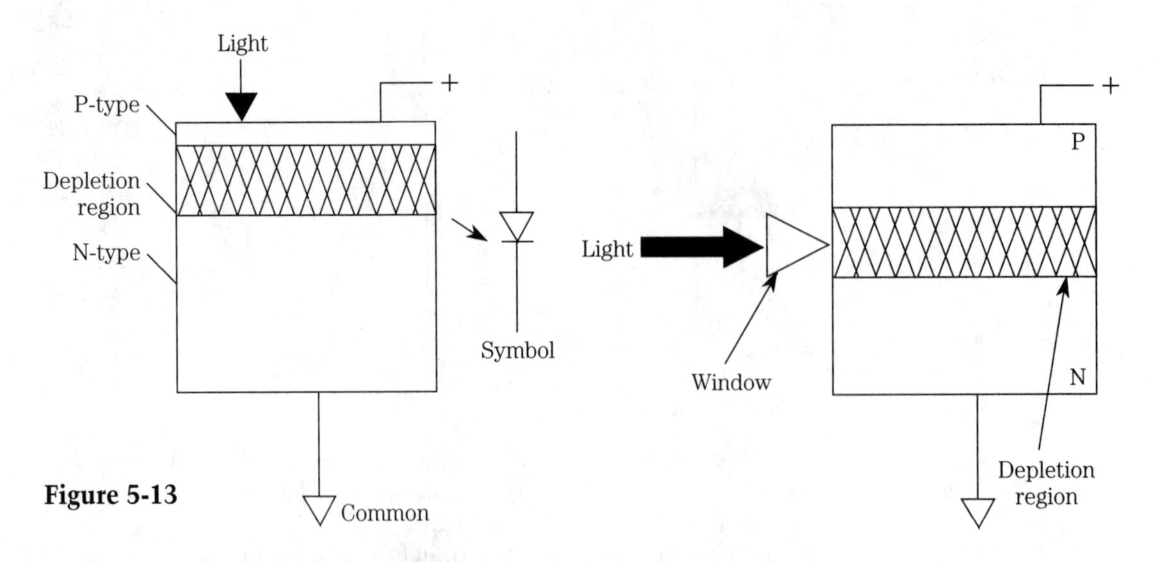

Figure 5-13

diodes are used to detect infrared energy from television remote controls. PIN diodes are also used at microwave frequencies, but in that case they do not have the light-transmitting windows.

Hot-carrier diodes

The concept of hot-carrier diodes is shown in Fig. 5-14. The P-type material *usually* used for the anode in junction diodes has been replaced by metal. There is practically no depletion region between the N-type cathode material and the metal anode.

The obvious advantage of this construction is that there is no junction voltage to be overcome before current starts to flow. Thus, as shown in the characteristic curve in Fig. 5-14, the forward current starts nearly at the point where forward voltage starts. This is especially important in low-voltage circuits.

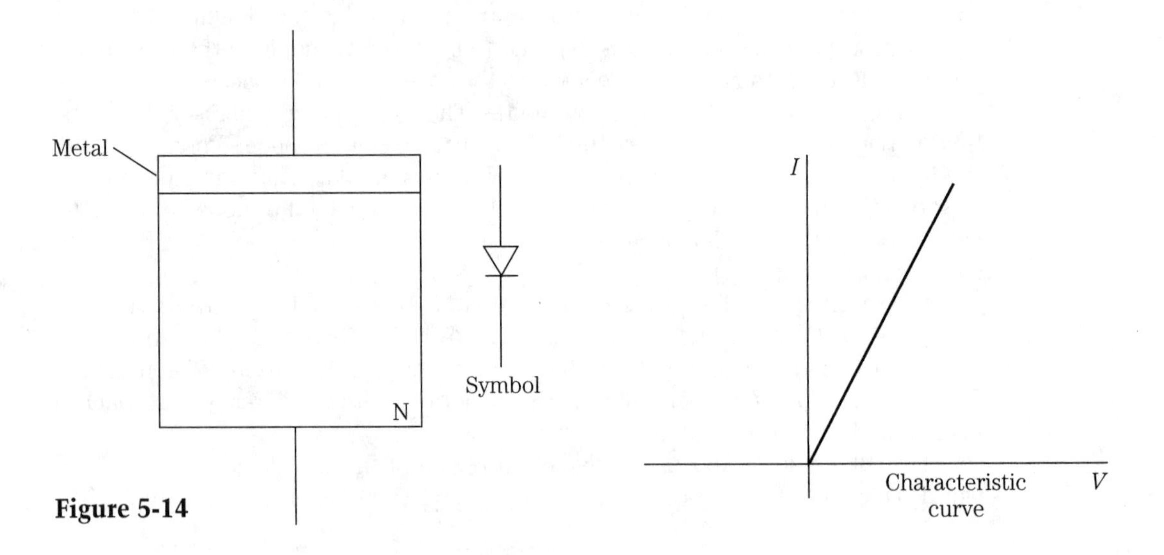

Figure 5-14

Figure 5-15 shows the symbol and characteristics of a point-contact diode. Because it uses a metal-to-semiconductor interface, it can be classified as an example of a hot-carrier type. The difference is in the construction, as shown in the illustration.

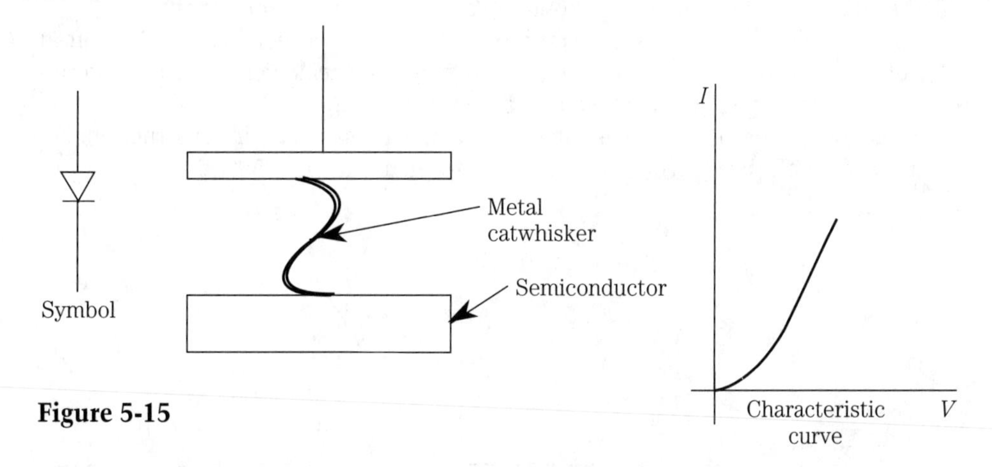

Figure 5-15

Because the anode is made with a piece of wire, there is very little area to contribute to the junction capacitance. That prevents high-frequency signals from sneaking through the interface without being rectified. This type of diode is used as a detector in radio and television receivers.

Three-layer and four-layer diodes

Consider the characteristic curve of the neon lamp shown in Fig. 5-16. It will not conduct until the applied voltage reaches a specified value called the breakover voltage (V_{bo}).

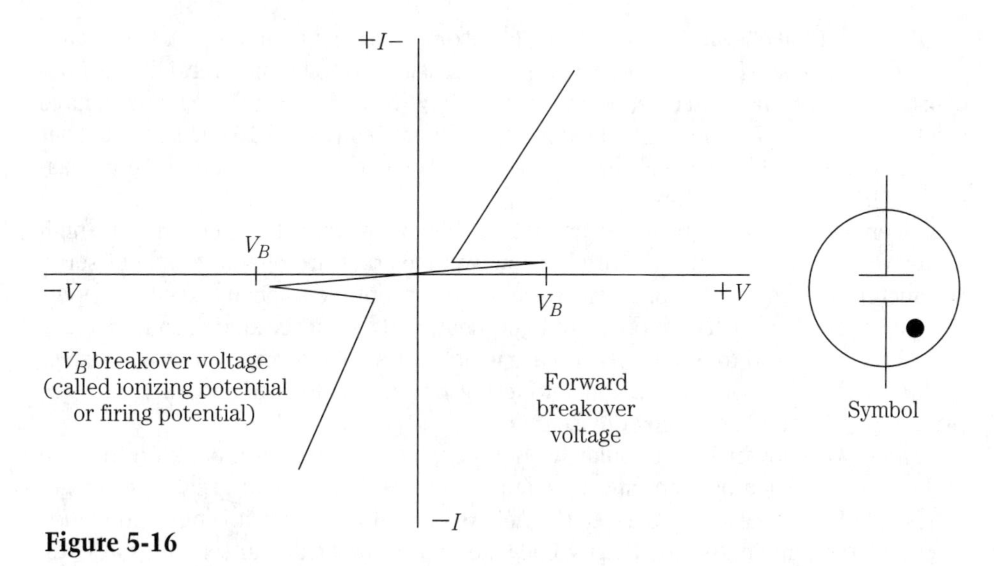

Figure 5-16

In the construction of the neon lamp, the manufacturers have (in some types) inserted a very small amount of radioactive material, which serves to guarantee that the lamp will start to conduct at a precise voltage because its radiation keeps the gas partially ionized. Without that radioactive material, the lamp starting voltage can be influenced by light, heat, and just about any other kind of radiant energy.

Now consider the characteristic curve of the three-layer diode (Fig. 5-17), often referred to as a *diac*. Except for value of voltage at the breakover point, the characteristic curve is very similar to the curve of the neon lamp.

The major difference between diacs and neon lamps is in their construction. Diacs are made with three layers of semiconductor material (Fig. 5-17).

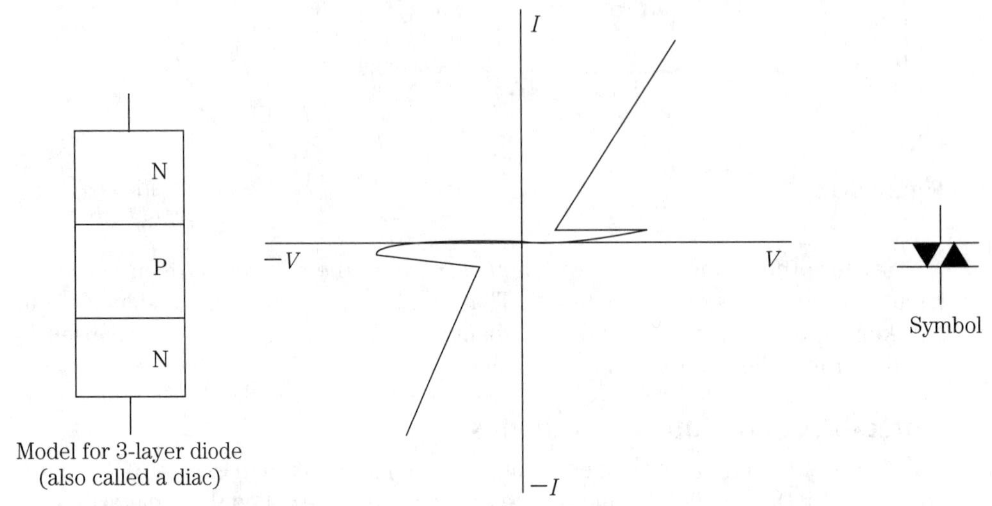

Figure 5-17

The essential characteristics of four-layer diodes are given in Fig. 5-18. They are also called *Shottky diodes*. These diodes are considered to be unilateral devices because they have a favored direction of current flow. Neon lamps can be made to have a similar characteristic curve by shaping the electrodes (Fig. 5-16). Current in that device is unilateral because electrons flow from the cathode to the anode more easily than from anode to cathode.

Neon devices are not nearly as popular as they were at one time, because the high breakover voltage does not fit with the low voltages that are characteristic of semiconductor devices. Also, neon devices are more susceptible to changes in their operating characteristics when exposed to light, heat, and any other kind of radiation.

Three-layer and four-layer diodes are usually classified as *thyristors* because of their well-defined conducting/nonconducting operation. In other words, they are primarily solid-state switching devices.

Note: We know it isn't popular to include vacuum tubes in a book on modern technology. As soon as transistors became popular, there was a great rush to get tubes out of textbooks. However, in the field of communications, it is not a good idea to completely eliminate tube theory because high-power transmitter tubes are still

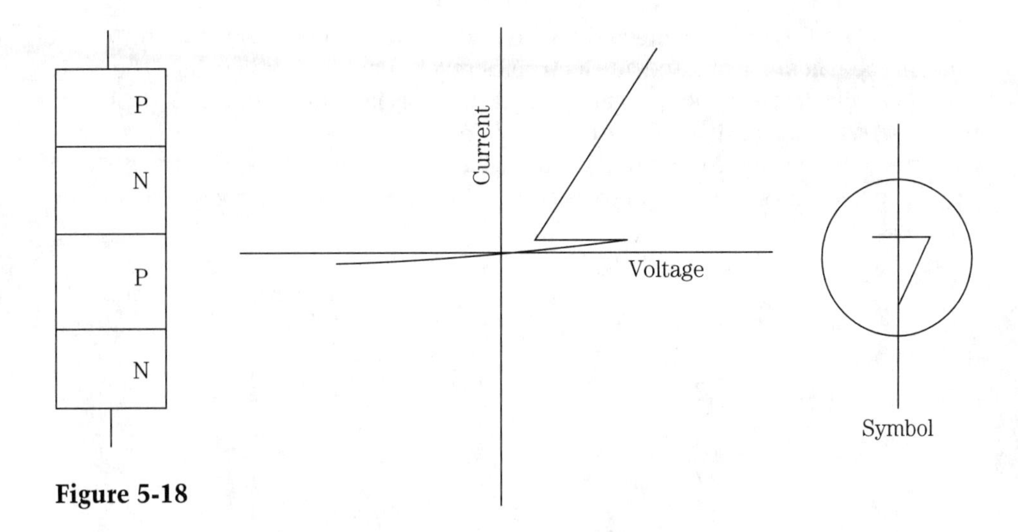

Figure 5-18

the predominant amplifier devices. Having said that, please know the main emphasis in this book is on solid-state components, circuits, and systems.

Vacuum tube amplifying devices

An important feature of many electronic circuits and systems is the ability to amplify a weak signal into a signal with a relatively high voltage or power. Tubes and various types of transistors are used for that purpose.

We start this section on amplifying devices with the study of vacuum tubes. Designers have often had to devise clever, and sometimes complicated, circuitry to get (bipolar) transistors to act like tubes.

Here is an example: vacuum tube devices have a high input impedance, which means you don't have to supply signal *power* (I^2R) to get the tube to amplify. In order to get a high input impedance for a (bipolar) transistor, it has been necessary to use an auxiliary circuit, like the *bootstrap circuit*, to simulate the high input impedance. (That circuit is covered in this book.)

The relatively low input impedance, and the fact that the (bipolar) transistor requires dc standby power to its input circuitry, is a disadvantage that has nagged designers for years. Their methods of circumventing the problem sometimes make the transistor circuit difficult to understand when you are primarily interested in amplification. The black box approach, covered later in this chapter, strips away the complicated circuitry.

Three-terminal vacuum tube amplifying devices

Important note: Many of the features of tubes discussed in this section apply equally to solid-state amplifying devices. This is especially true of field-effect transistors. Do not skip this material in the mistaken belief that it does not apply to modern equipment!

When Lee DeForest connected a wire grid between the cathode and plate of a vacuum tube diode, he started an electronics revolution that is still going. The basic concept of his device, which he called the *audion*, is illustrated in Fig. 5-19. Today, the newer versions of that tube are called *triodes* because of their three-electrode construction. The electrodes are the plate (P), the control grid (G) and the cathode (K). (The necessary filament is not usually included as a tube electrode in schematic drawings.)

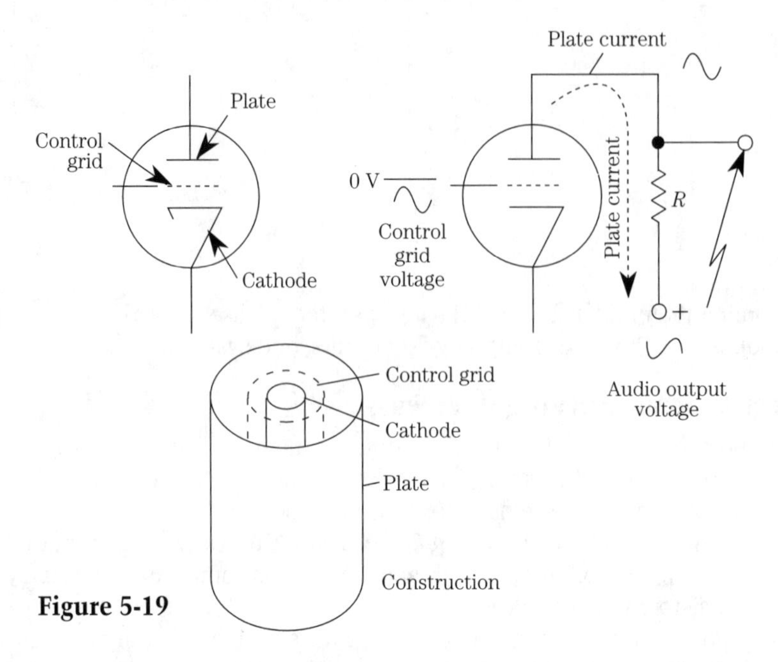

Figure 5-19

The principle of operation for the audion (and triode) is basically the same concept that applies to all amplifying devices; that is, a small-signal input results in a large-signal output.

The control grid is usually located closer to the cathode than to the plate because at that point the energy of the electrons is low. Therefore, a small amount of negative voltage on the grid has a great effect on the number of electrons per second that reach the plate.

Similar charges repel, so the negative grid voltage prevents some of the electrons from reaching the plate. The more negative the control grid, the fewer the number of electrons that can reach the plate. Saying it another way, a small change in the negative grid voltage will have a great effect on the plate current.

Voltage amplification

If a low-amplitude audio voltage is applied to the control grid, a high-amplitude audio plate current results. When that plate current flows through a resistor, a high-amplitude audio voltage is available at the plate (Fig. 5-19).

Observe that the control-grid signal voltage and the plate signal current are in phase. In other words, their maximum and minimum points occur at the same instant.

For the typical operation of a triode amplifier, the varying plate current flows through a resistor. That creates an audio voltage at the plate that has a higher amplitude than the audio voltage at the grid.

Because a low-amplitude audio input voltage produces a high audio output voltage, it is said that the tube "amplifies" the audio signal. That is not exactly correct, but the concept is popular and it is often used to explain audio voltage amplification.

In reality, the power supply sends current through the tube. The low-amplitude audio voltage on the grid can exert a great amount of control over that power supply current. Therefore, the tube doesn't actually amplify the signal. Rather, it uses the input signal to control the power supply current, which is where the amplification comes from.

Note carefully that the control grid voltage is not allowed to reach zero volts and it never goes positive. The waveforms used for comparing grid *voltage* and plate *current* in Fig. 5-19 are somewhat misleading; it is like comparing apples with oranges. Those two waveform amplitudes cannot be directly compared because one is current and the other is voltage. The current is drawn with a higher amplitude in that illustration only to emphasize the fact that a low-amplitude voltage controls a high-amplitude plate current.

Note that the control grid signal voltage and plate signal voltage shown in Fig. 5-19 are 180° out of phase, because the voltage across the load resistor (R_L) subtracts from the power supply voltage to give the voltage at the plate. For example, when the plate current is maximum, the drop across the load resistor is also maximum. When subtracted from the power supply voltage, that makes the voltage at the tube plate minimum.

Power amplification

The same concept of voltage amplification applies to the so-called audio power amplifier. Again, a relatively low amplitude audio voltage signal is applied to the grid, controlling the amount of current through the tube, and a high-power (I^2R) is delivered to the output circuit. Audio voltage amplifiers and audio power amplifiers do the same thing. One important operating difference is in the amount of current they are constructed to conduct.

A characteristic of tube power amplifiers is that they operate with a higher filament temperature, producing a higher current. If the grid of a power amplifier was located too close to the hot filament (or cathode), it would become heated and would likely emit electrons. That would be undesirable.

Locating the control grid a greater distance away from the cathode means that it has less control over the plate current so that the input signal voltage to the power amplifier must have a higher amplitude than the input to a voltage amplifier. *Power amplifiers in both tube and solid-state devices have a low-voltage amplification characteristic!*

Tube operation

Tube operation can be discussed by using the tube symbols. However, remember that the symbol gives no information about how the tube is actually constructed. For example, the symbol for the triode and the construction of most triodes are both shown in Fig. 5-19. As shown in the illustration, the cathode, grid, and plate are concentric in construction, but the symbol does not indicate that.

The grid of the triode is normally negative with respect to the cathode and plate. The negative grid dc voltage is called the *bias*. If the grid is allowed to become positive with respect to the cathode it will draw current. That would be destructive in most cases because the grids are made of small-diameter wire. In one application, called *grid leak bias*, the grid is permitted to become positive for a short duration.

Bias methods

There are six methods of obtaining the required negative dc grid bias. Keep this discussion in mind when you review the other amplifying devices. With minor changes, most of these methods of bias are used for solid-state amplifying devices. The method of obtaining bias tells a lot about what the amplifying device is doing in a circuit (Fig. 5-20).

When discussing the polarity of voltages around any amplifying device, the dc input electrode is the cathode, and it is considered to be at 0 Vdc. That is true regardless of what polarity of voltage is on that electrode in a circuit. For example, in the following discussion, the cathode is considered to be at 0 V even though it might be positive in an amplifying circuit. The six methods of biasing a tube are shown in Fig. 5-20.

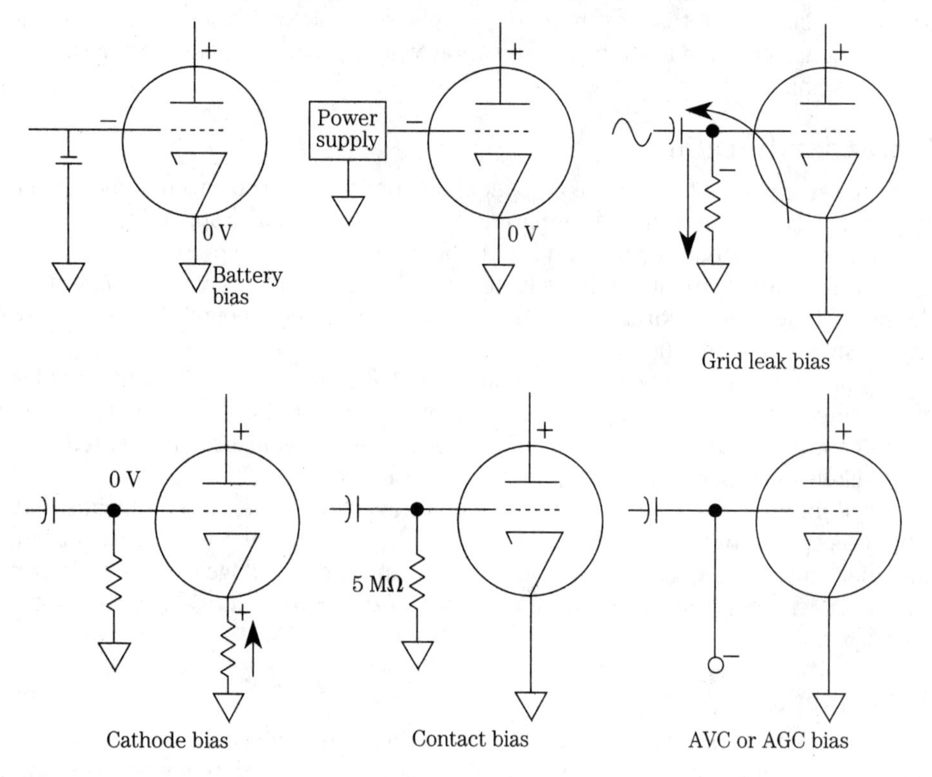

Figure 5-20

Battery bias

A battery or cell can be used to establish the required negative bias for tube operation. Usually, a cell is used. (A battery is a combination of cells.) The bias source is called a "C cell" or "C battery." The cell or battery does not have to supply current because in most circuits grid current is not allowed to flow. Therefore, C cells and batteries last a long time.

Power supply bias

The negative bias for the control grid can be obtained directly from a power supply, usually called the *C supply*. This method is popular in high-priced audio systems and measuring instruments. In those applications, it is important to keep the source of grid bias independent of the signal.

Cathode bias

This method is also called *self bias*. Electron flow through the tube also flows through the cathode resistor, making the cathode positive with respect to ground. At the same time, the grid is grounded through the grid resistor. Therefore, the cathode is positive with respect to the grid, which is the same as saying the grid is negative with respect to the cathode.

Contact bias

The grid of a tube has physical dimensions. Some electrons strike the grid when they are on their way to the plate. Those electrons must return to the cathode. Although the return current is very low, returning to the cathode through a high resistance will make the grid sufficiently negative to bias the tube. This only works for tubes with a very high voltage gain.

AVC or AGC bias

With this type of bias the receiver circuitry rectifies and filters the signal to produce the required negative bias voltage. The rectified signal is usually obtained from the detector stage. When the signal amplitude increases, the negative bias also increases, reducing the gain of the receiver. The effect is to regulate the receiver gain so that the output is not subject to fading as the signal strength varies.

Grid leak bias

This method is also called *signal bias*. For a very short part of the input signal, the grid is made positive and current flows from the grid into the coupling capacitor. For the balance of the signal period, electrons are driven off the capacitor through the grid resistor, making the grid negative.

An important characteristic of this type of bias is that when the signal is lost, the bias is lost. That could cause the tube to conduct heavily and it can be destroyed. To avoid that possibility, a low value of resistance is sometimes placed in the cathode circuit. That resistor acts like a fuse, and it burns out if the signal is lost (resistors are cheaper than tubes). The resistor in the cathode will supply sufficient bias to protect the tube.

A short review of methods used for obtaining electrons

Before discussing tetrodes and pentodes, you should review some of the methods used to obtain electrons from a material. This is an important subject because electronics is a technology in which electrons are put to work. The methods used to get the electrons helps in understanding electronic devices that control them.

Two methods of getting the electrons that are used for control are to use a battery or a generator. Those methods are usually thought of being chemical or electromechanical, rather than electronic.

There are at least six methods of getting "free" electrons that can be used for electronic control. These methods are used in many devices besides vacuum tubes. Here are the six methods:

Thermionic emission

When certain metals are heated to a sufficiently high temperature, electrons are emitted from their surface. To be useful in electronics, the metals must reach the temperature where they emit electrons before they reach their melting point.

The ability of the material to emit electrons is measured by its *work function*. Basically, the work function of a material is a measure of how much work is required to get an electron emitted from its surface. Materials with a low work function emit electrons better than those with a high work function.

Thorium oxide is an example of a material with a low work function, so it is an efficient electron emitter. That is why the cathodes of some types of vacuum tubes are coated with thorium oxide.

Field emission

If a sufficiently-high voltage is placed near the surface of some materials, that voltage will pull electrons off their surface. The procedure is called *field emission*. It is used to produce electron emission from the surface of the cathode in a neon lamp.

Secondary emission

High-speed electrons slamming onto the surface of some conductors causes secondary electrons to be knocked off their surface. (This is a watered-down theory, but it is sufficiently accurate for a description of secondary emission in electronic devices.)

Photoemission

Certain materials, such as selenium, emit electrons when they are struck by light. This is the method used for obtaining electron emission in some types of photoelectric devices.

Friction

Electrons can be easily rubbed off the surface of some insulating materials. The examples often used are: running a comb through hair or rubbing a cat's fur on a day when the air is dry.

Electrostatic voltages are produced by the friction method. This is a very important method of generating a very high voltage in electronics. For example, this method is used in electrostatic generators for generating thousands and even millions of volts of static electricity.

Photon bombardment

This method of freeing electrons was discussed in regard to the operation of PIN diodes. Please review that material.

Tetrodes, pentodes, and beam power tubes

The triode works very well at audio frequencies. At RF frequencies, however, its value as an amplifier drops rapidly as the frequency increases.

The most important problem at high frequencies is *interelectrode capacitance.* Remember that any time two metal surfaces are separated by a dielectric, a capacitor is formed.

Interelectrode capacitance

As shown with broken lines in Fig. 5-21, there is an interelectrode capacitance between the cathode and plate, cathode and control grid, and control grid and plate. The most important are those between the grid and plate.

The capacitance between the control grid and cathode permits some of the high-frequency input signal to be grounded. That is the same as lowering the amplitude of the input signal. The result is a reduced amplified signal at the output.

The capacitance between the control grid and plate permits a high-frequency signal to go from the plate to the control grid. When the amplifier output signal is

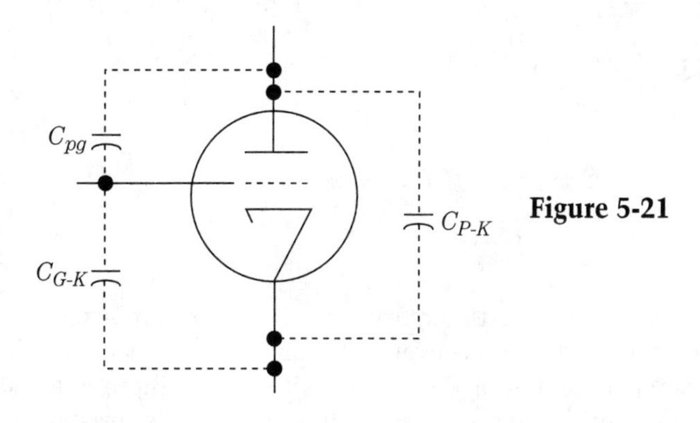

Figure 5-21

taken from the plate and delivered through C_{pg} to the grid, that feedback signal can have one of two effects:

- reduction of the input signal (because the grid and plate signals are out of phase), or
- oscillation at the input when the grid circuit contains an LC circuit.

In either case, the feedback is undesirable, and some preventative measure must be taken if high-frequency amplification is to take place. For example, neutralization can be used in the high-frequency amplifier. That is an amplifier circuit innovation for both tubes and transistors.

Two other methods of preventing plate-to-control grid capacitive feedback are to use a *tetrode* or *pentode* tube.

Tetrodes and pentodes

Figure 5-22 shows the schematic symbols used for the two tubes. Tetrodes are four-electrode tubes and pentodes are five-electrode tubes. The filament is not normally considered to be an active element in tubes; an exception to that is in tubes where the filament also serves as the cathode.

The screen grid of the tetrode is operated at the signal ground potential. This is a case where the screen is at a highly-positive dc voltage with respect to the cathode, but it is at ground potential for the signal (Fig. 5-22).

Always distinguish between the dc voltages and ac voltages in all amplifying devices. Keep in mind that the cathode is considered to be at 0 V and serves as a reference to the voltages on the other electrodes. A similar tradition is also used for transistors.

The screen grids of the tetrode and pentode serve as a shield between the grid and plate and, therefore, they eliminate the undesirable effects of grid-to-plate in-

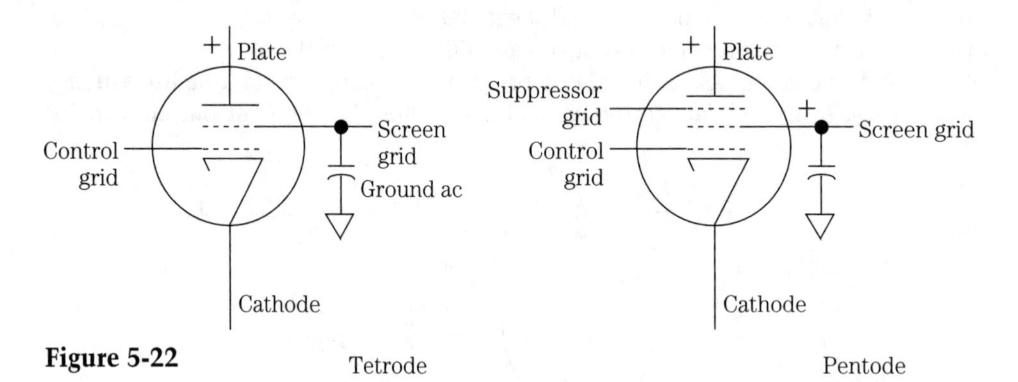

Figure 5-22

terelectrode capacitance. For this reason, pentodes are preferred over triodes for high-frequency amplification. However, neutralized triodes have less inherent noise.

The characteristic curve of the tetrode shows that there is a "negative resistance" region. Of course, there is no such thing as negative resistance. The name is derived from the fact that the plate current decreases in a range of increasing plate voltages. By comparison, an increase in voltage across a positive resistance would cause an increase in current.

In the tetrode, the positive screen grid along with the positive plate cause a high acceleration of electrons. The electrons moving from the cathode to the plate are accelerated to the level where they produce secondary emission when they bounce off the plate.

Electrons going from the cathode to the plate have a high energy (because of their high speed); therefore, they are not greatly attracted to the positive screen. However, secondary electrons from the plate have a low energy, and they can easily be attracted to the nearby positive screen grid. Electrons going to the screen grid subtract from the plate current, so the plate current decreases in the negative resistance region.

When the plate voltage becomes sufficiently high to attract most of the secondary electrons back to the plate, the negative resistance effect is eliminated and the screen current decreases.

Secondary electrons can be eliminated by adding a negative suppressor grid between the screen grid and plate. That is how the pentode works (Fig. 5-22). The negative *suppressor grid*, (SUG), repels the secondary electrons back to the plate so they cannot reach the screen grid. That, in turn, eliminates the negative resistance characteristic that is present in the tetrode.

Beam power tubes

In a beam power tube, a high-density electron stream is focused on the plate. That electron stream repels secondary electrons so that they cannot reach the screen grid. The advantage of this method is that it requires fewer electrodes than the pentode.

Noise

Noise is a problem with most amplifying devices. The noise generated by the amplifying device can mask the desired signal whenever the signal has a low amplitude. The types of noises discussed in this section are present in both tube and semiconductor devices.

Partition noise

Every time an electron passes an electrode in a vacuum tube (or transistor), there is a possibility that it might go to that electrode instead of going to the output.

For example, suppose that a few electrons strike the control grid at one instant. That subtracts from the plate current at that instant. In the next instant, many electrons, or no electrons, or a few electrons will go to the control grid (or screen grid or suppressor grid). The result is small variations in plate current, which represents a noise voltage when the plate current flows through the plate resistor.

In a tetrode or pentode, there are more electrodes than in a triode, so tetrodes and pentodes are noisier than triodes. All of the grid electrodes in a vacuum tube contribute to noise.

The screen grid in tetrodes and pentodes greatly reduces plate-to-control grid capacitance by acting as a grounded shield, but there is a penalty of adding noise to the desired signal.

Flicker noise

Electrons do not leave the cathode in a smooth manner. At any instant, there are more or less electrons leaving than left at the previous instant or leave in the next instant. That means the electrons move through the tube and arrive at the plate at an uneven manner. The result is a noise signal mixed with the plate current. When that noise current flows through the resistor in the plate circuit, it creates a noise voltage.

In addition to these tube noises, there can also be noise generated by the external amplifier circuitry. If the tube circuitry is made up of resistors, there is sure to be semiconductor noise added to the signal. This is especially true of resistors connected at the signal input circuit of a high-frequency amplifier where the resistor noise will be amplified.

Configurations

Three terminals are utilized in most amplifying devices. However, there are only two signals to consider: The input and the output signals. It follows that one of the electrodes of the amplifying devices must be common to both the input and output signal. This is true for all three-terminal amplifying devices.

Although it is true that one of the electrodes must be common to the input and output signal, the dc operating voltages do not necessarily have the same common point as the signal. Do not confuse the signal configuration with the dc configuration on the tube electrodes. An example was given in the discussion of the screen grid. The screen grid is at a positive dc level with respect to the cathode, but at the same time, it is at 0 V with respect to the signal.

Figure 5-23 shows the three possible configurations for a triode. Remember their names and their characteristics. Those characteristics are the same for tetrode and pentode tubes, and for solid-state amplifying devices. The characteristics are summarized in Table 5-1.

Table 5-1

Common emitter	Best gain
Common grid	Good current gain and voltage gain
Common plate	Good current gain No voltage gain

In the follower configuration, the plate is at signal ground.

In an amplifier, *voltage gain* (A_V) equals the output signal voltage (V_O) divided by the input signal voltage (V_{in}).

$A_v = V_o \div V_{in}$ (There are no units of measurement for A_v.)

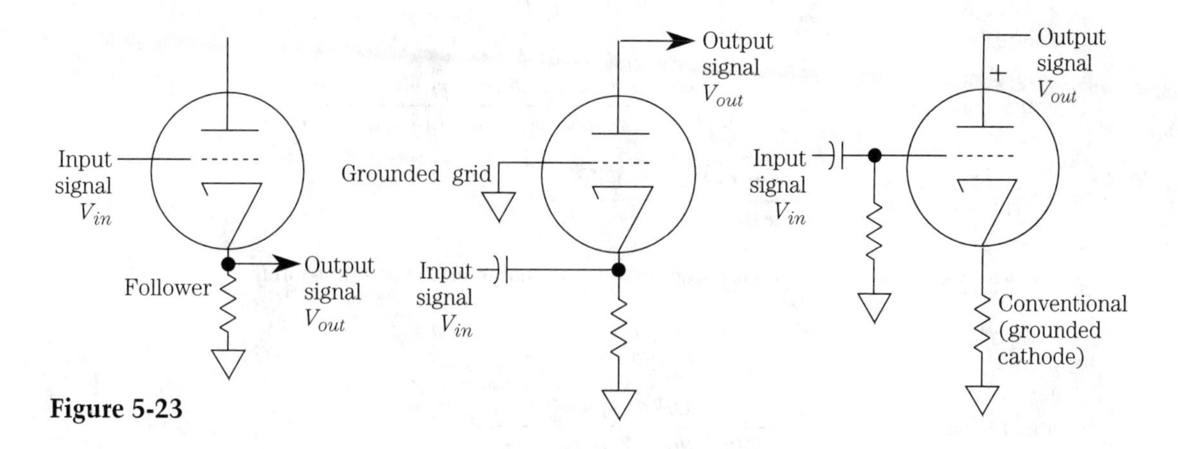

Figure 5-23

Voltage gain is less than one for all three-terminal amplifying devices in the follower configuration. Another way of saying that is the output signal voltage amplitude is lower than the input signal voltage amplitude.

Grounded grid circuits are useful at high frequencies because the control electrode acts as a shield to isolate the input and output signals. In other words, with this kind of configuration capacitive coupling between the input and output signals is minimized.

The grounded cathode circuit is by far the most popular configuration. It gives a high voltage gain. Refer again to Table 5-1.

Tube measurements

There are four important tube measurements shown in Fig. 5-24. Although they are shown with triodes, they apply equally to tetrodes and pentodes.

You should memorize the measurements shown in Fig. 5-24.

The Bipolar Junction Transistors (BJT)

Figure 5-25 shows the construction and symbols for NPN and PNP bipolar junction transistors. It has become fashionable to refer to *B*ipolar *J*unction *T*ransistors as BJTs. We will follow that tradition in this book. The three leads on the BJT go to the emitter (E), base (B), and collector (C). (For the NPN transistor the arrow is *N*ot *P*ointing i*N*.)

In most cases, we will concentrate on the NPN BJT. To get the PNP version, simply reverse the voltage polarities.

The term *bipolar* refers to the fact that there are two different types of charge carriers used in the BJT. They are electrons and holes. Electrons are negative charge carriers and holes are positive charge carriers. As a convenient model, it is acceptable to consider the holes as being positive electrons with a slightly greater mass. The holes do not move through P-type material quite as easily as electrons move through N-type material.

It is important to observe the arrows on the symbols in Fig. 5-25. For all semiconductor devices, the arrow always points toward an N-type material and away

* Four tube characteristics

* Transconductance (g_m) = $\dfrac{\textit{Small change in plate current}}{\textit{Small change in grid voltage that produced it}}$

* Voltage gain (A_v) = $\dfrac{\textit{Output signal voltage}}{\textit{Input signal voltage}}$

Strictly speaking, A_v is not a tube characteristic, it is characteristic of a tube circuit

* Plate resistance (R_p) *Change in plate current divided the related change in plate current*

Plate efficiency = $\dfrac{\textit{Output power}}{\textit{Plate voltage} \times \textit{plate current}}$

Again, this is very much a characteristic of the tube circuit

* Amplification factor (μ) = $\dfrac{\textit{Change in plate volts}}{\textit{Change in grid volts}}$

Figure 5-24

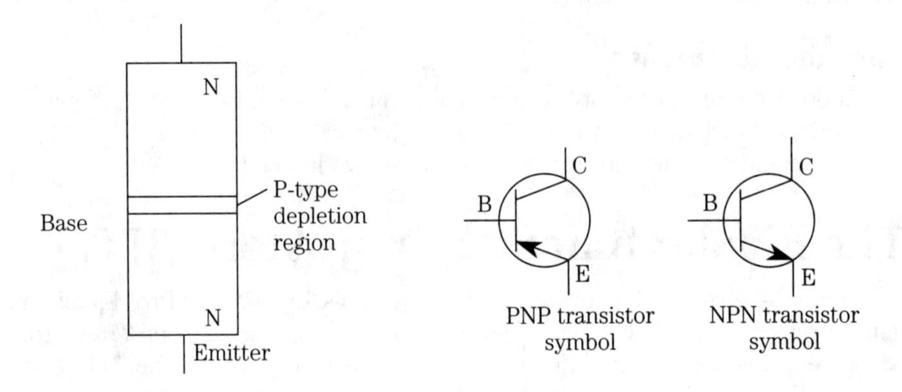

Figure 5-25

from a P-type material. Another way of saying that is the arrow always points in the direction of *conventional* current flow. Conventional current flows from positive to negative. (Remember that zener diodes are designed to operate with a reverse current. So, in the case of zeners, the arrow will point in the direction of electron current flow when that diode is connected in circuits.) The rule can be stated another way: The arrows on symbols point against the direction of electron flow.

In the vacuum tube devices, a voltage on the control grid determines the amount of plate current, so it is called a *voltage-operated device*. In the BJT, it is the input base current that determines the output collector current; therefore, the BJT is a current-operated device.

In the normal operation of a BJT, most of the current entering the base region from the emitter goes to the collector. The electron current distribution is shown in Fig. 5-26. The base current is only about 5% of the emitter current. The remainder of the emitter current goes to the collector.

Figure 5-26 also shows the two-diode representation that is sometimes given for a BJT. Do not be confused when you see it; you cannot make a transistor with two diodes! The only thing you can learn from that representation is that in normal operation, the emitter-base junction is forward biased and the base-collector junction is reverse biased.

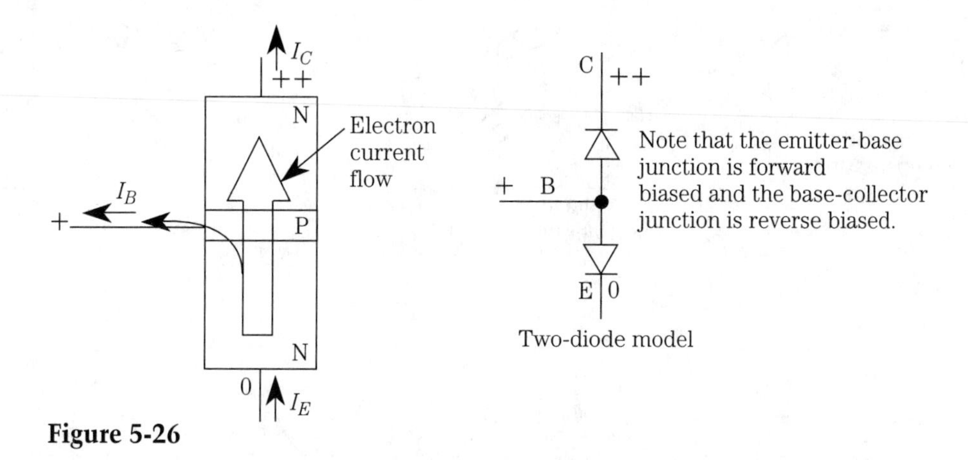

Figure 5-26

The collector terminal is more positive than the base terminal in most operations, so the base is negative with respect to the collector. Saying it another way, the base-collector junction of NPN and PNP transistors is normally reverse biased. Both voltages are positive with respect to the emitter in an NPN BJT.

Mathematically, the emitter, base, and collector currents in the BJT are related by the simple equation:

$$I_E = I_B + I_C$$

That current relationship is actually a statement of Kirchhoff's junction current law.

Refer to the NPN BJT in Fig. 5-26. Amplification in this device occurs because a small change in base current results in a larger change in collector current. Therefore, a low amplitude input signal current results in a high-amplitude collector current. The same thing is true for the PNP transistor.

BJT bias circuits

In the study of vacuum tube circuitry, you reviewed the fact that there are six methods of obtaining the necessary negative grid bias, and noted that some of those methods of bias can be used for bipolar transistors. The most important difference is in the polarity of the voltages required.

Figure 5-27 shows the most popular methods of biasing NPN transistors. (The same methods are used to bias the PNP types, but the polarities are reversed.) Each method will now be reviewed using the NPN type as examples. You might see different names for these methods of bias in other books and in magazine articles, but you will be able to recognize them from your knowledge of the transistors.

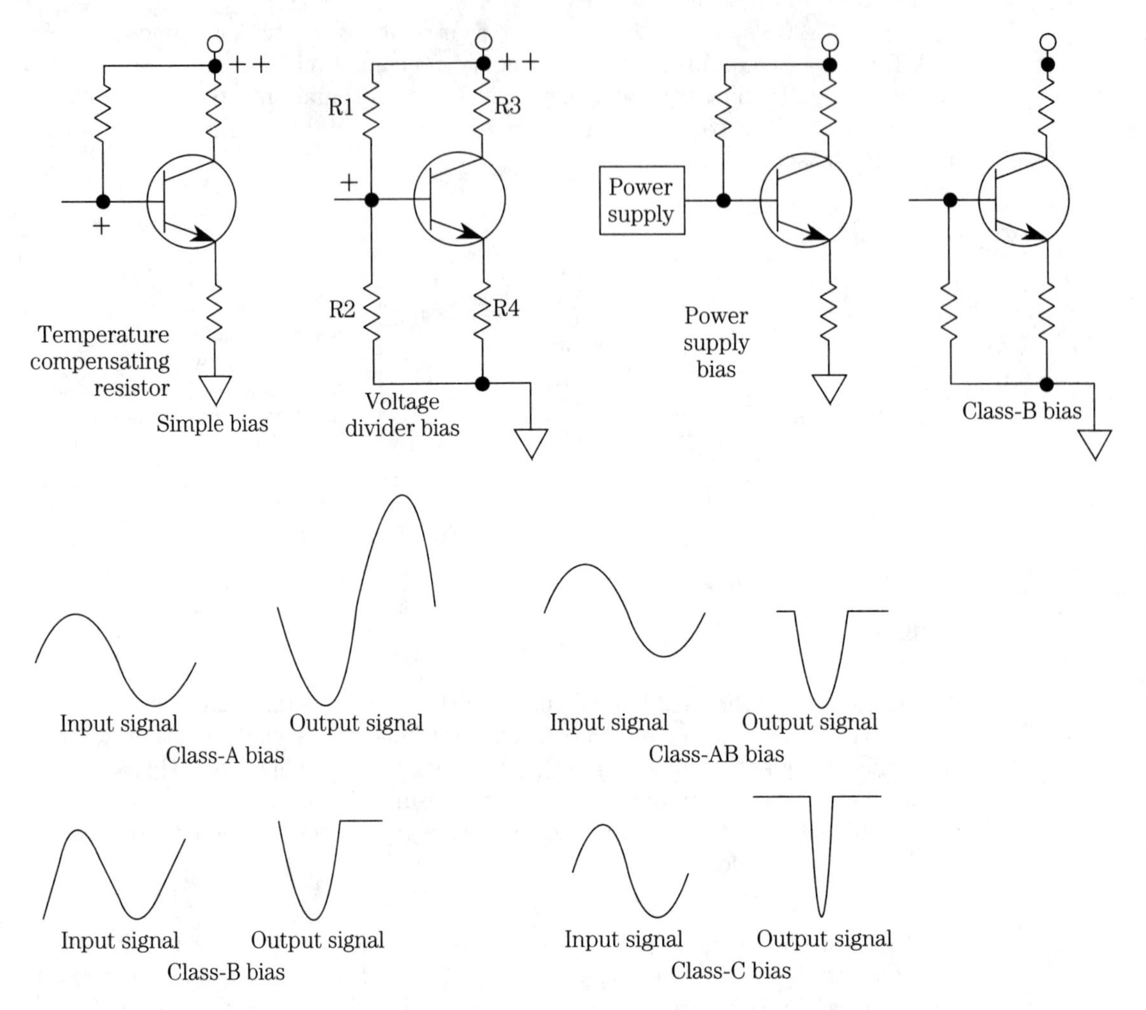

Figure 5-27

Simple bias

Refer again to Fig. 5-27. A single resistor is used to drop the positive supply voltage to a lower value of positive base voltage. The collector and the base both get their dc operating voltages from the same positive voltage source. The collector voltage is more positive than the base voltage. Another way of saying that is the base is negative with respect to the collector.

The base resistor normally is many times the resistance of the collector resistor. That is necessary because the value of base current is very low compared to the collector current, so a higher resistance is needed to get the required voltage drop.

If, for any reason, the base voltage becomes higher than the collector voltage, the *transistor will be quickly destroyed!* The collector voltage of an NPN transistor must be maintained more positive than the base voltage in normal operation of the transistor! For a PNP transistor, the collector voltage must be more negative than the base voltage.

The resistances of the semiconductor materials used for making N-type and P-type materials are greatly dependent upon their temperatures. Whenever their temperatures increase, their resistances are reduced and the current through them increases. That is why the simple bias circuit is seldom used without an emitter resistor. See the temperature-stabilized simple bias circuit in Fig. 5-27.

Suppose for some reason that the ambient (surrounding) temperature of the simple bias transistor circuit increases. That causes the temperature of the transistor to increase and lowers the resistance of the N-type and P-type materials. The result will be an increase in emitter-to-base current. The increased emitter-base current causes a higher current through the transistor, which causes its temperature to increase further. Again, the emitter-base current will increase and there will be an even greater increase in the transistor's temperature; its resistance decreases further and its emitter-base current increases further. The transistor is eventually destroyed by this upward spiral of temperature and current.

The destructive current rise and temperature rise just described is called a *thermal runaway*. It takes a little time to tell about the series of destructive events, but it happens very quickly and the transistor is rapidly destroyed.

The resistor in the emitter circuit of the temperature compensated simple bias circuit prevents the thermal runaway. When the transistor current starts to rise because of an increase in temperature, the voltage drop across the emitter resistor increases. That reduces the forward bias on the emitter-base junction and decreases the collector current, so the thermal runaway can't get started.

The emitter resistor is called a *temperature-compensating resistor*. In a vacuum tube circuit, a resistor in the cathode circuit is used to self-bias the tube. Do not confuse the purpose of the emitter resistor in the emitter circuits of Fig. 5-27 with the cathode resistor in vacuum tube circuits. The emitter resistor is not used for biasing the transistor. Its purpose is to provide temperature stabilization for the transistor circuit.

In the voltage divided bias circuit, there is a pull-down resistor (R2) between the base and common. That resistor makes the circuit more stable compared with the simple bias circuits. It works by holding down the dc base voltage and current to a value nearly independent of ambient temperature.

In some voltage divider circuits you may see a diode or thermistor in series with R2 to provide additional temperature stability. The diode has the same type of semiconductor junction as the transistor emitter-base junction. Any ambient temperature increase that affects the transistor emitter-base junction will produce the same effect in the diode, meaning that the forward resistance of the diode is also reduced. The base bias is automatically adjusted to compensate for any ambient temperature change.

If a thermistor is used in place of R2, an increase in ambient temperature will cause the thermistor resistance to decrease. That, in turn, will reduce the dc base bias and lower the base voltage and current. So, the tendency of the emitter-base current to increase with an increase in ambient temperature is offset by the decrease in base voltage caused by the diode or thermistor.

With power supply bias, a separate bias supply is used for the base bias. This method is used in more expensive equipment and in some types of regulated power supplies.

As with the vacuum tube circuits, an AVC/AGC bias voltage can be used in conjunction with the forward bias provided by the base resistor.

The circuit called class-B bias in Fig. 5-27 will not provide dc forward bias on the transistor base. Classes of bias are discussed in the next section. On the positive half cycles of the input signal, the emitter-base junction is forward biased and current flows through the transistor. On the negative half cycles of input signal, the emitter-base junction is reverse biased and no base current flows. Without a base current the transistor collector current is cut off. The overall effect of the class-B bias circuit is to provide rectification of the output signal.

To be absolutely accurate, the so-called "class-B bias" circuit does not bias the transistor for class-B operation. Instead, it is a form of class-AB bias, but it is usually called *class-B bias*.

When comparing the transistor bias methods with those used in vacuum tube circuits, note that there is no equivalent to grid-leak bias or contact bias.

A short review of classes of operation

Simplified versions of the classes of operation reviewed in Fig. 5-28.

Class A bias—The output current flows during a complete cycle of input signal.

Class B bias—The output current flows for 50% of an input cycle.

Class AB bias—The output current flows for nearly half cycle of input signal.

Class C bias—The output current flows for considerably less than 50% of input cycle.

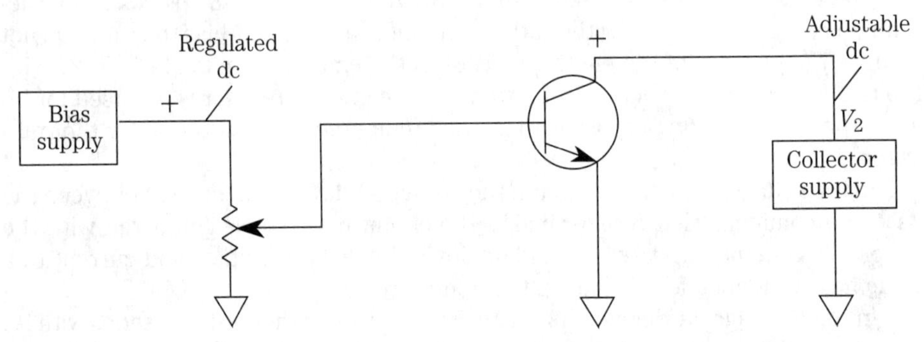

Figure 5-28

Parameters and abbreviations for bipolar transistors

The military specification for bipolar transistors lists about 250 parameters and abbreviations for bipolar transistors. You are not expected to memorize all of them. There are, however, a few that would be worthwhile to know. They are summarized in Table 5-2. A few explanations of those listed will be helpful.

Table 5-2

BV_{CEO} Breakdown voltage (collector-to-emitter with base open).

BV_{CBO} Breakdown voltage (collector-to-base with emitter open).

Av = Voltage gain

Ai = Current gain

Ap = Power gain

$$\alpha_{DC} = \frac{I_C}{I_E} \approx h_{FB}$$

$$\beta_{DC} = \frac{I_C}{I_B} \approx h_{FE}$$

$$\alpha_{ac} = \frac{\Delta I_C}{\Delta I_E} \approx h_{fb}$$

$$\beta_{ac} = \frac{\Delta I_C}{\Delta I_B} \approx h_{fe}$$

≈ means "approximately equal to"

The parameters are used in evaluating transistors and for design work. Once you review the scheme of the parameter subscripts, it is much easier to recall their meanings.

Two very important parameters are alpha (α) and beta (β). They are not listed by those abbreviations in Table 5-2, but you must know them.

The abbreviations h_{FB} and h_{fb} are called α (alpha) in many publications and specification sheets. The h_{FB} and h_{fb} hybrid parameters are based upon a hybrid four-terminal network with the transistor in the common base configuration. The subscript letters mean *forward base*. The capital letters mean dc or RMS values and lowercase letters mean ac or varying values.

Here is the way α_{DC} is obtained:

$$\alpha_{DC} = h_{FB} = I_C/I_E$$

The letters I_C and I_E mean collector current and emitter current. When the measurements are made the base is grounded.

Here is the way β_{DC} is obtained:

$\beta_{DC} = h_{fe} = I_C/I_B$ (FE) means *forward emitter*

For α_{ac} and β_{ac}, the subscripts are lowercase.

$$\alpha_{ac} = h_{fb} = i_c/i_e$$

The value of α is always less than 1.

$$\beta_{ac} = h_{fe} = i_c/i_b$$

The value of β is always greater than 1. The measurement of β is often used to give a quick evaluation of a transistor's condition.

The capital letters are used for leakage current (for example: I_{CBO}) and breakdown voltage (for example: BV_{CEO}) follow the same rule. The first two letters tell where the measurement is taken and the third tells what is done with the electrode not in the measurement.

Examples

I_{CBO} means the collector-to-base leakage current with the emitter open. This is a reverse current.

Refer back to the two-diode illustration and note that current which flows from collector to base is a reverse current. There should be no reverse current flowing from collector to base because that junction is normally reverse biased. However, no PN junction is perfect. The reverse collector-to-base current will subtract from the desired collector current. That, in turn, reduces the gain of the transistor.

BV_{CEO} means the breakdown voltage from collector to emitter with the base open. Instead of O for open, the third subscript can be S meaning shorted to ground.

The more exact measurements for α and β take leakage currents into consideration. The equations are:

$$\alpha = h_{FB} = (I_C - I_{CBO})/I_E$$

$$\beta = h_{FE} = (I_C - I_{CBO})/I_B$$

Physically, α is a measure of how much emitter current becomes collector current in the common base configuration. Also, β is the current gain, and it is an indirect measurement of how much influence the base current has on the collector current. Ideally, the value of beta should be high for a transistor that is being used as a voltage amplifier.

Another important equation gives the relationship between alpha and beta. If you know one of those parameters, the other one can be derived by algebraic manipulation. The equations are:

alpha = beta/(1 + beta) and

beta = alpha/(1 – alpha)

Alpha and beta cutoff frequencies

Important parameters not given in Table 5-1 are the alpha and beta cutoff frequencies. Both the alpha and the beta drop off at high frequencies.

You will see the values 70.7% and –3 dB used to mark high-frequency points on the bandwidth graphs of circuits. Those points have also been used to mark the frequency where alpha and beta drop from their maximum, or midfrequency, gain to 70.7 (3 dB) of maximum.

Maximum gain is often identified as the gain at 1000 hertz even though that is nowhere near the midpoint of the frequency range.

Using these reference points, alpha and beta cutoff can be defined as follows:

- *Alpha Cutoff Frequency*—The point on a graph of alpha vs. frequency where alpha drops to 70.7% (–3 dB) of its maximum gain.
- *Beta Cutoff Frequency*—The point on a graph of beta vs. frequency where beta drops to 70.7% (–3 dB) of its maximum gain.

Although those points are well-established as high-frequency cutoff marks, a transistor manufacturer is more likely to define the beta cutoff frequency as the point where beta drops to unity (1.0). When you are reading specification charts, it is always important to determine how the manufacturer defines a parameter.

Gain bandwidth product

Load lines for bipolar transistors

The method of plotting a load line for a diode was discussed earlier in this chapter. This is a continuation of that discussion using a bipolar transistor in place of the diode.

In order to plot any load line, it is necessary to have a characteristic curve for the device. The curve(s) for bipolar transistors can be obtained in a number of ways. One method is shown in Fig. 5-28. A dc voltage (V_1) is used to supply the base-to-emitter voltage(V_{BE}) of the transistor. A different dc voltage (V_2) is used to supply the collector-to-emitter voltage (V_{CE}).

The base voltage is set to get the desired base current. Then, the collector voltage is varied through a range that corresponds to the manufacturers specifications for the transistor. After the curve for that base current is drawn, the base voltage is set to the next value and the procedure is repeated. The result is a family of curves. An example is shown in Fig. 5-29.

A second method of getting a family of characteristic curves is shown in Fig. 5-30. In this case the base voltage is adjusted one step at a time. Each step repre-

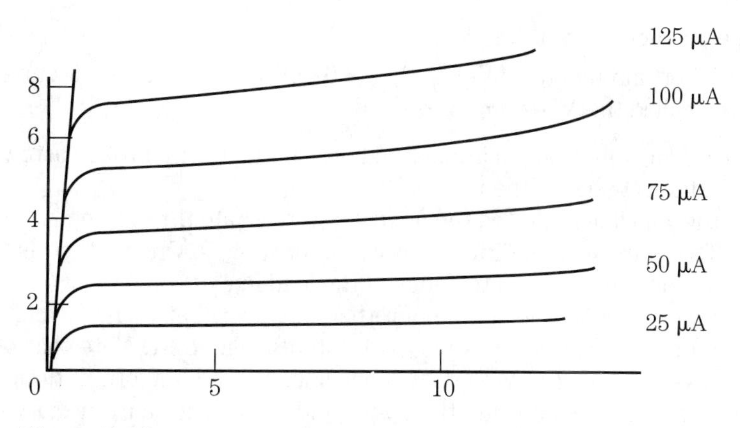

Figure 5-29

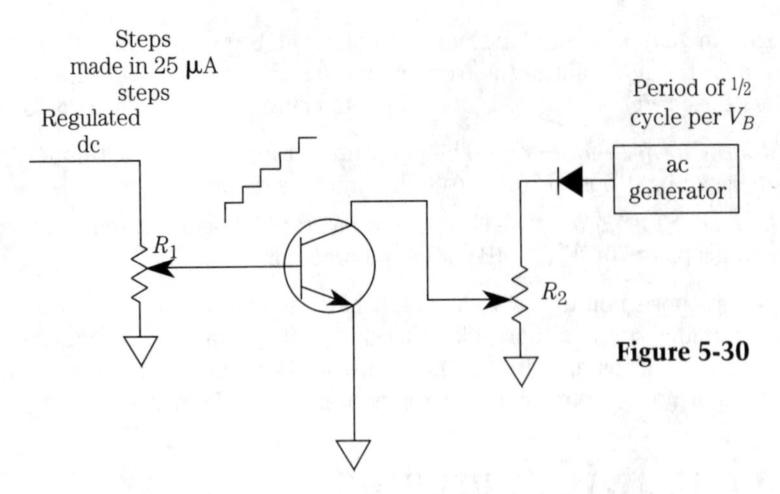

Figure 5-30

sents a base voltage needed to get a desired base current. For each base current, the collector voltage is swept through a range to get the collector curve for that base current. Each curve can be displayed on an oscilloscope.

Observe that the collector voltage is swept by a rectified sine wave. So, the collector voltage actually goes from a minimum value to maximum value and back to the minimum value for each step of base voltage. That works very well as long as there are no reactive components in the collector circuitry. [With reactive components, the curve does not trace (left-to-right) from 0 V to 12 V and retrace (right-to-left) from 12 V to 0 V along the same line.]

The method just described, graphs a family of characteristic curves by plotting one curve at a time. Commercial curve tracers can plot the complete family of curves as an oscilloscope display.

Figure 5-31 shows how a load line is plotted on a family of curves. The circuit in the illustration shows the transistor connected to a Thevenin voltage and Thevenin resistance. That Thevenin generator is the equivalent of the circuit the transistor is normally connected into.

As with the diode in the circuit the two points that determine the load line are:

- The open circuit voltage (V_{TH}) = 10 V.
- The short-circuit current (V_{TH}/R_{TH}) = 6 mA. (The short-circuit current is often called the *Norton current*.)

The 10-V and 6-mA points are marked on the graph with broken arrows and the load line is drawn between them.

Once the load line is plotted the next step is to locate the quiescent point (Q) on that line. There are several different ways of locating Q. One method is to go to a point that is half way between the ends of the load line.

A better method is to locate a point that is halfway between 0 V and the maximum value of V_{CE}. That is the way Q is located in Fig. 5-31. Note that Q is about halfway between 0 V and 10 V on the voltage line. Another important thing to note is that it is nearly halfway between the 25-μA and 75-μA base current curves. That is important for obtaining low distortion in a transistor amplifier.

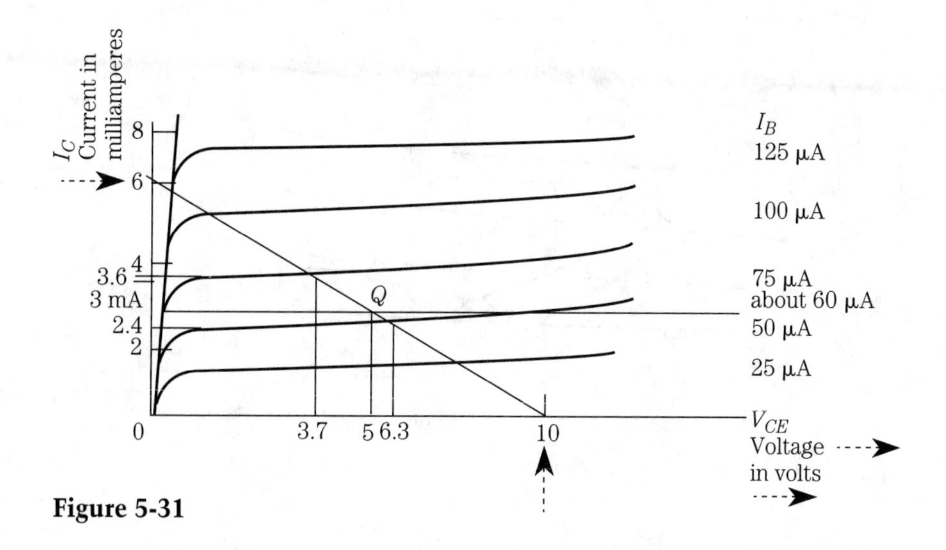

Figure 5-31

From Fig. 5-31, you can see that with no input signal (at the *Q* point) the base current is about 60 mA, the collector-to-emitter voltage is 5 V, and the collector current is about 3 mA.

When an input signal swings the base current between 25 mA and 75 mA, the collector current swings between (about) 2.4 mA and (about) 3.6 mA. At the same time, the collector voltage swings between (about) 3.7 V and (approximately) 6.3 V. The ranges are shown on the curve in Fig. 5-31.

The maximum allowable transistor dissipation

One of the parameters that is usually supplied by the transistor manufacturer is the maximum power the transistor is allowed to dissipate. That parameter makes it possible to plot a power dissipation curve on the family of curves.

Assume that the manufacturer has specified that the transistor having the family of curves shown in Fig. 5-32 cannot dissipate more than 18 mW. The equation for dc power dissipation is:

$$P = VI$$

Solving that equation for *V* gives:

$$V = P/I = 18/I$$

The procedure is to pick a value of *I*, and solve for the corresponding value of *V* using the equation just given. For example, when $I = 6$ mA, the value of *V* is:

$$V = 18 \text{ mW}/6 \text{ mA} = 3 \text{ V}$$

The point on the curve that corresponds to 3 V and 6 mA (usually identified as 3,6) is marked on the curve with a #1.

The procedure is repeated for additional values of *I* and corresponding values of *V*. When the points are connected, the result is the maximum power curve of Fig. 5-32.

The maximum power curve is shown with the load line. As you can see, the power curve does not cross the load line. For a class A-amplifier, the power curve

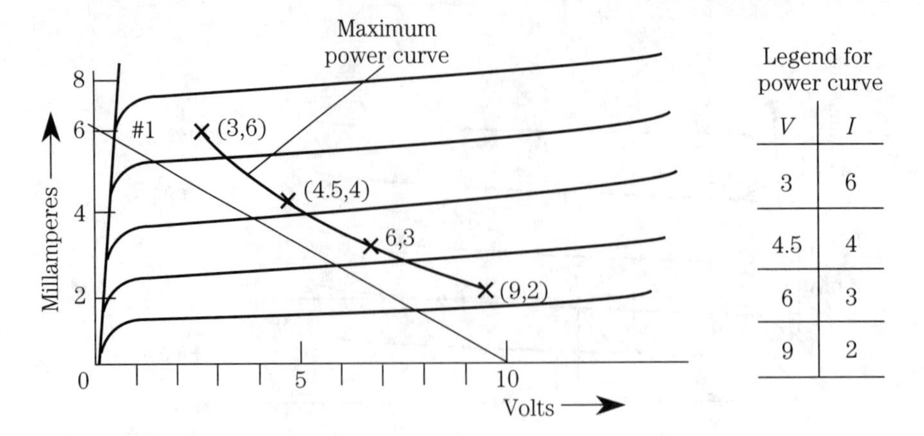

Figure 5-32

must not cross the load line. If it does cross the load line, the circuit resistances and/or voltages must be changed, so there is no point where they cross.

The optimum operation occurs when the power line is tangent to the load line. If the load line cuts through the power curve the transistor is overloaded.

Using the characteristic curves, the load line, and the power curve as just described makes it an easy matter to design transistor amplifier circuits. However, there is a disadvantage to that design procedure. If the family of curves is obtained with a specific transistor, it might turn out that the transistor used is not typical of all transistors having the same identification number. Obviously, a number of transistor curves must be averaged with a number of transistors in order to get a typical curve.

Field-Effect Transistor (FETs)

Field-effect transistors were invented before bipolar transistors. However, it was not possible to put them into production because the necessary materials were not available at the time.

Once the field-effect transistors became mass production items, they began to overtake the bipolar transistors in a number of applications. There are two good reasons for the popularity of FETs.

- No power is required by the bias circuitry.

FETs are voltage-operated. In that respect, they are very similar to vacuum tube triodes.

By way of comparison, bipolar transistors are current operated devices. They must be supplied with current in their bias circuit, and that current represents power (I^2R) loss.

So, when bipolar transistors are in a dc quiescent standby condition, they are dissipating power in their base circuits. That isn't too important when you have one or two transistors, but consider applications where there are 5000 transistors in a system. In these cases, the total standby power dissipation affects battery life and cost of operation.

- Another advantage of the field-effect transistor is that it does not require input signal power for operation. Unlike the BJT, the FET requires only signal voltage for its operation. That is very important; the signal source does not have to supply power to a FET amplifier. In text books this advantage is described by saying the FET "has a high input impedance."

Junction Field-Effect Transistors (JFETs)

Figure 5-33 shows a model and a symbol for both types of JFETs. In the upper illustration a P-type material is shown embedded in an N-type material. The P-type material is called the *gate* and the N-type material is called the *channel*. The device illustrated is called an *N-channel JFET*.

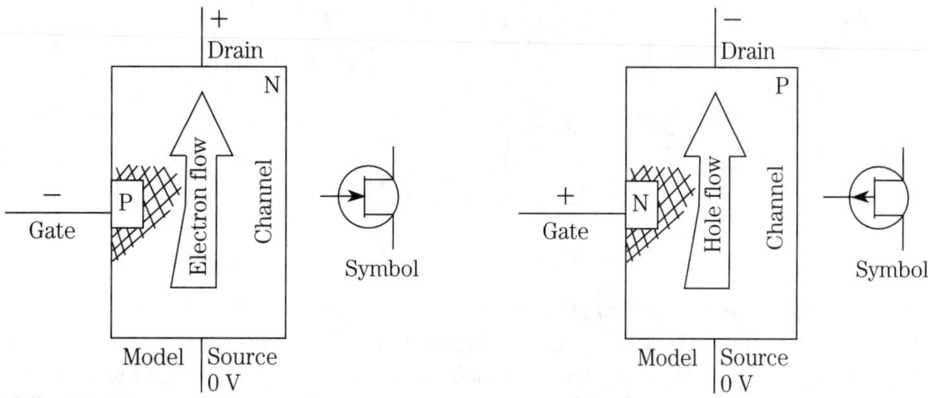

Figure 5-33

We will concentrate on N-channel FETs in this discussion. However, you are expected to understand the operation and the symbols for P-channel FETs. You will be tested on the P-channel devices in the review tests. To answer those questions simply reverse the polarities of the voltages on the electrodes from what they are on the N-channel devices.

The three electrodes on the JFET are called *source*, *gate,* and *drain*, and are marked on the illustrations. The names are descriptive of their function.

Observe the polarities of the voltages on the electrodes. As usual, the terminal where the charge carriers enter, that is, the source, is marked 0 V. The control electrode is the gate. In an N-channel JFET, it is normally negative with respect to the source. The place where the charge carriers go is the drain, and is positive with respect to the source in the N-channel device.

Note that the polarities of the voltages on the N-channel JFET are the same as the polarities of voltages on a triode vacuum tube. You will find that if you have a knowledge of tube operation you can readily understand the operation of field-effect transistors in circuits.

Referring to the N-channel JFET, if the gate is negative and the drain is positive there is a reverse bias between the gate and the drain. Also, there is a reverse bias between the gate and source. That reverse bias causes a depletion region around the gate, and is represented by the shaded area in Fig. 5-33.

Because of the reverse bias, there is no gate current flowing in the normal operation of a JFET. That is why no input power is required for this device. In order for electrical power to be dissipated, there must be current flow.

In Fig. 5-34, the operation of an N-channel JFET is illustrated. When compared with the model shown in Fig. 5-33, the gate has an additional negative bias that is indicated with two or three negative signs.

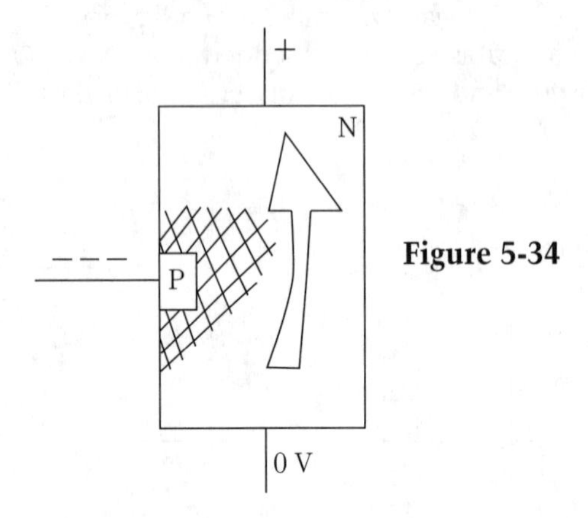

Figure 5-34

Increasing the negative bias increases the size of the depletion region; at the same time it reduces the size of the path for electron flow. Note that the more negative the gate voltage, the more restricted the channel becomes. Saying it another way, increasing the negative gate voltage decreases the drain current, which is another similarity between the N-channel JFET and the triode vacuum tube. In both devices, an increase in the negative voltage on the control electrode reduces the output current. The fundamental theory of operation for both devices is that a small change in voltage on the control electrode produces a relatively large change in output current.

In a P-channel device (Fig. 5-33) the gate must be positive with respect to the emitter, and the drain must be negative. As with the N-channel device, there is a depletion region around the gate. In this case, making the gate more positive increases the size of the depletion region and, therefore, decreases the current through the channel. Current through the channel is considered to be positive hole flow.

The tradition is carried on for the direction of arrows on schematic symbols. Note that the arrow on the N-channel JFET points toward the part of the symbol that represents the N-channel. Also, the arrow points away from the P-material in the P-channel in the JFET symbol.

Earlier in the chapter it was stated that the arrow on semiconductor symbols points in the direction of conventional current flow. For field-effect transistors, that statement has to be modified. In this case, the arrow points in the direction of current flow *if that current was allowed to flow!* For the JFET, the depletion region prevents current flow. For types of FETs that are reviewed next, there is an insulation region that prevents the conventional current flow.

Figure 5-35 shows the methods of biasing a JFET. They are basically the same methods used for triode tubes.

By far, the most common methods of biasing JFETs and triodes is the automatic bias, or self bias. With this method, there is a source-to-drain current that flows through the source resistor. That produces a voltage drop that makes the source

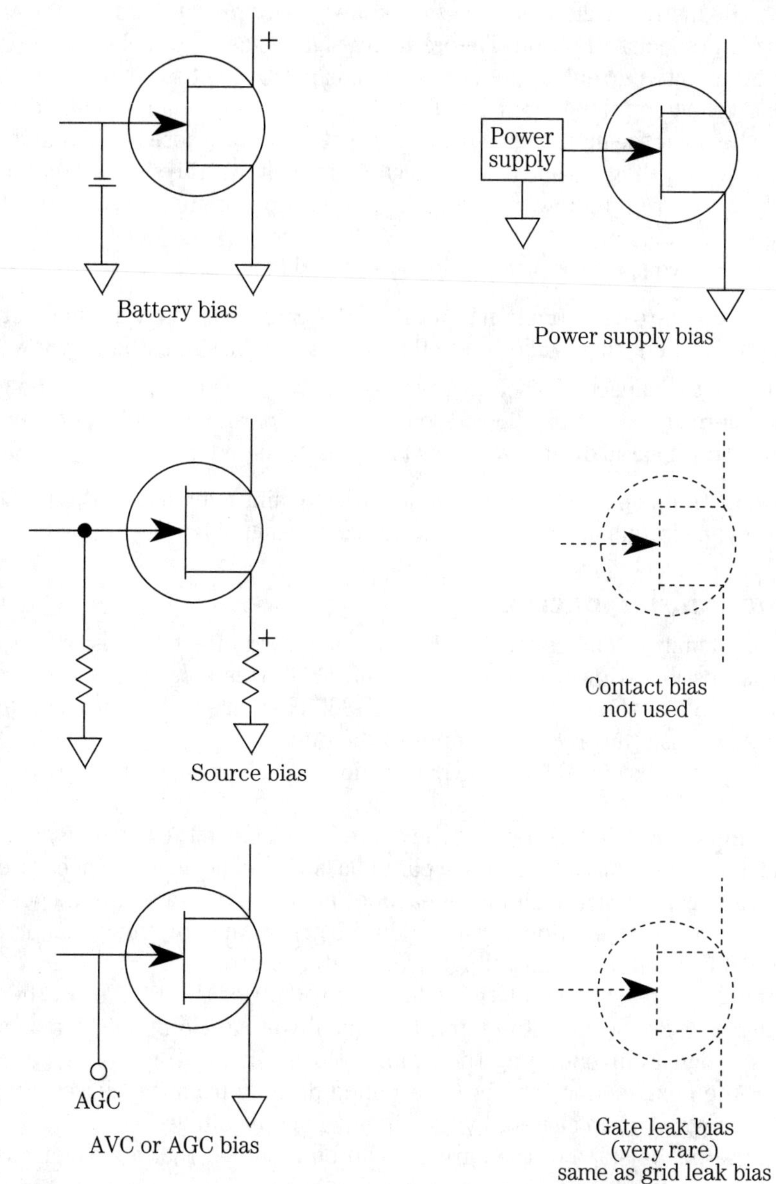

Figure 5-35

positive with respect to common. Because there is no dc current in the gate circuit, the dc gate voltage is at common (or ground) potential.

Remember: In order to get a voltage across a resistor it is necessary to have a current flowing through it. (That statement disregards resistor noise voltage.)

With the gate at 0 V and the drain at a positive voltage, it follows that the gate is negative with respect to the source. That is what you would expect for the polarities marked on the N-channel JFET.

The remaining methods of bias are the same as was discussed for the triode. However, the method called *signal bias* (known as *grid-leak bias* in tube circuits) requires some special attention. Theoretically, signal bias is possible with a JFET because there is gate current whenever the gate is positive with respect to the source. A positive voltage on the gate means that the source-to-gate junction is forward biased, so whenever the input signal makes the gate positive there is gate current flow. When the signal voltage drops below the gate bias voltage, the signal forces a capacitor discharge current to flow through the gate resistor forcing the gate voltage to go negative.

There are two special considerations for signal bias:

- The source-to-gate junction in most JFETs cannot handle very much current, so the current that is allowed to flow with signal bias must be very low.
- During the capacitor discharge period, the reverse voltage across the gate-to-source must not be allowed to exceed the breakdown voltage of that junction. That allowable reverse voltage is specified by the manufacturer.

The constraints placed on JFET signal bias explains why that method is not popular. However, circuits have been able to utilize the signal for bias.

Forward transfer function

Using a family of curves for the JFET (or any field-effect transistor) a load line analysis is possible. However, with tubes and FETs it is more convenient to use a transfer curve like the one shown in Fig. 5-36. They are called *forward transfer functions* because they relate the input to the output.

The lines are extended from the flat portion of each curve in the family. That flat portion is called the *constant-current region.*

Returning to Fig. 5-36, note that the curve is entirely in the negative gate region. The transfer curve is actually part of a parabola, so it can be analyzed accurately with mathematics. Notice that the input signal does not drive the gate into the positive region. This is class-A operation. As shown in the graph, the output signal current can be obtained if the input signal voltage is accurately plotted.

Nothing can be said about the voltage gain (A_V) of the FET from the signals shown in Fig. 5-36, because the input is a signal voltage along the V_G axis and the output is a signal current along the I_D axis. However, the transconductance (discussed in the next section) can be determined directly from the curves. Simply divide the change in drain current by the change in gate voltage.

Because of the slope of the curve, it can be assumed that a smaller change in gate voltage can produce a larger drain current change.

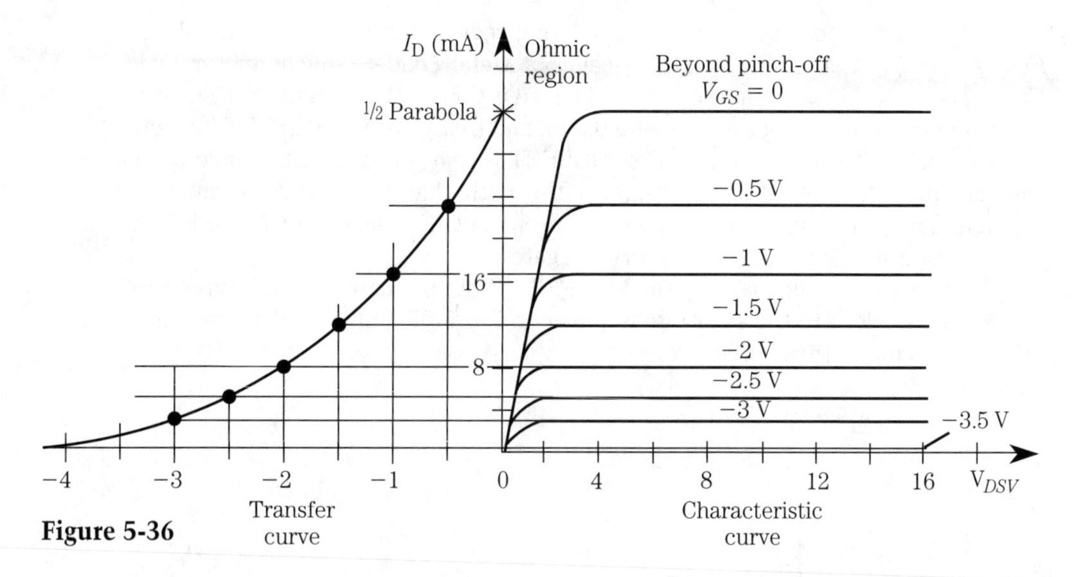

Figure 5-36

To summarize the operation of the field-effect transistor, a small input signal voltage controls a relatively large change in the output current. You will remember that the same thing was said of the triode tube.

JFET transconductance

The parameter that relates the input signal-voltage amplitude to the output signal current is called *transconductance*.

Mathematically:

Transconductance = change in drain current/change in gate voltage

This is often written as:

$$g_m = \Delta I_D \Delta V_{GS}$$

Where g_m is the transconductance as measured in Siemens.

At one time, conductance and transconductance was measured in mhos. That unit is still around, but Siemens is now preferred. The equation is for the transconductance measured across the FET from gate to drain. (A similar parameter is given for vacuum tubes.)

MOSFETs

Remember that an excessive gate current will produce internal heat in the JFET that can cause its destruction. Circuit designers must always be careful not to let that happen, but it is difficult when the input signal amplitude varies over a wide range.

To eliminate that problem, an insulating region has been placed around the gate (Fig. 5-27). The N-channel *depletion MOSFET* is similar to the N-channel JFET except for the insulating region in the MOSFET; its purpose is to prevent the flow of gate current when the gate-to-source junction is forward biased.

A name given to a FET with an insulating region was IGFET, *I*nsulated *G*ate *F*ield-*E*ffect *T*ransistor, but it is now called a *M*etal *O*xide *S*emiconductor *F*ield-*E*ffect *T*ransistor. (The name *IGFET* is still used in some other countries.) The insulating region around the gate is a metal oxide, and that gives the MOSFET its name.

The reason it is called a depletion MOSFET is that a (negative) voltage on the gate of an N-channel MOSFET depletes the path that the electrons can move through. That, in turn, decreases the drain current. Enhancement MOSFETs increase the path with an increase in forward gate bias.

The substrate is the material the MOSFET is built upon, usually a P-type material. A separate lead to the substrate is shown in Fig. 5-37, but the substrate is most often connected to the source internally.

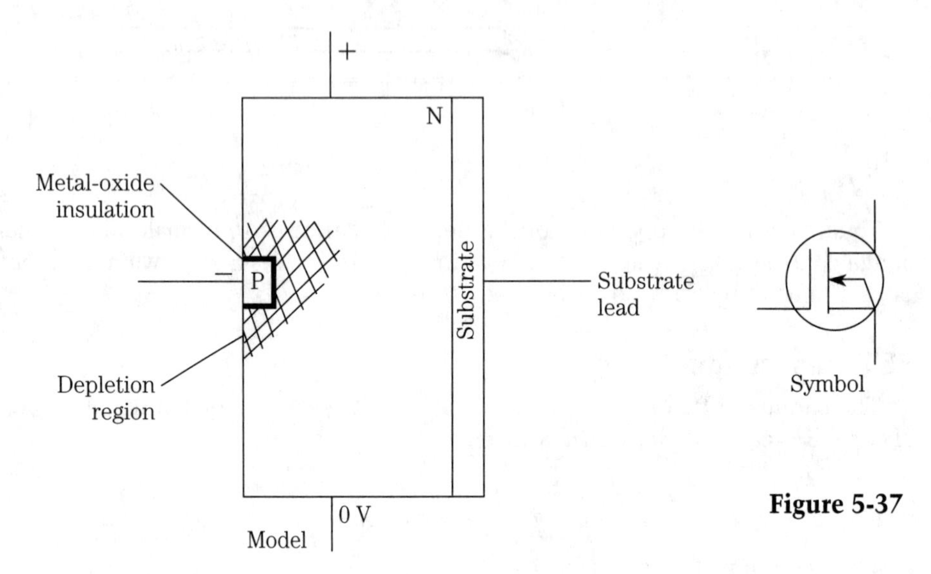

Figure 5-37

When depletion MOSFETs were first introduced, they were a nightmare to technicians. Any static electricity that touched the gate lead destroyed the insulation around the gate and made the device useless. The static electricity could be on a technician's hand, on the tip of a soldering iron, on its shipping container, or any number of other places.

The MOSFET was shipped with a small-diameter wire wrapped around the leads. The procedure was to insert the leads into the socket, then pull the wire off. That took a considerable amount of dexterity.

Today, the MOSFETs have been provided with internal zener diodes to prevent a static voltage buildup, as shown schematically in Fig. 5-38. However, despite the zener protection, the manufacturers suggest that technicians proceed with caution when handling MOSFETs.

Assume the gate of a depletion region is open (not connected to anything). If you connect an ohmmeter across the depletion MOSFET, between the source and drain, it will show resistance. It is, after all, a semiconductor. That leads to an im-

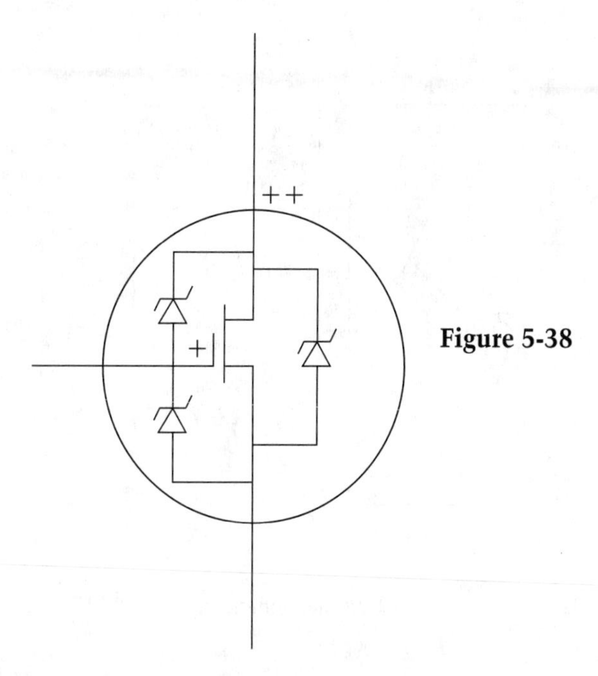

Figure 5-38

portant word of caution. Without gate bias to limit the current, you must be careful not to connect a voltage across the depletion MOSFET that can destroy the channel with an excessively high current!

The fact that there is no gate current in a MOSFET circuit means there is no partition noise. That is why you see depletion MOSFETs used extensively in radio frequency amplifier circuits. Noise in the RF circuits causes hissing noises in the audio output, and causes snow in video output circuits.

The enhancement MOSFET

The model and the symbol for enhancement MOSFETs is shown in Fig. 5-39. In this case, the device has a depletion region that reaches completely across the channel. However, it is a convenient model that truly describes the action of the device.

No current can flow from source to drain without a positive gate bias that reduces the depletion region. As shown in the illustration, the operating polarities on the N-channel version of the enhancement MOSFET shows a positive gate as well as a positive drain.

As shown in Fig. 5-39, the depletion region fills the place used in the depletion MOSFET for the channel. Because it is a semiconductor with a relatively high resistance, it prevents current from flowing in the absence of a gate voltage. An ohmeter connected between the gate and drain of an enhancement MOSFET should show a high resistance when the gate lead is open.

A positive voltage on the gate of the enhancement MOSFET draws electrons from the substrate, and that creates the N-type channel needed for source-to-drain current flow. The more positive the drain, the greater the channel. Saying it another way, increasing the positive gate voltage increases the drain current.

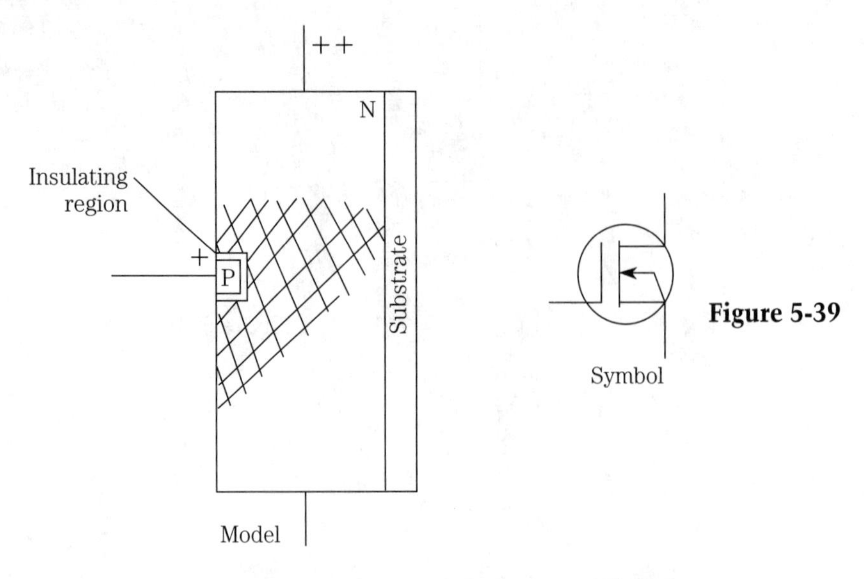

Figure 5-39

N-Channel enhancement MOSFET

If you compare the polarities of voltages on the N-channel enhancement MOSFET with the operating voltages on an NPN bipolar transistor, you will see they are the same.

Even though the gate is positive in the enhancement MOSFET, no current flows in the gate output. Therefore, the device has a high input impedance like the JFET and the depletion MOSFET.

The biases on the depletion MOSFET will not work on an enhancement MOSFET because they would put a negative voltage on the gate, and that would drive the negative charge carriers in the substrate away from the desired channel position. However, the types of biases for an NPN bipolar transistor will work.

MOSFETs can be analyzed with a family of curves and a load line. However, the use of transfer function curves is preferred. Most of the commercial curve tracers in use today can draw both families of curves and transfer curves.

A word of caution: technicians sometimes get complacent about touching transistor circuits because they usually operate at low voltages. However, there are some important exceptions. Enhancement MOSFETs are often used in transmitter circuits where the drain voltage is as high as 600 volts! There are also some stacked bipolar circuits that operate well above 100 volts! Be careful when working in high-voltage semiconductor circuits!

Special semiconductor devices

For many years, MOSFETs were used only in low-power circuits. The problem was that any heat generated at the internal PN junction was trapped inside the device, and the buildup of heat destroyed the MOSFET.

Near the end of the 1980s, a specially fabricated MOSFET was introduced. It is called a *VFET* and it can be used in high-power circuits. Figure 5-40 shows the model. This power FET has replaced bipolar transistors as audio power amplifiers and other applications where a power amplifying device is needed. There is no special symbol for a VFET.

In a standard amplifier configuration, a MOSFET voltage amplifier works with a drain load resistor. However, there are special series combinations of MOSFETs in which one serves as an amplifier and the second one serves as a resistive load (Fig. 5-40). The combination is called *CMOS* (pronounced "Sea Moss"), which is a Complementary Metal Oxide Semiconductor.

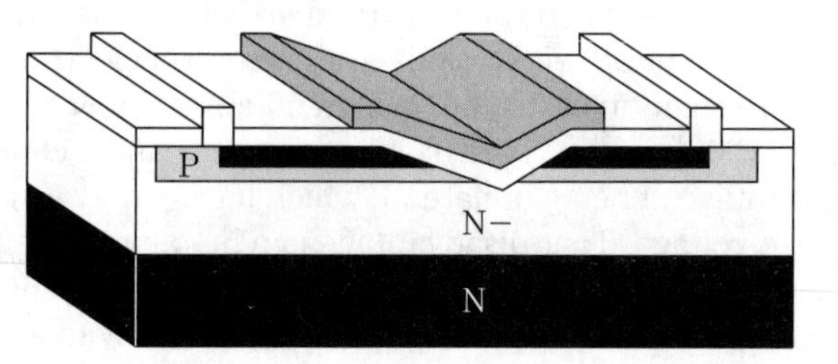

Figure 5-40

As shown in the illustration, one MOSFET is an N-channel type and the other is a P-channel type. That explains the name "complementary." One advantage of this combination is that both MOSFETs can be made on an integrated circuit at the same time.

Bipolar power transistors can be made to control large currents and, therefore, large amounts of power. The disadvantage of those power BJTs is that their gate circuit requires relatively high amounts of power.

Figure 5-40 shows a BJT circuit used for getting high power control without the need for a high base current, called a *Darlington Amplifier* or a *Darlington Pair*. Darlingtons can be made with separate transistors, but you will usually find them in a single power transistor package. Because the combination has three leads to the outside world, the package looks no different from a single power transistor. They are usually represented on schematics with a standard BJT symbol.

Darlington pairs have a high input impedance compared to a single power transistor. Also, they have a high current gain (beta); in fact, the beta of the combination is equal to the product of the individual betas. Because the transistors are usually matched, the Darlington amplifier is sometimes called a *beta-squared* (β^2) *amplifier*.

The BIFET amplifying device shown in Fig. 5-40 has the FET advantage of high input impedance along with the high current advantage of the BJT.

SCR and triac thyristors

There are two important switching devices that are examples of three-terminal devices. They are: the *silicon controlled rectifier (SCR)* and the *triac*. The symbols for both are shown in Fig. 5-41.

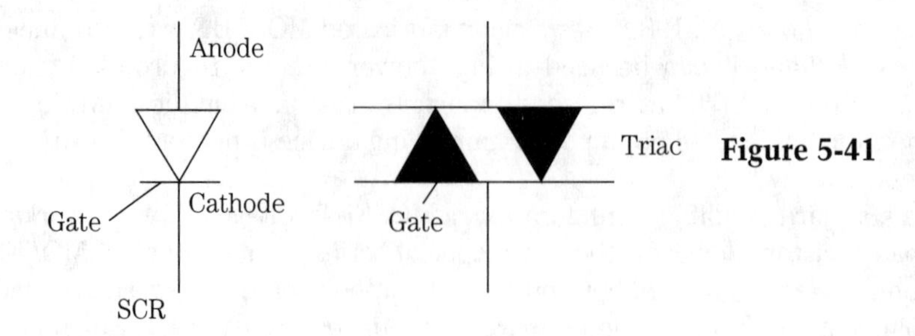

Figure 5-41

The SCR is a semiconductor replacement of a thyratron or gas-filled triode tube. Both devices are very rapid switches, and can be turned on with a pulse to their control electrode. Neither can be shut off with a signal to the control electrode.

SCRs and thyratrons are sometimes used to switch off a portion of an ac voltage waveform. In that way, the RMS portion value of the waveform can be controlled.

SCRs have the disadvantage of being unilateral, which means they can only control one half of a sine-wave voltage. That disadvantage can be taken care of by connecting them back-to-back as shown in Fig. 5-42. A triac is the same thing, except that the two SCRs are built into the same package. As with the individual SCRs, the triacs cannot be turned off with a signal to the gate.

SCRs and triacs are the subject of some review questions. Be sure you know the purpose of the snubber circuit, shown in Fig. 5-42. It prevents the SCR from being turned on by a high reverse voltage during the nonconducting half cycle. That can happen when the load is inductive and there is a high counter voltage.

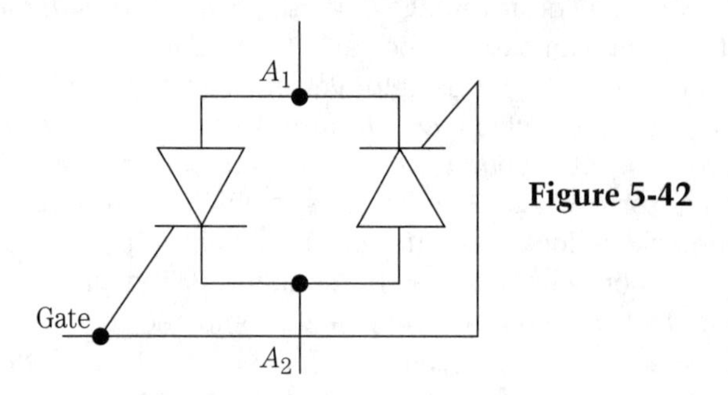

Figure 5-42

Programmed review

Start with Block No. 1. Pick the answer that you believe is correct. Go to the next block and check your answer. All answers are in italics. There is only one choice for each block. There is some material in this section that was not covered in the chapter.

Block 1

A resistor is connected across the antenna terminals of a high-gain receiver that is operating at room temperature. Which of the following statement is true?

A. There is a noise voltage injected into the receiver.
B. A noise voltage is not injected into the receiver, so nothing happens.

Block 2

The correct answer is A. The same thing happens when an antenna is connected to the antenna terminals because of the antenna resistance.

Here is your next question: Resistor noise is:

A. white noise. B. blue noise.

Block 3

The correct answer is A. There is no type of noise called *blue noise.*

Here is your next question: What is the name of the component in Fig. 5-43?

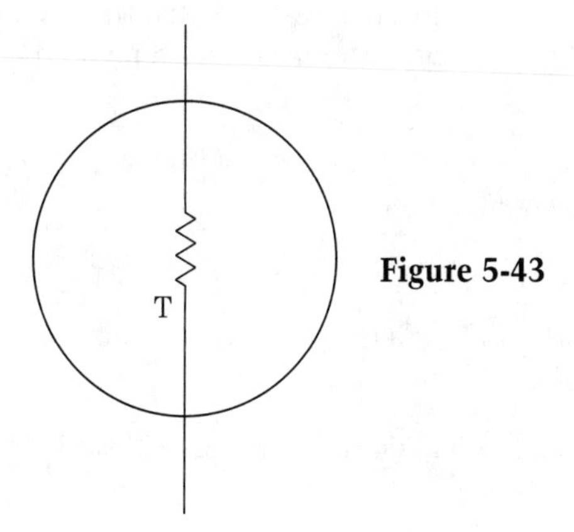

Figure 5-43

Block 4

The correct answer is thermistor. There are two types of thermistors: NTC and PTC. Thermistors with a negative temperature coefficient (NTC) decrease in resistance when their temperature is increased. Thermistors with a positive temperature coefficient (PTC) increase in resistance when their temperature is increased. For the application in Fig. 5-50, an NTC thermistor must be used so it matches the temperature coefficient of the transistor emitter-base junction.

Here is your next question: A series of electrical measurements or values that determine the characteristics or behavior of a device in a circuit are called a ____________.

Block 5

The correct answer is parameter. As a model, think of a parameter as being the thing you choose to make the answer come outright. For example, if you want a circle with an area of 10 square inches the parameter is the radius (or diameter). Likewise, if you want to make a capacitor that has a capacitance of 0.1 μF, the parameters are the area of one plate, the thickness of the dielectric, and the dielectric constant.

Here is your next question: For a certain BJT, the collector current is 20 mA and the base current is 1000 nanoamperes. What is the value of emitter current?

Block 6

The correct answer is 21 mA. The emitter current is the sum of the base and collector currents. Note that 1000 nanoamperes is 1 mA.

Here is your next question: In an amplifier, the emitter-base junction of a BJT is:

A. reverse biased. B. forward biased.

Block 7

The correct answer is B. The question assumes that the amplifier is class A. That is the classification most often used. If the amplifier is operated class B, there is no forward bias.

Here is your next question: What is the name of the field-effect transistor that is designed to work as a power amplifier?

Block 8

The correct answer is VFET. That is a generic term. Some manufacturers have given them different names.

Here is your next question: What do the letters CMOS stand for?

Block 9

The correct answer is Complementary Metal Oxide Semiconductor. The word complementary means that an N-channel and a P-channel field-effect transistor are used together to complement each other. Metal oxide refers to the insulating material around the gate. The combination is a semiconductor device. This is a very popular method of making integrated circuits.

Here is your next question: The BJT combination in Fig. 5-44 is called a ____________ amplifier. It has a ________ (high)(low) input impedance.

Block 10

The correct answers are Darlington pair and *high.* The combination has a beta that is equal to the product of the beta of each transistor. If they are matched transistors the beta of the combination is beta^2.

Here is your next question: Diodes are connected in parallel to get ______________.

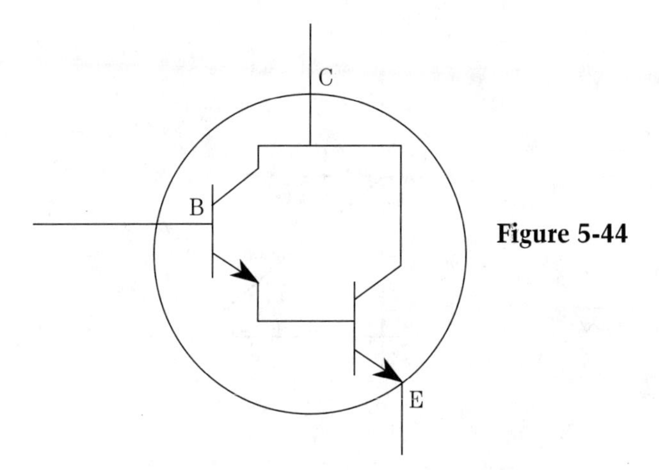

Figure 5-44

Block 11

The correct answer is: a higher current rating than possible for one of the diodes in the parallel combination. The current rating of a parallel combination of diodes is equal to the sum of the current ratings of each diode in the parallel combination. The peak inverse voltage (PIV) rating is equal to the lowest PIV rating in the parallel combination.

Here is your next question: Diodes are connected in series to:

Block 12

The correct answer is: a higher PIV rating than possible for one of the diodes in the series combination. The PIV rating of the combination is equal to the sum of the PIV ratings of each diode in the series combination. The Peak Inverse Voltage rating of the series combination is equal to the lowest PIV rating in the combination of diodes.

Here is your next question: Which of the following diodes must be operated with a reverse current?

A. Varactor diode. B. Zener diode.

Block 13

The correct answer is A. Both must be operated with a reverse voltage, but only the zener diode is operated with a reverse current.

Here is your next question: Assume the drain of the MOSFET in Fig. 5-45 has the correct operating voltage. The drain current can be increased by moving the arm of the variable resistor toward:

A. X. B. Y.

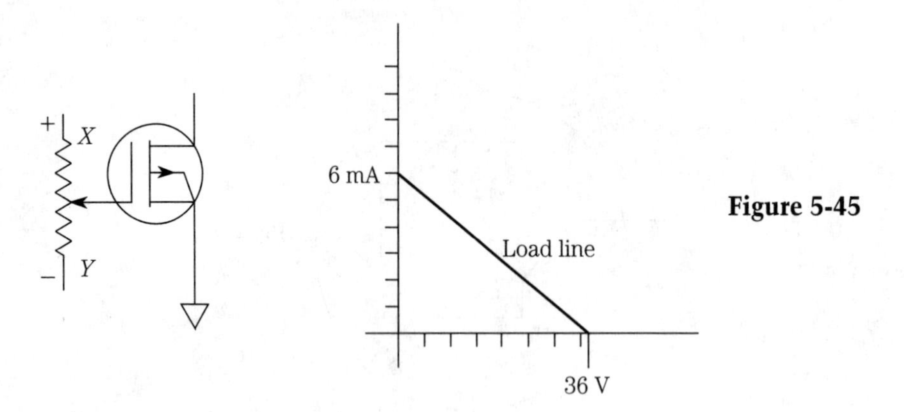

Figure 5-45

Block 14

The correct answer is B. It is a P-channel MOSFET and it requires a negative polarity on the gate and drain. For the position shown in the illustration, there is 0 V on the gate and the MOSFET cannot conduct. Moving the arm of the switch toward Y pulls holes out of the substrate and creates a channel for drain current.

Here is your next question: Write an equation for JFET transconductance.

Block 15

The correct answer is:

$$9m = I_C/V_{GS}$$

Be sure you know the meaning of transconductance and how it is calculated.

Here is your next question: The load line in Fig. 5-45 is drawn on the basis of a Thevenin generator. What is the internal resistance of that generator?

Block 16

The correct answer is 6000 Ω. The point on the current axis is 6 mA and the point on the voltage axis is 36 V. They are the Norton current and the Thevenin voltage. The question is asking for the Thevenin resistance:

$$R_{TH} = V_{TH}/I_N = 36 \text{ V}/6 \text{ mA} = 6000 \text{ V}$$

Here is your next question: Energy by virtue of motion is called ____________ energy.

Block 17

The correct answer is kinetic energy. An arrow moves with great speed so it has a high kinetic energy. If it strikes a wood target, its kinetic energy changes to heat energy.

Here is your next question: An avalanching current occurs in a ________________.

A. neon lamp. B. silicon diode.

Block 18

The correct answer is A. If you had trouble with this question, review the operation of the neon lamp as it relates to Fig. 5-11.

Here is your next question: Another name for a LAD is ____________.

Block 19

The correct answer is Light-Activated Diode *or photodiode.* The semiconductor devices that radiate light or utilize light are called *optoelectronic devices.*

Here is your next question: What voltage polarity is required at point X in Fig. 5-46?

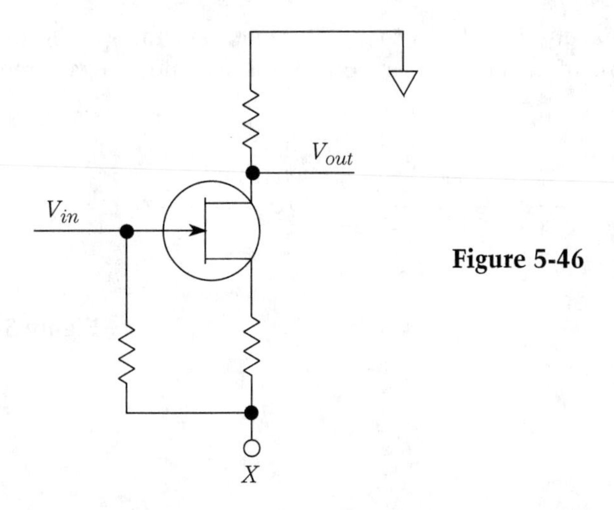

Figure 5-46

Block 20

The correct answer is negative. For an N-channel JFET, the source is negative with respect to the drain.

Here is your next question: The circuit in Fig. 5-46 is in a:

A. common drain configuration.
B. common gate configuration.
C. common source configuration.

Block 21

The correct answer is C. If you answered A, you were tricked by the fact that the drain is connected through a resistor to common. Remember this very important rule: the configuration is determined by where the signal enters and where the signal leaves. The remaining point determines the configuration.

In Fig. 5-46, the signal enters the gate and leaves the source so that it is a common drain circuit configuration.

Here is your next question: Give two reasons for the popularity of MOSFETs.

Block 22

The text emphasized that field-effect transistors do not require signal power or dc standby power. A very important advantage of MOSFETs and other field-effect transistors is their low noise. There is no partition noise, so their high-frequency noise is low. You will often see MOSFETs in RF and IF amplifier circuits.

Here is your next question: Write the conjugate of $3 - j5$.

Block 23

The correct answer is $3 + j5$. To write the conjugate, retain the values, but change the sign of the j term.

Here is your next question: Refer to the Thevenin Generator in Fig. 5-47. What is the rectangular form of the impedance required for maximum power transfer.

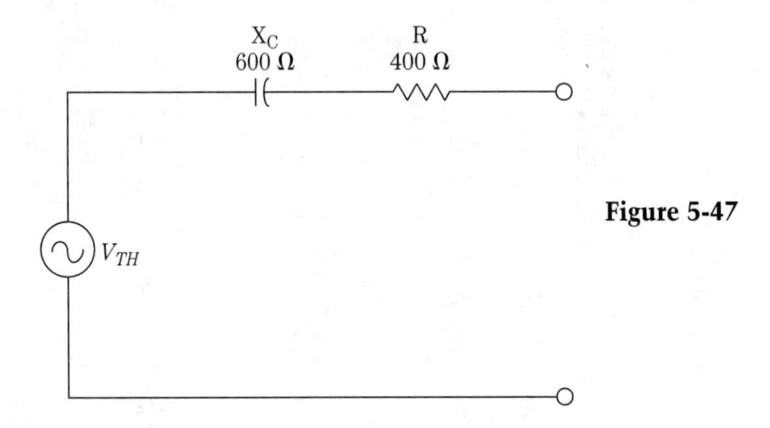

Figure 5-47

Block 24

The correct answer is $400 + j600$. In order to get maximum power out to the load, it is necessary for the phase angle between the voltage and current to be 0°. In other words, the inductive reactance and capacitive reactance be equal. Also, the internal resistance must equal the load resistance. All of the conditions are met when the load impedance is the conjugate of the internal impedance.

Here is your next question: A screen grid resistor burns out after smoking. First thing to check would be:

A. Short in screen grid bypass capacitor.
B. An open turn in plate coil.
C. Short in B+ bypass capacitor.
D. Open grid leak resistor.

Block 25

The correct answer is A.

Here is your next question: When a vacuum tube operates at VHF or higher as compared to lower frequencies:

A. Transit time of electrons becomes important.
B. It is necessary to make larger components.
C. It is necessary to increase grid spacing.
D. Only a pentode is satisfactory.

Block 26

The correct answer is A.

Here is your next question: What is the photoconductive effect?

A. The conversion of photon energy to electromotive energy.
B. The increased conductivity of an illuminated semiconductor junction.
C. The conversion of electromotive energy to photon energy.
D. The decreased conductivity of an illuminated semiconductor junction.

Block 27

The correct answer is B.

Here is your next question: What happens to photoconductive material when light shines on it?

A. The conductivity of the material increases.
B. The conductivity of the material decreases.
C. The conductivity of the material stays the same.
D. The conductivity of the material becomes temperature dependent.

Block 28

The correct answer is A.

Here is your next question: What happens to the resistance of a photoconductive material when light shines on it?

A. It increases.
B. It becomes temperature dependent.
C. It stays the same.
D. It decreases.

Block 29

The correct answer is D.

Here is your next question: What happens to the conductivity of a semiconductor junction when it is illuminated?

A. It stays the same.
B. It becomes temperature dependent.
C. It increases.
D. It decreases.

Block 30

The correct answer is C.

Here is your next question: What is an optocoupler?

A. An LCD in a phototransistor.

B. A frequency-modulated helium-neon laser.
C. An amplitude-modulated helium-neon laser.
D. An LED and a phototransistor.

Block 31

The correct answer is D.

Here is your next question: What is an optoisolator?

A. An LED and a phototransistor.
B. A P-N junction that develops an excess positive charge when exposed to light.
C. An LED and a capacitor.
D. An LED and a solar cell.

Block 32

The correct answer is A.

Here is your next question: What is an optical shaft encoder?

A. An array of optocouplers chopped by a stationary wheel.
B. An array of optocouplers whose light transmission path is controlled by a rotating wheel.
C. An array of optocouplers whose propagation velocity is controlled by a stationary wheel.
D. An array of optocouplers whose propagation velocity is controlled by a rotating wheel.

Block 33

The correct answer is B.

Here is your next question: What does the photoconductive effect in crystalline solids produce a noticeable change in?

A. The capacitance of the solid.
B. The inductance of the solid.
C. The specific gravity of the solid.
D. The resistance of the solid.

Block 34

The correct answer is D.

Here is your next question: Structurally, what are the two main categories of semiconductor diodes?

A. Junction and point contact.
B. Electrolytic and junction.
C. Electrolytic and point contact.
D. Vacuum and point contact.

Block 35

The correct answer is A.

Here is your next question: What are the two primary classifications of Zener diodes?

A. Hot carrier and tunnel. C. Voltage regulator and voltage reference.
B. Varactor and rectifying. D. Forward and reversed biased.

Block 36

The correct answer is C.

Here is your next question: What is the principal characteristic of a Zener diode?

A. A constant current under conditions of varying voltage.
B. A constant voltage under conditions of varying current.
C. A negative resistance region.
D. An internal capacitance that varies with the applied voltage.

Block 37

The correct answer is B.

Here is your next question: What is the range of voltage ratings available in Zener diodes?

A. 2.4 V to 200 V. B. 1.2 V to 7 V. C. 3 V to 2000 V. D. 1.2 V to 5.6 V.

Block 38

The correct answer is A.
Here is your next question: What is the principal characteristic of a tunnel diode?

A. A high forward resistance. C. A negative resistance region.
B. A very high PIV. D. A high forward current rating.

Block 39

The correct answer is C.

Here is your next question: What special type of diode is capable of both amplification and oscillation?

A. Point contact diodes. B. Zener diodes. C. Tunnel diodes. D. Junction diodes.

Block 40

The correct answer is C.

Here is your next question: What type of semiconductor diode varies its internal capacitance as the voltage applied to its terminals varies?

A. A varactor diode. C. A silicon-controlled rectifier.
B. A tunnel diode. D. A Zener diode.

Block 41

The correct answer is A.

Here is your next question: What is the principal characteristic of a varactor diode?

A. It has a constant voltage under conditions of varying current.
B. Its internal capacitance varies with the applied voltage.
C. It has a negative resistance region.
D. It has a very high PIV.

Block 42

The correct answer is B.

Here is your next question: What is a common use of a varactor diode?

A. As a constant current source.
B. As a constant voltage source.
C. As a voltage controlled inductance.
D. As a voltage controlled capacitance.

Block 43

The correct answer is D.

Here is your next question: What is a common use of a hot-carrier diode?

A. As balanced mixers in SSB generation.
B. As a variable capacitance in an automatic frequency control circuit.
C. As a constant voltage reference in a power supply.
D. As VHF and UHF mixers and detectors.

Block 44

The correct answer is D.

Here is your next question: What limits the maximum forward current in a junction diode?

A. The peak inverse voltage.
B. The junction temperature.
C. The forward voltage.
D. The back EMF.

Block 45

The correct answer is B.

Here is your next question: How are junction diodes rated?

A. Maximum forward current and capacitance.
B. Maximum reverse current and PIV.
C. Maximum reverse current and capacitance.
D. Maximum forward current and PIV.

Block 46

The correct answer is D.

Here is your next question: What is a common use for point-contact diodes?

A. As a constant-current source.
B. As a constant-voltage source.
C. As an RF detector.
D. As a high-voltage rectifier.

Block 47

The correct answer is C.

Here is your next question: What type of diode is made of a metal whisker touching a very small semiconductor die?

A. Zener diode. B. Varactor diode. C. Junction diode. D. Point contact diode.

Block 48

The correct answer is D.

Here is your next question: What is one common use for PIN diodes?

A. As a constant-current source. C. As an RF switch.
B. As a constant-voltage source. D. As a high-voltage rectifier.

Block 49

The correct answer is C.

Here is your next question: What special type of diode is often used in RF switches, attenuators, and various types of phase shifting devices?

A. Tunnel diodes. B. Varactor diodes. C. PIN diodes. D. Junction diodes.

Block 50

The correct answer is C.

Here is your next question: What are the three terminals of a bipolar transistor?

A. Cathode, plate, and grid. C. Gate, source, and sink.
B. Base, collector, and emitter. D. Input, output, and ground.

Block 51

The correct answer is B.

Here is your next question: What is the meaning of the term *alpha* with regard to bipolar transistors?

A. The change of collector current with respect to base current.
B. The change of base current with respect to collector current.
C. The change of collector current with respect to emitter current.
D. The change of collector current with respect to gate current.

Block 52

The correct answer is C.

Here is your next question: What is the term used to express the ratio of change in dc collector current to a change in emitter current in a bipolar transistor?

A. Gamma. B. Epsilon. C. Alpha. D. Beta.

Block 53

The correct answer is C.

Here is your next question: What is the meaning of the term *beta* with regard to bipolar transistors?

A. The change of collector current with respect to base current.
B. The change of base current with respect to emitter current.
C. The change of collector current with respect to emitter current.
D. The change in base current with respect to gate current.

Block 54

The correct answer is A.

Here is your next question: What is the term used to express the ratio of change in the dc collector current to a change in base current in a bipolar transistor?

A. Alpha. B. Beta. C. Gamma. D. Delta.

Block 55

The correct answer is B.

Here is your next question: What is the meaning of the term alpha cutoff frequency with regard to bipolar transistors?

A. The practical lower frequency limit of a transistor in common emitter configuration.
B. The practical upper frequency limit of a transistor in common base configuration.
C. The practical lower frequency limit of a transistor in common base configuration.
D. The practical upper frequency limit of a transistor in common emitter configuration.

Block 56

The correct answer is B.

Here is your next question: What is the term used to express that frequency at which the grounded base current gain has decreased to 0.7 of the gain obtainable at 1 kHz in a transistor?

A. Corner frequency.
B. Alpha cutoff frequency.
C. Beta cutoff frequency.
D. Alpha rejection frequency.

Block 57

The correct answer is B.

Here is your next question: What is the meaning of the term beta cutoff frequency with regard to a bipolar transistor?

A. That frequency at which the grounded base current gain has decreased to 0.7 of that obtainable at 1 kHz in a transistor.
B. That frequency at which the grounded emitter current gain has decreased to 0.7 of that obtainable at 1 kHz in a transistor.

C. That frequency at which the grounded collector current gain has decreased to 0.7 of that obtainable at 1 kHz in a transistor.
D. That frequency at which the grounded gate current gain has decreased to 0.7 of that obtainable at 1 kHz in a transistor.

Block 58

The correct answer is B.

Here is your next question: What is the meaning of the term *transition region* with regard to a transistor?

A. An area of low charge density around the P-N junction.
B. The area of maximum P-type charge.
C. The area of maximum N-type charge.
D. The point where wire leads are connected to the P- or N-type material.

Block 59

The correct answer is A.

Here is your next question: What does it mean for a transistor to be fully saturated?

A. The collector current is at its maximum value.
B. The collector current is at its minimum value.
C. The transistor's alpha is at its maximum value.
D. The transistor's beta is at its maximum value.

Block 60

The correct answer is A.

Here is your next question: What does it mean for a transistor to be cut off?

A. There is no base current.
B. The transistor is at its operating point.
C. No current flows from emitter to collector.
D. Maximum current flows from emitter to collector.

Block 61

The correct answer is C.

Here is your next question: What are the elements of a unijunction transistor?

A. Base 1, base 2, and emitter. C. Gate, base 1, and base 2.
B. Gate, cathode, and anode. D. Gate, source, and sink.

Block 62

The correct answer is A.

Here is your next question: For best efficiency and stability, where on the load line should a solid-state power amplifier be operated?

A. Just below the saturation point. C. At the saturation point.
B. Just above the saturation point. D. At 1.414 times the saturation point.

Block 63

The correct answer is A.

Here is your next question: What two elements widely used in semiconductor devices exhibit both metallic and nonmetallic characteristics?

A. Silicon and gold. C. Galena and germanium.
B. Silicon and germanium. D. Galena and bismuth.

Block 64

The correct answer is B.

Here is your next question: What are the three terminals of an SCR?

A. Anode, cathode, and gate. C. Base, collector, and emitter.
B. Gate, source, and sink. D. Gate, base 1, and base 2.

Block 65

The correct answer is A.

Here is your next question: What are the two stable operating conditions of an SCR?

A. Conducting and nonconducting.
B. Oscillating and quiescent.
C. Forward conducting and reverse conducting.
D. NPN conduction and PNP conduction.

Block 66

The correct answer is A.

Here is your next question: When an SCR is in the triggered or on condition, its electrical characteristics are similar to what other solid-state device (as measured between its cathode and anode)?

A. The junction diode. C. The hot-carrier diode.
B. The tunnel diode. D. The varactor diode.

Block 67

The correct answer is A.

Here is your next question: Under what operating condition does an SCR exhibit electrical characteristics similar to a forward-biased silicon rectifier?

A. During a switching transition. C. When it is gated "off."
B. When it is used as a detector. D. When it is gated "on."

Block 68

The correct answer is D.

Here is your next question: What is the transistor called that is fabricated as two complementary SCRs in parallel with a common gate terminal?

A. Triac. B. Bilateral SCR. C. Unijunction transistor. D. Field-effect transistor.

Block 69

The correct answer is A.

Here is your next question: What are the three terminals of a Triac?

A. Emitter, base 1, and base 2. C. Base, emitter, and collector.
B. Gate, anode 1, and anode 2. D. Gate, source, and sink.

Block 70

The correct answer is B.

Here is your next question: What is the normal operating voltage and current for a light-emitting diode?

A. 60 V and 20 mA. B. 5 V and 50 mA. C. 1.7 V and 20 mA. D. 0.7 V and 60 mA.

Block 71

The correct answer is C.

Here is your next question: What type of bias is required for an LED to produce luminescence?

A. Reverse bias. B. Forward bias. C. Zero bias. D. Inductive bias.

Block 72

The correct answer is B.

Here is your next question: What are the advantages of using an LED?

A. Low-power consumption and long life.
B. High lumens per cm per cm and low-power consumption.
C. High lumens per cm per cm and low-voltage requirement.
D. A current flows when the device is exposed to a light source.

Block 73

The correct answer is A.

Here is your next question: What colors are available in LEDs?

A. Yellow, blue, red, and brown. C. Violet, blue, orange, and red.
B. Red, violet, yellow, and peach. D. Red, green, orange, and yellow.

Block 74

The correct answer is D.

Here is your next question: How can a neon lamp be used to check for the presence of RF?

A. A neon lamp will go out in the presence of RF.
B. A neon lamp will change color in the presence of RF.
C. A neon lamp will light only in the presence of very low frequency RF.
D. A neon lamp will light in the presence of RF.

Block 75

The correct answer is D.

Here is your next question: What is an enhancement-mode FET?

A. An FET with a channel that blocks voltage through the gate.
B. An FET with a channel that allows a current when the gate voltage is zero.
C. An FET without a channel to hinder current through the gate.
D. An FET without a channel; no current occurs with zero gate voltage.

Block 76

The correct answer is D.

Here is your next question: What is a depletion-mode FET?

A. An FET that has a channel with no gate voltage applied; a current flows with zero gate voltage.
B. An FET that has a channel that blocks current when the gate voltage is zero.
C. An FET without a channel; no current flows with zero gate voltage.
D. An FET without a channel to hinder current through the gate.

Block 77

The correct answer is A.

Here is your next question: Why do many MOSFET devices have built-in gate-protective Zener diodes?

A. The gate-protective Zener diode provides a voltage reference to provide the correct amount of reverse-bias gate voltage.
B. The gate-protective Zener diode protects the substrate from excessive voltages.
C. The gate-protective Zener diode keeps the gate voltage within specifications to prevent the device from overheating.
D. The gate-protective Zener diode prevents the gate insulation from being punctured by small static charges or excessive voltages.

Block 78

The correct answer is D.

Here is your next question: What do the initials CMOS stand for?

A. Common mode oscillating system.
B. Complementary mica-oxide silicon.
C. Complementary metal-oxide semiconductor.
D. Complementary metal-oxide substrate.

Block 79

The correct answer is C.

Here is your next question: Why are special precautions necessary in handling FET and CMOS devices?

A. They are susceptible to damage from static charges.
B. They have fragile leads that might break off.
C. They have micro-welded semiconductor junctions that are susceptible to breakage.
D. They are light sensitive.

Block 80

The correct answer is A.

Here is your next question: How does the input impedance of a field-effect transistor compare with that of a bipolar transistor?

A. One cannot compare input impedance without first knowing the supply voltage.
B. An FET has low input impedance; a bipolar transistor has high input impedance.
C. The input impedance of FETs and bipolar transistors is the same.
D. An FET has high input impedance; a bipolar transistor has low input impedance.

Block 81

The correct answer is D.

Here is your next question: What are the three terminals of a field-effect transistor?

A. Gate 1, gate 2, drain.
B. Emitter, base, collector.
C. Emitter, base 1, base 2.
D. Gate, drain, source.

Block 82

The correct answer is D.

Here is your next question: What are the two basic types of junction field-effect transistors?

A. N-channel and P-channel.
B. High power and low power.
C. MOSFET and GaAsFET.
D. Silicon FET and germanium FET.

Block 83

The correct answer is A.

Here is your next question: What is the voltage drop across R1? (Please refer to Fig. 5-48.)

A. 9 V. B. 7 V. C. 5 V. D. 3 V.

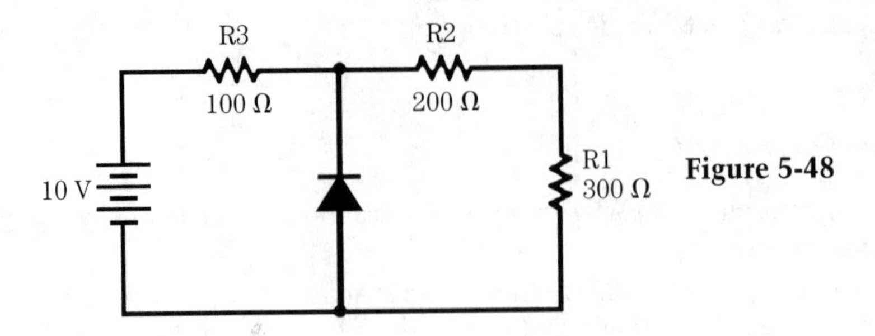

Figure 5-48

Block 84

The correct answer is C.

Here is your next question: What is the voltage drop across R1? (Please refer to Fig. 5-49.)

A. 1.2 V. B. 2.4 V. C. 3.7 V. D. 9 V.

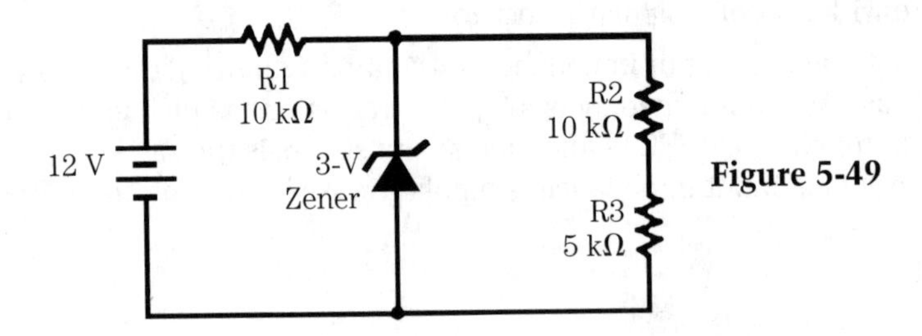

Figure 5-49

Block 85

The correct answer is D.

Here is your next question: When S1 is closed, light L1 and L2 go on. What is the condition of both lamps when both S1 and S2 are closed? (Please refer to Fig. 5-50.)

A. Both lamps stay on. C. Both lamps turn off.
B. L1 turns off; L2 stays on. D. L1 stays on; L2 turns off.

Block 86

The correct answer is D.

Here is your next question: If S1 is closed, both lamps light, what happens when S1 and S2 are closed? (Please refer to Fig. 5-51.)

A. L1 and L2 are off. C. L1 is off and L2 is on.
B. L1 is on and L2 is flashing. D. L1 is on and L2 is off.

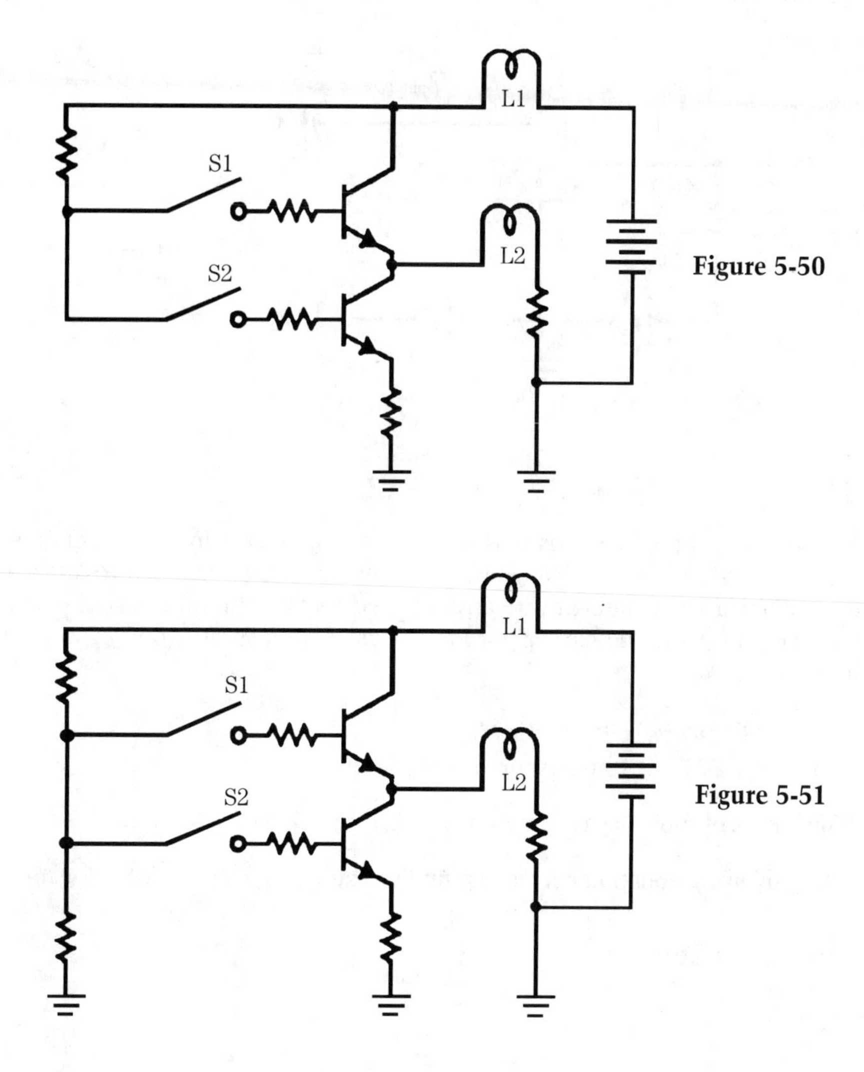

Figure 5-50

Figure 5-51

Block 87

The correct answer is D.

Here is your next question: What change is needed in order to correct the grounded emitter amplifier shown? (Please refer to Fig. 5-52.)

A. No change is necessary.
B. Polarities of emitter-base battery should be reversed.
C. Polarities of collector-base battery should be reversed.
D. Point A should be replaced with a low-value capacitor.

Block 88

The correct answer is A.

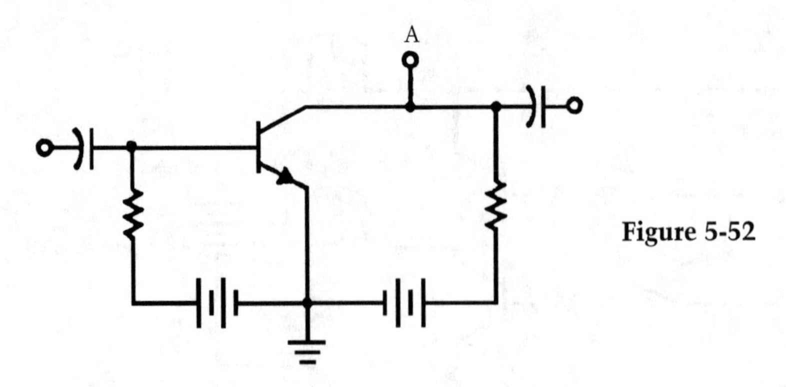

Figure 5-52

Quiz

You will likely find questions in this quiz (and all quizzes in this book) on subjects not covered in the text. They are on subjects you are expected to know as a technician. If you are unfamiliar with any of the subjects in this quiz, take the time to review that material in the earlier portion of this chapter. Write your answers on a separate sheet.

1. An emitter resistor is used for:

   A. biasing a BJT. B. temperature stabilization.

2. Which type of tube bias is sometimes called "signal bias?"

3. Is the following equation correct for finding the voltage across R2 in Fig. 5-53?

$$V_2 = R_2/(R_1 + R_2)$$

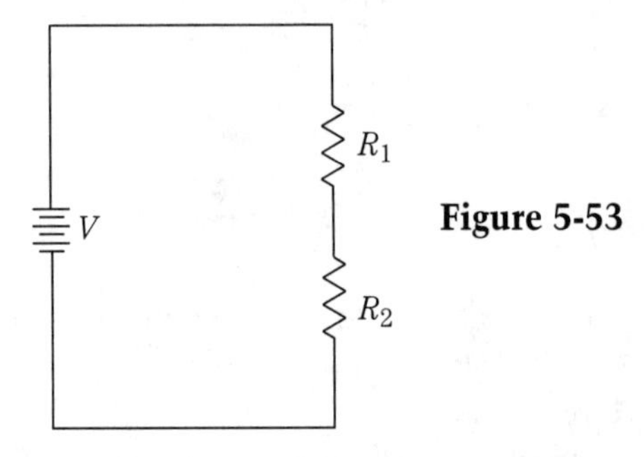

Figure 5-53

4. Write the meaning of the letters in the parameter BV_{ceo}.

5. Write the equation for finding the beta of a transistor when its alpha is known.

6. A line is drawn between the open-circuit voltage and the short-circuit current on a JFET characteristic curve. What is that line called?

7. Which of the following is an advantage of a MOSFET over a BJT in an RF amplifier circuit?

   A. Low-voltage operation. B. Low noise.

8. What polarity of voltage is required at point "X" in the amplifier circuit of Fig. 5-54?

9. What is the configuration of the circuit in Fig. 5-54?

   A. Common source. B. Common gate. C. Common drain.

10. What is the purpose of R2 in the circuit of Fig. 5-54?

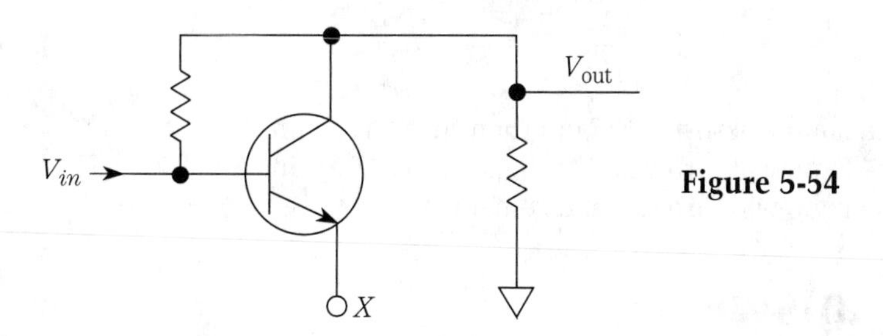

Figure 5-54

11. Why are diodes sometimes connected in series?

12. The voltage gain of an emitter follower circuit is:

    A. high. B. low.

13. The current gain of an emitter follower circuit is:

    A. high. B. low.

14. Which type of amplifier circuit can be used to match a high input impedance to a low output impedance?

15. When a voltage is connected across a Barium Titanate crystal, it changes shape (flexes). What is the name of this effect?

16. Which of the following is an advantage of connecting two transistors in the same package to make a Darlington amplifier?

    A. Low internal heat. B. High beta.

17. Which type of bias requires operation beyond the cutoff region?

18. Which type of diode has a negative resistance region in its characteristic curve?

19. For a JFET a change in drain current divided by a change in the gate voltage is called what?

20. Refer to the circuit in Fig. 5-55. Assume the SCR is not conducting. The switch is momentarily closed and then opened. Which of the following is correct?

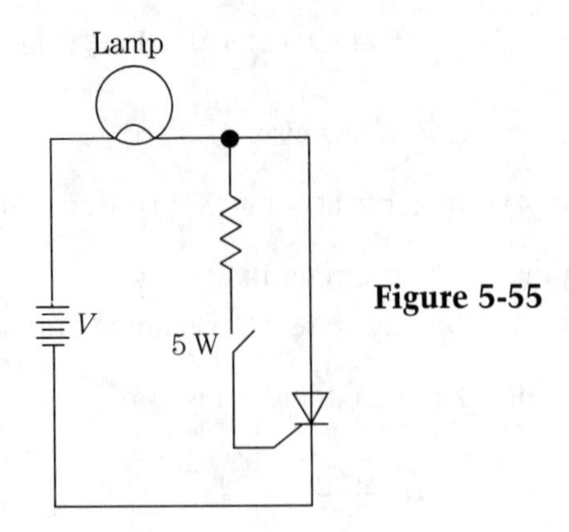

Figure 5-55

A. The lamp will turn OFF and then turn ON again.
B. The lamp will turn ON and then turn OFF again.
C. The lamp will turn ON and remain ON.

Quiz answers

1. A.
2. Grid leak bias.
3. No!

$$V_2 = (V)[R_2/(R_1 + R_2)]$$

4. *B*ack *V*oltage, *c*ollector-to-*e*mitter with the base *o*pen (*Note:* back voltage means reverse voltage.)
5. beta = alpha/(1–alpha).
6. Load line.
7. B.
8. Negative.
9. A.
10. To obtain gate bias.
11. A higher peak inverse voltage rating.
12. Low (It is always less than 1.0).
13. High (That is true for all follower circuits).
14. A follower circuit.

15. Piezoelectric.

16. B (In fact, high internal heat is sometimes listed as a disadvantage of the Darlington amplifier).

17. Class C bias.

18. Tunnel diode.

19. Transconductance.

20. C (Once the SCR conducts, the gate has no control over its operation).

6
CHAPTER

Transmitter and receiver programmed reviews
Questions related to transmitters and receivers

Technicians usually do well with this subject.

Start with Block 1. Pick the answer you believe is correct. Go to the next block and check your answer. All answers are in italics. There is only one choice for each block.

Block 1

Here is your first question: What is a maritime mobile repeater station?

A. A fixed land station used to extend the communications range of ship and coast stations.
B. An automatic on-board radio station which facilitates the transmissions of safety communications aboard ship.
C. A mobile radio station that links two or more public coast stations.
D. A one-way low-power communications system used in the maneuvering of vessels.

Block 2

The correct answer is A.

Here is your next question: Who has ultimate control of service at a ship's radio station?

A. The master of the ship.
B. A holder of a First Class Radiotelegraph Certificate with a six months service endorsement.
C. The Radio Office-in-Charge authorized by the captain of the vessel.
D. An appointed licensed radio operator who agrees to comply with all Radio Regulations in force.

Block 3

The correct answer is A.

Here is your next question: What transmitting equipment is authorized for use by a station in the maritime services?

A. Transmitters that haven't been certified by the manufacturer for maritime use.
B. Unless specifically accepted, only transmitter type accepted by the Federal Communications Commission for Part 80 operations.
C. Equipment that has been inspected and approved by the U. S. Coast Guard.
D. Transceivers and transmitters that meet all ITU specifications for use in maritime mobile service.

Block 4

The correct answer is B.

Here is your next question: What is the antenna requirement of a radiotelephone installation aboard a passenger vessel?

A. The antenna must be located a minimum of 15 meters from the radiotelegraph antenna.
B. An emergency reserve antenna system must be provided for communications on 156.8 MHz.
C. The antenna must be vertically polarized and as nondirectional and efficient as is practicable for the transmission and reception of ground waves over seawater.
D. All antennas must be tested and the operational results logged at least once during each voyage.

Block 5

The correct answer is C.

Here is your next question: What is the average range of VHF marine transmissions?

A. 150 miles. B. 50 miles. C. 20 miles. D. 10 miles.

Block 6

The correct answer is C.

Here is your next question: VHF ship station transmitters must have the capability of reducing carrier power to:

A. 1 W. B. 10 W. C. 25 W. D. 50 W.

Block 7

The correct answer is A.

Here is your next question: If your transmitter is producing spurious harmonics or is operating at a deviation from the technical requirements of the station authorization:

A. continue operating until returning to port.
B. repair problem within 24 hours.
C. cease transmission.
D. reduce power immediately.

Block 8

The correct answer is C.

Here is your next question: What is facsimile?

A. The transmission of characters by radioteletype that form a picture when printed.
B. The transmission of still pictures by slow-scan television.
C. The transmission of video by television.
D. The transmission of printed pictures for permanent display on paper.

Block 9

The correct answer is D.

Here is your next question: What is the modern standard scan rate for a facsimile picture transmitted by a radio station?

A. The modern standard is 240 lines per minute.
B. The modern standard is 50 lines per minute.
C. The modern standard is 150 lines per second.
D. The modern standard is 60 lines per second.

Block 10

The correct answer is A.

Here is your next question: What is the approximate transmission time for a facsimile picture transmitted by a radio station?

A. Approximately 6 minutes per frame at 240 lines per minute.
B. Approximately 3.3 minutes per frame at 240 lines per minute.
C. Approximately 6 seconds per frame at 240 lines per minute.
D. $\frac{1}{60}$ second per frame at 240 lines per minute.

Block 11

The correct answer is B.

Here is your next question: What is the term for the transmission of printed pictures by radio for the purpose of a permanent display?

A. Television. B. Facsimile. C. Xerography. D. ACSSB.

Block 12

The correct answer is B.

Here is your next question: In facsimile, how are variations in picture brightness and darkness converted into voltage variations?

A. With an LED.
B. With a Hall-effect transistor.
C. With a photodetector.
D. With an optoisolator.

Block 13

The correct answer is C.

Here is your next question: A 25 MHz amplitude modulated transmitter's actual carrier frequency is 25.00025 MHz without modulation and is 24.99950 MHz when modulated. What statement is true?

A. If the allowed frequency tolerance is 0.001%, this is an illegal transmission.
B. If the allowed frequency tolerance is 0.002%, this is an illegal transmission.
C. Modulation should not change carrier frequency.
D. If the authorized frequency tolerance is 0.005% for the 25-MHz band this transmitter is operating legally.

Block 14

The correct answer is D.

Here is your next question: What ferrite device can be used instead of a duplexer to isolate a microwave transmitter and receiver when both are connected to the same antenna?

A. Isolator. B. Circulator. C. Magnetron. D. Simplex.

Block 15

The correct answer is B.

Here is your next question: How does the bandwidth of the transmitted signal affect selective fading?

A. It is more pronounced at wide bandwidths.
B. It is more pronounced at narrow bandwidths.
C. It is equally pronounced at both narrow and wide bandwidths.
D. The receiver bandwidth determines the selective fading effect.

Block 16

The correct answer is A.

Here is your next question: What is transequatorial propagation?

A. Propagation between two points at approximately the same distance north and south of the magnetic equator.
B. Propagation between two points on the magnetic equator.
C. Propagation between two continents by way of ducts along the magnetic equator.
D. Propagation between any two stations at the same latitude.

Block 17

The correct answer is A.

Here is your next question: What is the maximum range for signals using transequatorial propagation?

A. About 1,000 miles. C. About 5,000 miles.
B. About 2,500 miles. D. About 7,500 miles.

Block 18

The correct answer is C.

Here is your next question: What is the best time of day for transequatorial propagation?

A. Morning. C. Afternoon or early evening.
B. Noon. D. Transequatorial propagation only works at night.

Block 19

The correct answer is C.

Here is your next question: The average range for VHF communications is:

A. 5 miles. B. 15 miles. C. 30 miles. D. 100 miles.

Block 20

The correct answer is C.

Here is your next question: The band of frequencies least susceptible to atmospheric noise and interference is:

A. 30 to 300 kHz. C. 3 to 30 MHz.
B. 300 to 3000 kHz. D. 300 to 3000 MHz.

Block 21

The correct answer is D.

Here is your next question: What is the most the actual transmitter frequency could differ from a reading of 156,520,000 Hz on a frequency counter with a time base accuracy of ±0.1 ppm?

A. 15.652 Hz. B. 0.1 MHz. C. 1.4652 Hz. D. 1.5652 kHz.

Block 22

The correct answer is A.

Here is your next question: What is the most the actual transmitter frequency could differ from a reading of 156,520,000 Hz on a frequency counter with a time base accuracy of ±10 ppm?

A. 146.42 Hz. B. 10 Hz. C. 156.52 kHz. D. 1565.20 Hz.

Block 23

The correct answer is D.

Here is your next question: What is the most the actual transmitter frequency could differ from a reading of 462,100,000 Hz on a frequency counter with a time base accuracy of ±1.0 ppm?

A. 46.21 MHz. B. 10 Hz. C. 1.0 MHz. D. 462.1 Hz.

Block 24

The correct answer is D.

Here is your next question: What is the most the actual transmitter frequency could differ from a reading of 462,100,000 Hz on a frequency counter with a time base accuracy of ±0.1 ppm?

A. 46.21 Hz. B. 0.1 MHz. C. 462.1 Hz. D. 0.2 MHz.

Block 25

The correct answer is A.

Here is your next question: What is the most the actual transmitter frequency could differ from a reading of 462,100,000 Hz on a frequency counter with a time base accuracy of ±10 ppm?

A. 10 MHz. B. 10 Hz. C. 4621 Hz. D. 462.1 Hz.

Block 26

The correct answer is C.

Here is your next question: What is the name of the condition that occurs when the signals of two transmitters in close proximity mix together in one or both of their final amplifiers, and unwanted signals at the sum and difference frequencies of the original transmissions are generated?

A. Amplifier desensitization.
B. Neutralization.
C. Adjacent channel interference.
D. Intermodulation interference.

Block 27

The correct answer is D.

Here is your next question: How does intermodulation interference between two transmitters usually occur?

A. When the signals from the transmitters are reflected out of phase from airplanes passing overhead.
B. When they are in close proximity and the signals mix in one or both of their final amplifiers.

C. When they are in close proximity and the signals cause feedback in one or both of their final amplifiers.
D. When the signals from the transmitters are reflected in phase from the airplanes passing overhead.

Block 28

The correct answer is B.

Here is your next question: How can intermodulation interference between two transmitters in close proximity often be reduced or eliminated?

A. By using a class-C final amplifier with high driving power.
B. By installing a terminated circulator or ferrite isolator in the feed line to the transmitter and duplexer.
C. By installing a band-pass filter in the antenna feed line.
D. By installing a low-pass filter in the antenna feed line.

Block 29

The correct answer is B.

Here is your next question: How can even-order harmonics be reduced or prevented in transmitter amplifier design?

A. By using a push-push amplifier. C. By operating class-C.
B. By using a push-pull amplifier. D. By operating class-AB.

Block 30

The correct answer is B.

Here is your next question: What is the term used to refer to the condition where the signals from a very strong station are superimposed on other signals being received?

A. Intermodulation distortion. C. Receiver quieting.
B. Cross-modulation interference. D. Capture effect.

Block 31

The correct answer is B.

Here is your next question: If there are too many harmonics from a transmitter, check the:

A. coupling. B. tuning of circuits. C. shielding. D. any of the above.

Block 32

The correct answer is D.

Here is your next question: What is the effective radiated power of a repeater with 50-W transmitter power output, 4-dB feedline loss, 3-dB duplexer and circulator loss, and 6-dB antenna gain?

A. 158 W, assuming the antenna gain is referenced to a half-wave dipole.
B. 39.7 W, assuming the antenna gain is referenced to a half-wave dipole.
C. 251 W, assuming the antenna gain is referenced to a half-wave dipole.
D. 69.9 W, assuming the antenna gain is referenced to a half-wave dipole.

Block 33

The correct answer is B.

Here is your next question: What is the effective radiated power of a repeater with 50-W transmitter power output, 5-dB feedline loss, 4-dB duplexer and circulator loss, and 7-dB antenna gain?

A. 300 W, assuming the antenna gain is referenced to a half-wave dipole.
B. 315 W, assuming the antenna gain is referenced to a half-wave dipole.
C. 31.5 W, assuming the antenna gain is referenced to a half-wave dipole.
D. 69.9 W, assuming the antenna gain is referenced to a half-wave dipole.

Block 34

The correct answer is C.

Here is your next question: What is the effective radiated power of a repeater with 75-W transmitter power output, 4-dB feedline loss, 3-dB duplexer and circulator loss, and 10-dB antenna gain?

A. 600 W, assuming the antenna gain is referenced to a half-wave dipole.
B. 75 W, assuming the antenna gain is referenced to a half-wave dipole.
C. 18.75 W, assuming the antenna gain is referenced to a half-wave dipole.
D. 150 W, assuming the antenna gain is referenced to a half-wave dipole.

Block 35

The correct answer is D.

Here is your next question: What is the effective radiated power of a repeater with 75-W transmitter power output, 5-dB feedline loss, 4-dB duplexer and circulator loss, and 6-dB antenna gain?

A. 37.6 W, assuming the antenna gain is referenced to a half-wave dipole.
B. 237 W, assuming the antenna gain is referenced to a half-wave dipole.
C. 150 W, assuming the antenna gain is referenced to a half-wave dipole.
D. 23.7 W, assuming the antenna gain is referenced to a half-wave dipole.

Block 36

The correct answer is A.

Here is your next question: What is the effective radiated power of a repeater with 100-W transmitter power output, 4-dB feedline loss, 3-dB duplexer and circulator loss, and 7-dB antenna gain?

A. 631 W, assuming the antenna gain is referenced to a half-wave dipole.

B. 400 W, assuming the antenna gain is referenced to a half-wave dipole.
C. 25 W, assuming the antenna gain is referenced to a half-wave dipole.
D. 100 W, assuming the antenna gain is referenced to a half-wave dipole.

Block 37

The correct answer is D.

Here is your next question: What is the effective radiated power of a repeater with 100-W transmitter power output, 5-dB feedline loss, 4-dB duplexer and circulator loss, and 10-dB antenna gain?

A. 800 W, assuming the antenna gain is referenced to a half-wave dipole.
B. 126 W, assuming the antenna gain is referenced to a half-wave dipole.
C. 12.5 W, assuming the antenna gain is referenced to a half-wave dipole.
D. 1260 W, assuming the antenna gain is referenced to a half-wave dipole.

Block 38

The correct answer is B.

Here is your next question: What is the effective radiated power of a repeater with 120-W transmitter power output, 5-dB feedline loss, 4-dB duplexer and circulator loss, and 6-dB antenna gain?

A. 601 W, assuming the antenna gain is referenced to a half-wave dipole.
B. 240 W, assuming the antenna gain is referenced to a half-wave dipole.
C. 60 W, assuming the antenna gain is referenced to a half-wave dipole.
D. 379 W, assuming the antenna gain is referenced to a half-wave dipole.

Block 39

The correct answer is C.

Here is your next question: What is the effective radiated power of a repeater with 150-W transmitter power output, 4-dB feedline loss, 3-dB duplexer and circulator loss, and 7-dB antenna gain?

A. 946 W, assuming the antenna gain is referenced to a half-wave dipole.
B. 37.5 W, assuming the antenna gain is referenced to a half-wave dipole.
C. 600 W, assuming the antenna gain is referenced to a half-wave dipole.
D. 150 W, assuming the antenna gain is referenced to a half-wave dipole.

Block 40

The correct answer is D.

Here is your next question: What is the effective radiated power of a repeater with 200-W transmitter power output, 4-dB feedline loss, 4-dB duplexer and circulator loss, and 10-dB antenna gain?

A. 317 W, assuming the antenna gain is referenced to a half-wave dipole.
B. 2000 W, assuming the antenna gain is referenced to a half-wave dipole.

C. 126 W, assuming the antenna gain is referenced to a half-wave dipole.
D. 260 W, assuming the antenna gain is referenced to a half-wave dipole.

Block 41

The correct answer is A.

Here is your next question: What is the effective radiated power of a repeater with 200-W transmitter power output, 4-dB feedline loss, 3-dB duplexer and circulator loss, and 6-dB antenna gain?

A. 252 W, assuming the antenna gain is referenced to a half-wave dipole.
B. 63.2 W, assuming the antenna gain is referenced to a half-wave dipole.
C. 632 W, assuming the antenna gain is referenced to a half-wave dipole.
D 159 W, assuming the antenna gain is referenced to a half-wave dipole.

Block 42

The correct answer is D.

Here is your next question: What is an optoisolator?

A. An LED and a phototransistor.
B. A P-N junction that develops an excess positive charge when exposed to light.
C. An LED and a capacitor.
D. An LED and a solar cell.

Block 43

The correct answer is A.

Here is your next question: What is an optical shaft encoder?

A. An array of optocouplers chopped by a stationary wheel.
B. An array of optocouplers whose light transmission path is controlled by a rotating wheel.
C. An array of optocouplers whose propagation velocity is controlled by a stationary wheel.
D. An array of optocouplers whose propagation velocity is controlled by a rotating wheel.

Block 44

The correct answer is B.

Here is your next question: What does the photoconductive effect in crystalline solids produce a noticeable change in?

A. The capacitance of the solid.
B. The inductance of the solid.
C. The specific gravity of the solid.
D. The resistance of the solid.

Block 45

The correct answer is D.

Here is your next question: What is the meaning of the term *time constant of an RC circuit*?

A. The time required to charge the capacitor in the circuit to 36.8% of the supply voltage.
B. The time required to charge the capacitor in the circuit to 36.8% of the supply current.
C. The time required to charge the capacitor in the circuit to 63.2% of the supply current.
D. The time required to charge the capacitor in the circuit to 63.2% of the supply voltage.

Block 46

The correct answer is D.

Here is your next question: What is the meaning of the phrase *time constant of an RL circuit*?

A. The time required for the current in the circuit to build up to 36.8% of the maximum value.
B. The time required for the voltage in the circuit to build up to 63.2% of the maximum value.
C. The time required for the current in the circuit to build up to 63.2% of the maximum value.
D. The time required for the voltage in the circuit to build up to 36.8% of the maximum value.

Block 47

The correct answer is C.

Here is your next question: What is the term for the time required for the capacitor in an RC circuit to be charged to 63.2% of the supply voltage?

A. An exponential rate of one.
B. One time constant.
C. One exponential period.
D. A time factor of one.

Block 48

The correct answer is B.

Here is your next question: What is the term for the time required for the current in an RL circuit to build up to 63.2% of the maximum value?

A. One time constant.
B. An exponential period of one.
C. A time factor of one.
D. One exponential rate.

Block 49

The correct answer is A.

Here is your next question: What is emission A3C?

A. Facsimile. B. RTTY. C. ATV. D. Slow-scan TV.

Block 50

The correct answer is A.

Here is your next question: What type of emission is produced when an amplitude modulated transmitter is modulated by a facsimile signal?

A. A3F. B. A3C. C. F3F. D. F3C.

Block 51

The correct answer is B.

Here is your next question: What is facsimile?

A. The transmission of tone-modulated telegraphy.
B. The transmission of a pattern of printed characters designed to form a picture.
C. The transmission of printed pictures by electrical means.
D. The transmission of moving pictures by electrical means.

Block 52

The correct answer is C.

Here is your next question: What is emission F3C?

A. Voice transmission. B. Slow-scan TV. C. RTTY. D. Facsimile.

Block 53

The correct answer is D.

Here is your next question: What type of emission is produced when a frequency modulated transmitter is modulated by a facsimile signal?

A. F3C. B. A3C. C. F3F. D. A3F.

Block 54

The correct answer is A.

Here is your next question: What is emission A3F?

A. RTTY. B. Television. C. SSB. D. Modulated CW.

Block 55

The correct answer is B.

Here is your next question: What type of emission is produced when an amplitude modulated transmitter is modulated by a television signal?

A. F3F. B. A3F. C. A3C. D. F3C.

Block 56

The correct answer is B.

Here is your next question: What is emission F3F?

A. Modulated CW. B. Facsimile. C. RTTY. D. Television.

Block 57

The correct answer is D.

Here is your next question: What type of emission is produced when a frequency modulated transmitter is modulated by a television signal?

A. A3F. B. A3C. C. F3F. D. F3C.

Block 58

The correct answer is C.

Here is your next question: How can an FM-phone signal be produced?

A. By modulating the supply voltage to a class-B amplifier.
B. By modulating the supply voltage to a class-C amplifier.
C. By using a reactance modulator on an oscillator.
D. By using a balanced modulator on an oscillator.

Block 59

The correct answer is C.

Here is your next question: How can a double-sideband phone signal be produced?

A. By using a reactance modulator on an oscillator.
B. By varying the voltage to the varactor in an oscillator circuit.
C. By using a phase detector, oscillator, and filter in a feedback loop.
D. By modulating the plate supply voltage to a class-C amplifier.

Block 60

The correct answer is D.

Here is your next question: How can a single-sideband phone signal be produced?

A. By producing a double-sideband signal with a balanced modulator and then removing the unwanted sideband by filtering.
B. By producing a double-sideband signal with a balanced modulator and then removing the unwanted sideband by heterodyning.
C. By producing a double-sideband signal with a balanced modulator and then removing the unwanted sideband by mixing.
D. By producing a double-sideband signal with a balanced modulator and then removing the unwanted sideband by neutralization.

Block 61

The correct answer is A.

Here is your next question: For many types of voices, what is the ratio of PEP-to-average power during a modulation peak in a single-sideband phone signal?

A. Approximately 1.0 to 1. C. Approximately 2.5 to 1.
B. Approximately 25 to 1. D. Approximately 100 to 1.

Block 62

The correct answer is C.

Here is your next question: In a single-sideband phone signal, what determines the PEP-to-average power ratio?

A. The frequency of the modulating signal. C. The speech characteristics.
B. The degree of carrier suppression. D. The amplifier power.

Block 63

The correct answer is C.

Here is your next question: What is the approximate dc input power to a class-B RF power amplifier stage in an FM-phone transmitter when the PEP output is 1500 W?

A. Approximately 900 W. C. Approximately 2500 W.
B. Approximately 1765 W. D. Approximately 3000 W.

Block 64

The correct answer is C.

Here is your next question: What is the approximate dc input power to a class-C RF power amplifier stage in an RTTY transmitter when the PEP output power is 1000 W?

A. Approximately 850 W. C. Approximately 1667 W.
B. Approximately 1250 W. D. Approximately 2000 W.

Block 65

The correct answer is B.

Here is your next question: What is the type of modulation in which the modulating signal varies the duration of the transmitted pulse?

A. Amplitude modulation. C. Pulse-width modulation.
B. Frequency modulation. D. Pulse-height modulation.

Block 66

The correct answer is C.

Here is your next question: In a pulse-position modulation system, what parameter does the modulating signal vary?

A. The number of pulses per second.
B. Both the frequency and amplitude of the pulses.
C. The duration of the pulses.
D. The time at which each pulse occurs.

Block 67

The correct answer is D.

Here is your next question: Why is the transmitter peak power in a pulse modulation system much greater than its average power?

A. The signal duty cycle is less than 100%.
B. The signal reaches peak amplitude only when voice modulated.
C. The signal reaches peak amplitude only when voltage spikes are generated within the modulator.
D. The signal reaches peak amplitude only when the pulses are also amplitude modulated.

Block 68

The correct answer is A.

Here is your next question: What is one way that voice is transmitted in a pulse-width modulation system?

A. A standard pulse is varied in amplitude by an amount depending on the voice waveform at that instant.
B. The position of a standard pulse is varied by an amount depending on the voice waveform at that instant.
C. A standard pulse is varied in duration by an amount depending on the voice waveform at that instant.
D. The number of standard pulses per second varies depending on the voice waveform at that instant.

Block 69

The correct answer is C.

Here is your next question: What is amplitude compandored single sideband?

A. Reception of single sideband with a conventional CW receiver.
B. Reception of single sideband with a conventional FM receiver.
C. Single sideband incorporating speech compression at the transmitter and speech expansion at the receiver.
D. Single sideband incorporating speech expansion at the transmitter and speech compression at the receiver.

Block 70

The correct answer is C.

Here is your next question: What is meant by compandoring?

A. Compressing speech at the transmitter and expanding it at the receiver.
B. Using an audio-frequency signal to produce pulse-length modulation.
C. Combining amplitude and frequency modulation to produce a single-sideband signal.
D. Detecting and demodulating a single-sideband signal by converting it to a pulse-modulated signal.

Block 71

The correct answer is A.

Here is your next question: What is the purpose of a pilot tone in an amplitude compandored single-sideband system?

A. It permits rapid tuning of a mobile receiver.
B. It replaces the suppressed carrier at the receiver.
C. It permits rapid change of frequency to escape high-powered interference.
D. It acts as a beacon to indicate the present propagation characteristic of the band.

Block 72

The correct answer is A.

Here is your next question: What is the approximate frequency of the pilot tone in an amplitude compandored single-sideband system?

A. 1 kHz. B. 5 MHz. C. 455 kHz. D. 3 kHz.

Block 73

The correct answer is D.

Here is your next question: How many voice transmissions can be packed into a given frequency band for amplitude-compandored single-sideband systems over conventional FM-phone systems?

A. 2. B. 4. C. 8. D. 16.

Block 74

The correct answer is B.

Here is your next question: What term describes a wide-bandwidth communications systems in which the RF carrier varies according to some predetermined sequence?

A. Amplitude compandored single sideband.
B. SITOR.
C. Time-domain frequency modulation.
D. Spread spectrum communication.

Block 75

The correct answer is D.

Questions related to receivers

Block 1

Here is your first next question: In what frequencies does the Communications Act require radio watches by compulsory radiotelephone stations?

A. Watches are required on 500 kHz and 2182 kHz.
B. Continuous watch is required on 2182 kHz only.
C. On all frequencies between 405 to 535 kHz, 1605 to 3500 kHz, and 156 to 162 MHz.
D. Watches are required on 2182 kHz and 156.800 MHz.

Block 2

The correct answer is D.

Here is your next question: What is the international VHF digital selective calling channel?

A. 2182 kHz. B. 156.35 MHz. C. 156.525 MHz. D. 500 kHz.

Block 3

The correct answer is C.

Here is your next question: What channel must compulsorily equipped vessels monitor at all times in the open sea?

A. Channel 8,156.4 MHz.
B. Channel 16,156.8 MHz.
C. Channel 22A,157.1 MHz.
D. Channel 6,156.3 MHz.

Block 4

The correct answer is B.

Here is your next question: Which VHF channel is used only for digital selective calling?

A. Channel 70. B. Channel 16. C. Channel 22A. D. Channel 6.

Block 5

The correct answer is A.

Here is your next question: As an alternative to keeping watch on a working frequency in the band 1600 to 4000 kHz, an operator must tune station receiver to monitor 2182 kHz:

A. at all times.
B. during distress calls only.
C. during daytime hours of service.
D. during the silence periods each hour.

Block 6

The correct answer is A.

Here is your next question: Two way communications with both stations operating on the same frequency is:

A. radiotelephone. B. duplex. C. simplex. D. multiplex.

Block 7

The correct answer is C.

Here is your next question: What is a selective fading effect?

A. A fading effect caused by small changes in beam heading at the receiving station.
B. A fading effect caused by phase differences between radio wave components of the same transmission, as experienced at the receiving station.
C. A fading effect caused by large changes in the height of the ionosphere, as experienced at the receiving station.
D. A fading effect caused by time differences between the receiving and transmitting stations.

Block 8

The correct answer is B.

Here is your next question: What is the propagation effect called when phase differences between radio wave components of the same transmission are experienced at the recovery station?

A. Faraday rotation. C. Selective fading.
B. Diversity reception. D. Phase shift.

Block 9

The correct answer is C.

Here is your next question: What is the major cause of selective fading?

A. Small changes in beam heading at the receiving station.
B. Large changes in the height of the ionosphere, as experienced at the receiving station.
C. Time differences between the receiving and transmitting stations.
D. Phase differences between radio wave components of the same transmission, as experienced at the receiving station.

Block 10

The correct answer is D.

Here is your next question: Which emission modes suffer the most from selective fading?

A. CW and SSB.
B. FM and double sideband AM.
C. SSB and image.
D. SSTV and CW.

Block 11

The correct answer is B.

Here is your next question: How does the bandwidth of the transmitted signal affect selective fading?

A. It is more pronounced at wide bandwidths.
B. It is more pronounced at narrow bandwidths.
C. It is equally pronounced at both narrow and wide bandwidths.
D. The receiver bandwidth determines the selective fading effect.

Block 12

The correct answer is A.

Here is your next question: The band of frequencies least susceptible to atmospheric noise and interference is:

A. 30 to 300 kHz.
B. 300 to 3000 kHz.
C. 3 to 30 MHz.
D. 300 to 3000 MHz.

Block 13

The correct answer is D.

Here is your next question: What is receiver desensitizing?

A. A burst of noise when the squelch is set too low.
B. A burst of noise when the squelch is set too high.
C. A reduction in receiver sensitivity because of a strong signal on a nearby frequency.
D. A reduction in receiver sensitivity when the AF gain control is turned down.

Block 14

The correct answer is C.

Here is your next question: What is the term used to refer to the reduction of receiver gain caused by the signals of a nearby station transmitting in the same frequency band?

A. Desensitizing.
B. Quieting.
C. Cross-modulation interference.
D. Squelch gain rollback.

Block 15

The correct answer is A.

Here is your next question: What is the term used to refer to a reduction in receiver sensitivity caused by unwanted high-level adjacent channel signals?

A. Intermodulation distortion.
B. Quieting.
C. Desensitizing.
D. Overloading.

Block 16

The correct answer is C.

Here is your next question: How can receiver desensitizing be reduced?

A. Ensure good RF shielding between the transmitter and receiver.
B. Increase the transmitter audio gain.
C. Decrease the receiver squelch gain.
D. Increase the receiver bandwidth.

Block 17

The correct answer is A.

Here is your next question: What is cross-modulation interference?

A. Interference between two transmitters of different modulation type.
B. Interference caused by audio rectification in the receiver preamp.
C. Harmonic distortion of the transmitted signal.
D. Modulation from an unwanted signal is heard in addition to the desired signal.

Block 18

The correct answer is D.

Here is your next question: What is the term used to refer to the condition where the signal from a very strong station are superimposed on other signals being received?

A. Intermodulation distortion.
B. Cross-modulation interference.
C. Receiver quieting.
D. Capture effect.

Block 19

The correct answer is B.

Here is your next question: How can cross-modulation in a receiver be reduced?

A. By installing a filter at the receiver.
B. By using a filter at the receiver.
C. By increasing the receiver's RF gain while decreasing the AF gain.
D. By adjusting the pass-band tuning.

Block 20

The correct answer is A.

Here is your next question: What is the result of cross-modulation?

A. A decrease in modulation level of transmitted signals.
B. Receiver quieting.
C. The modulation of an unwanted signal is heard on the desired signal.
D. Inverted sidebands in the final stage of the amplifier.

Block 21

The correct answer is C.

Here is your next question: What is the capture effect?

A. All signals on a frequency are demodulated by an FM receiver.
B. All signals on a frequency are demodulated by an AM receiver.
C. The loudest signal received is the only demodulated signal.
D. The weakest signal received is the only demodulated signal.

Block 22

The correct answer is C.

Here is your next question: What is the term used to refer to the reception blockage of one FM-phone signal by another FM-phone signal?

A. Desensitization.
B. Cross-modulation interference.
C. Capture effect.
D. Frequency discrimination.

Block 23

The correct answer is C.

Here is your next question: With which emission type is the capture-effect most pronounced?

A. FM. B. SSB. C. AM. D. CW.

Block 24

The correct answer is A.

Here is your next question: What is one of the most significant problems you might encounter when you try to receive signals with a mobile station?

A. Ignition noise.
B. Doppler shift.
C. Radar interference.
D. Mechanical vibrations.

Block 25

The correct answer is A.

Here is your question: What is the proper procedure for suppressing electrical noise in a mobile station?

A. Apply shielding and filtering where necessary.
B. Insulate all plain sheet metal surfaces from each other.
C. Apply anti-static spray liberally to all nonmetallic surfaces.
D. Install filter capacitors in series with all dc wiring.

Block 26

The correct answer is A.

Here is your next question: How can ferrite beads be used to suppress ignition noise?

A. Install them in the resistive high-voltage cable every two years.
B. Install them between the starter solenoid and the starter motor.
C. Install them in the primary and secondary ignition leads.
D. Install them in the antenna lead to the radio.

Block 27

The correct answer is C.

Here is your next question: How can conducted and radiated noise caused by an alternator be suppressed?

A. By installing filter capacitors in series with the dc power lead and by installing a blocking capacitor in the field lead.
B. By connecting the radio's power leads to the battery by the longest possible path and by installing a blocking capacitor in series with the positive lead.
C. By installing a high-pass filter in series with the radio's power lead to the vehicle's electrical system and by installing a low-pass filter in parallel with the field lead.
D. By connecting the radio's power leads directly to the battery and by installing coaxial capacitors in the alternator leads.

Block 28

The correct answer is D.

Here is your next question: What is a major cause of atmospheric static?

A. Sunspots. B. Thunderstorms. C. Airplanes. D. Meteor showers.

Block 29

The correct answer is B.

Here is your next question: How can you determine if a line-noise interference problem is being generated within a building?

A. Check the power-line voltage with a time-domain reflectometer.
B. Observe the ac waveform on an oscilloscope.
C. Turn off the main circuit breaker and listen on a battery-operated radio.
D. Observe the power-line voltage on a spectrum analyzer.

Block 30

The correct answer is C.

Here is your next question: What causes receiver desensitizing?

A. Audio gain adjusted too low.
B. Squelch gain adjusted too high.
C. The presence of a strong signal on a nearby frequency.
D. Squelch gain adjusted too low.

Block 31

The correct answer is C.

Here is your next question: How can alternator noise be minimized?

A. By connecting the radio's power leads to the battery by the longest possible path.
B. By connecting the radio's power lead to the battery by the shortest possible path.
C. By installing a high-pass filter in series with the radio's dc power lead to the vehicle's electrical system.
D. By installing filter capacitors in series with the dc power lead.

Block 32

The correct answer is B.

Here is your next question: In radio circuits, the component most apt to break down is the:

A. Resistor. B. Crystal. C. Transformer. D. Wiring.

Block 33

The correct answer is A.

Here is your next question: Motorboating (low-frequency oscillations) in an amplifier can be stopped by:

A. grounding the screen grid.
B. by passing the screen grid resistor with a 0.1-µF capacitor.
C. connecting a capacitor between the B+ lead and ground.
D. grounding the plate.

Block 34

The correct answer is C.

Here is your next question: Auto interference to radio reception can be eliminated by:

A. installing resistive spark plugs.
B. installing capacitive spark plugs.
C. installing resistors in series with the spark plugs.
D. installing two copper-braid ground strips.

Block 35

The correct answer is A.

Here is your next question: What would be the bandwidth of a good crystal-lattice bandpass filter for a single-sideband phone emission?

A. 6 kHz at –6 dB. B. 2.1 kHz at –6 dB. C. 500 Hz at –6 dB. D. 15 kHz at –6 dB.

Block 36

The correct answer is B.

Here is your next question: What would be the bandwidth of a good crystal-lattice bandpass filter for a double-sideband phone emission?

A. 1 kHz at –6 dB. B. 500 Hz at –6 dB. C. 6 kHz at –6 dB. D. 15 kHz at –6 dB.

Block 37

The correct answer is C.

Here is your next question: What is a crystal-lattice filter?

A. A power supply filter made with crisscrossed quartz crystals.
B. An audio filter made with four quartz crystals at 1-kHz intervals.
C. A filter with infinitely wide and shallow skirts made using quartz crystals.
D. A filter with narrow bandwidth and steep skirts made using quartz crystals.

Block 38

The correct answer is D.

Here is your next question: What technique can be used to construct low-cost, high-performance crystal-lattice filters?

A. Splitting and tumbling. C. Etching and splitting.
B. Tumbling and grinding. D. Etching and grinding.

Block 39

The correct answer is D.

Here is your next question: What is the input impedance of a theoretically ideal op amp?

A. 100 Ω. B. 1000 Ω. C. Very low. D. Very high.

Block 40

The correct answer is C.

Here is your next question: What is the output impedance of a theoretically ideal op amp?

A. Very low. B. Very high. C. 100 Ω. D. 1000 Ω.

Block 41

The correct answer is B.

Here is your next question: What is a phase-locked loop circuit?

A. An electronic servo loop consisting of a ratio detector, reactance modulator, and voltage-controlled oscillator.
B. An electronic circuit also known as a monostable multivibrator.

C. An electronic circuit consisting of a precision push-pull amplifier with a differential input.
D. An electronic servo loop consisting of a phase detector, a low-pass filter and voltage-controlled oscillator.

Block 42

The correct answer is A.

Here is your next question: What functions are performed by a phase-locked loop?

A. Wideband AF and RF power amplification.
B. Comparison of two digital input signals, digital pulse counter.
C. Photovoltaic conversion, optical coupling.
D. Frequency synthesis, FM demodulation.

Block 43

The correct answer is D.

Here is your next question: A circuit compares the output from a voltage-controlled oscillator and a frequency standard. The difference between the two frequencies produces an error voltage that changes the voltage-controlled oscillator frequency. What is the name of the circuit?

A. A doubly balanced mixer. C. A differential voltage amplifier.
B. A phase-locked loop. D. A variable frequency oscillator.

Block 44

The correct answer is C.

Here is your next question: What do the initials TTL stand for?

A. Resistor-transistor logic. C. Diode-transistor logic.
B. Transistor-transistor logic. D. Emitter-coupled logic.

Block 45

The correct answer is B.

Here is your next question: What is the recommended power supply voltage for TTL series integrated circuits?

A. 12.00 V. B. 50.00 V. C. 5.00 V. D. 13.60 V.

Block 46

The correct answer is A.

Here is your next question: What logic state do the inputs of a TTL device assume if they are left open?

A. A high logic state.

B. A low logic state.
C. The device becomes randomized and will not provide consistent high or low logic states.
D. Open inputs on a TTL device are ignored.

Block 47

The correct answer is B.

Here is your next question: What level of input voltage is high in a TTL device operating with a 5-V power supply?

A. 2.0 to 5.5 V. B. 1.5 to 3.0 V. C. 1.0 to 1.5 V. D. –5.0 to –2.0 V.

Block 48

The correct answer is C.

Here is your next question: What level of input voltage is low in a TTL device operating with a 5-V power supply?

A. –2.0 to –5.5 V. B. 2.0 to 5.5 V. C. –0.6 to 0.8 V. D. –0.8 to 0.4 V.

Block 49

The correct answer is C.

Here is your next question: Why do circuits containing TTL devices have several bypass capacitors per printed circuit board?

A. To prevent RFI to receivers.
B. To keep the switching noise within the circuit, thus eliminating RFI.
C. To filter out switching harmonics.
D. To prevent switching transients from appearing on the supply line.

Block 50

The correct answer is A.

Here is your next question: What is a CMOS IC?

A. A chip with only P-channel transistors.
B. A chip with P-channel and N-channel transistors.
C. A chip with only N-channel transistors.
D. A chip with only bipolar transistors.

Block 51

The correct answer is C.

Here is your next question: What is one major advantage of CMOS over other devices?

A. Small size.
B. Low current consumption.
C. Low cost.
D. Ease of circuit design.

Block 52

The correct answer is D.

Here is your next question: Why do CMOS digital integrated circuits have high immunity to noise on the input signal or power supply?

A. Larger bypass capacitors are used in CMOS circuit design.
B. The input switching threshold is about two times the power-supply voltage.
C. The input switching threshold is about one-half the power-supply voltage.
D. Input signals are stronger.

Block 53

The correct answer is B.

Here is your next question: What two factors determine the sensitivity of a receiver?

A. Dynamic range and third-order intercept.
B. Cost and availability.
C. Intermodulation distortion and dynamic range.
D. Bandwidth and noise figure.

Block 54

The correct answer is D.

Here is your next question: What is the limiting condition for sensitivity in a communications receiver?

A. The noise floor of the receiver.
B. The power-supply output ripple.
C. The two-tone intermodulation distortion.
D. The input impedance to the detector.

Block 55

The correct answer is A.

Here is your next question: What is the theoretical minimum noise floor of a receiver with a 400-Hz bandwidth?

A. –141 dBm. B. –148 dBm. C. –174 dBm. D. –180 dBm.

Block 56

The correct answer is B.

Here is your next question: How can selectivity be achieved in the front-end circuitry of a communications receiver?

A. By using an audio filter.
B. By using a preselector.
C. By using an additional RF amplifier stage.
D. By using an additional IF amplifier stage.

Block 57

The correct answer is B.

Here is your next question: A receiver selectivity of 2.4 kHz in the IF circuitry is optimum for what type of signals?

A. CW. B. SSB voice. C. Double-sideband AM voice. D. FSK RTTY.

Block 58

The correct answer is B.

Here is your next question: What occurs during CW reception if too narrow a filter bandwidth is used in the IF stage of a receiver?

A. Undesired signals will reach the audio stage. C. Cross-modulation distortion.
B. Output-offset overshoot. D. Filter ringing.

Block 59

The correct answer is D.

Here is your next question: A receiver selectivity of 10 kHz in the IF circuitry is optimum for what type of signals?

A. SSB voice. B. Double-sideband AM. C. CW. D. FSK RTTY.

Block 60

The correct answer is B.

Here is your next question: What degree of selectivity is desirable in the IF circuitry of a single-sideband phone receiver?

A. 1 kHz. B. 2.4 kHz. C. 4.2 kHz. D. 4.8 kHz.

Block 61

The correct answer is B.

Here is your next question: What is an undesirable effect of using too wide a filter bandwidth in the IF section of a receiver?

A. Output-offset overshoot. C. Thermal-noise distortion.
B. Undesired signals will reach the audio stage. D. Filter ringing.

Block 62

The correct answer is B.

Here is your next question: How should the filter bandwidth of a receiver IF section compare with the bandwidth of a received signal?

A. Filter bandwidth should be slightly greater than the received-signal bandwidth.
B. Filter bandwidth should be approximately half the received-signal bandwidth.
C. Filter bandwidth should be approximately two times the received-signal bandwidth.
D. Filter bandwidth should be approximately four times the received-signal bandwidth.

Block 63

The correct answer is A.

Here is your next question: What degree of selectivity is desirable in the IF circuitry of an FM-phone receiver?

A. 1 kHz. B. 2.4 kHz. C. 4.2 kHz. D. 15 kHz.

Block 64

The correct answer is D.

Here is your next question: How can selectivity be achieved in the IF circuitry of a communications receiver?

A. Incorporate a means of varying the supply voltage to the local oscillator circuitry.
B. Replace the standard JFET mixer with a bipolar transistor followed by a capacitor of the proper value.
C. Remove AGC action from the IF stage and confine it to the audio stage only.
D. Incorporate a high-*Q* filter.

Block 65

The correct answer is D.

Here is your next question: What is meant by the dynamic range of a communications receiver?

A. The number of kHz between the lowest and the highest frequency to which the receiver can be tuned.
B. The maximum possible undistorted audio output of the receiver, referenced to 1 mW.
C. The ratio between the minimum discernible signal and the largest tolerable signal without causing audible distortion products.
D. The difference between the lowest-frequency signal and the highest-frequency signal detectable without moving the tuning knob.

Block 66

The correct answer is C.

Here is your next question: What is the term for the ratio between the largest tolerable receiver input signal and the minimum discernible signal?

A. Intermodulation distortion. C. Noise figure.
B. Noise floor. D. Dynamic range.

Block 67

The correct answer is D.

Here is your next question: What type of problems are caused by poor dynamic range in a communications receiver?

A. Cross-modulation of the desired signal and desensitization from strong adjacent signals.
B. Oscillator instability requiring frequency retuning, and loss of ability to recover the opposite sideband, should it be transmitted.
C. Cross-modulation of the desired signal and insufficient audio power to operate the speaker.
D. Oscillator instability and severe audio distortion of all but the strongest received signals.

Block 68

The correct answer is A.

Here is your next question: The ability of a communications receiver to perform well in the presence of strong signals outside the band of interest is indicated by what parameter?

A. Noise figure. C. Signal-to-noise ratio.
B. Blocking dynamic range. D. Audio output.

Block 69

The correct answer is B.

Here is your next question: What is meant by the phrase *noise figure of a communications receiver*?

A. The level of noise entering the receiver from the antenna.
B. The relative strength of a received signal 3 kHz removed from the carrier frequency.
C. The level of noise generated in the front end and succeeding stages of a receiver.
D. The ability of a receiver to reject unwanted signals at frequencies close to the desired one.

Block 70

The correct answer is C.

Here is your next question: Which stage of a receiver primarily establishes its noise figure?

A. The audio stage. C. The RF stage.
B. The IF strip. D. The local oscillator.

Block 71

The correct answer is C.

Here is your next question: Where is the noise generated that primarily determines the signal-to-noise ratio in a VHF (150 MHz) marine-band receiver?

A. In the receiver front end.
C. In the atmosphere.
B. Man-made noise.
D. In the ionosphere.

Block 72

The correct answer is A.

7

CHAPTER

Digital basics

You should consider this to be required reading for all certification and license tests with the possible exception of the Marine Radio License.

Some introductory concepts

Figure 7-1 shows the input and output signals for an *inverter*, one of the logic gates that you will review in this chapter. The output signal is 180° out of phase with the input signal. At least, that is the *theory* of operation. In practice, it takes time for the signal to pass through the gate, so the phase difference is not exactly 180°. The delay might be only 20 ns, but that is enough to produce glitches in another circuit.

As an example, suppose the out-of-phase signals are delivered to a circuit that has no output unless both inputs are positive. This is shown in Fig. 7-2.

If the signals are really 180° out of phase, there will never be an output. But, if one of the input signals is obtained from an inverter, there is a very short period of time on each half cycle when both signals are positive. This will result in undesirable spikes, or glitches.

The short-duration pulses at the output of the device in Fig. 7-2 are examples of glitches. You might not be able to observe them with an oscilloscope unless it has a bandwidth rating of at least 20 MHz.

As with electronic circuits, pins of devices such as ICs are numbered counterclockwise as viewed from the top. There is always an identifier to tell you where pin #1 is. Notches or dots are often used with DIP packages, and tabs are used on TO packages. DIP packages are the dual in-line plastic packages like the one in Fig. 7-3. TO packages are round metal packages of the type often used for transistors.

The quickest and most convenient way to measure logic levels is to use a *logic probe*—not a voltmeter. Figure 7-4 shows how a logic probe is represented in this chapter. There are three logic levels that can be measured by the probe. As shown in Fig. 7-4, they are 0, 1, and pulse.

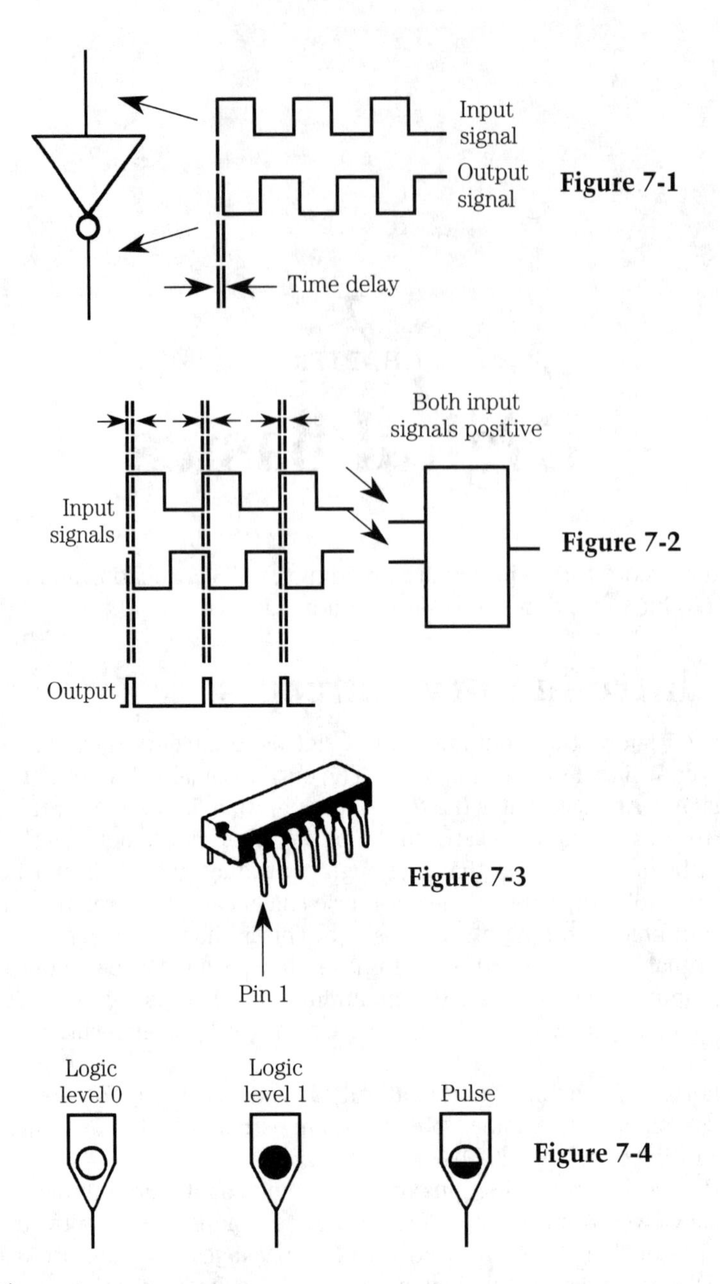

Figure 7-1

Figure 7-2

Figure 7-3

Figure 7-4

The basic gates

There are seven basic components that you will be working with: AND, OR, NOT, NAND, NOR, exclusive OR, and LOGIC COMPARATOR. When they are in integrated circuit form, they are usually called gates.

There are four important features of these basic components that you should know: the *schematic drawing symbol*, the *truth table*, a *circuit example*, and the *math symbol*. Memorize these features when they are given for each of the basic gates!

The AND gate

The first component to study is the AND gate. The four important features of this gate are given in Fig. 7-5. Three schematic symbols are given. The upper symbol is used most often in consumer electronics and industrial schematics. The lower symbol can be used on industrial diagrams. It would be a good idea to learn all types of symbols used for all gates.

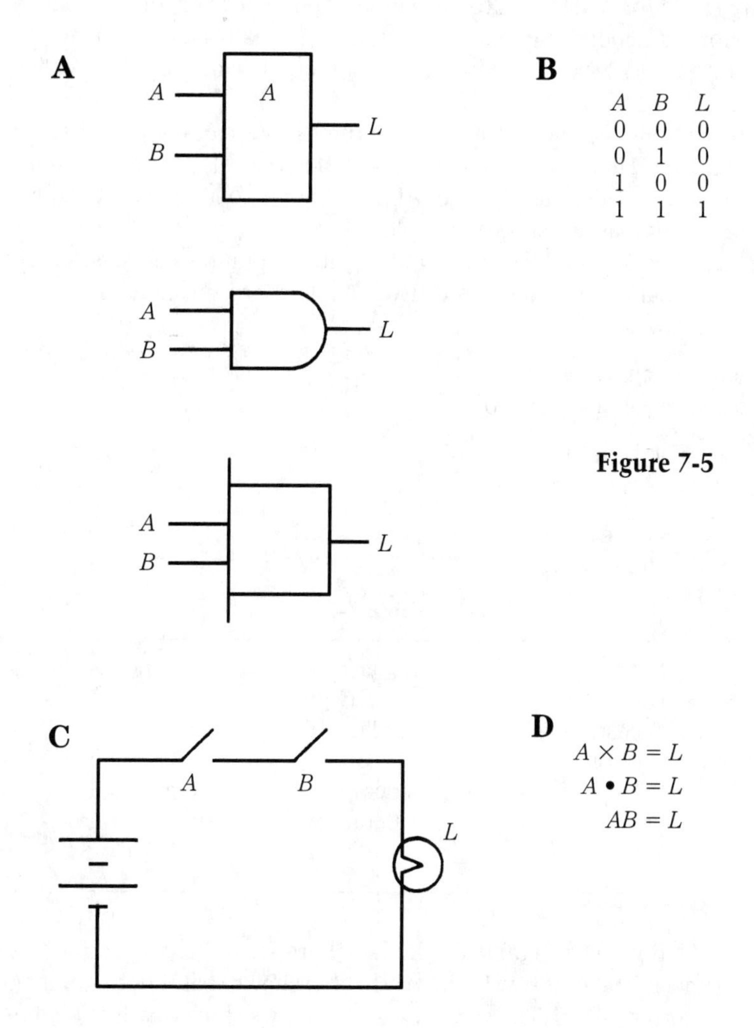

A	*B*	*L*
0	0	0
0	1	0
1	0	0
1	1	1

$A \times B = L$

$A \bullet B = L$

$AB = L$

Figure 7-5

There are only two inputs (*A* and *B*) and one output (*L*), so it is called a *two-input AND*. For all of the gates covered in this chapter, the input and output designations are arbitrarily chosen. Any other combination of letters or numbers can be used. More important, the input and output signals can represent some physical function. Here is an example:

INPUT #1—All Windows Closed = Logic 1
INPUT #2—All Doors Closed = Logic 1
OUTPUT—Green Light On = Logic 1

For the circuit shown in Fig. 7-5, switch A represents all windows closed and switch B represents all doors closed. The green light (L) is ON only when the two switches are closed.

It is possible for AND gates to have more inputs, but only one output terminal is possible. However, more than one connection can usually be made to an output. The number of inputs is called the *fan in*, and the number of outputs is called the *fan out*.

The truth table lists all of the possible things that can happen at the input and output terminals. There are only two possible logic levels for each terminal, represented by 0 and 1. Don't think of 0 and 1 as being numbers, they can represent anything. A few examples are shown in Table 7-1.

The logic levels (1 and 0) can be represented a number of ways. The usual way is to let 0 V equal logic 0, and a positive voltage represent equal logic 1. Here are some other examples:

Pressure = Logic Level 1
No Pressure = Logic Level 0

Light = Logic Level 1
No Light = Logic Level 0

Sound = Logic Level 1
No Sound = Logic Level 0

Table 7-1

Motor	Not running = 0	Running = 1
Lamp	Off = 0	On = 1
Tape recorder	Not recording = 0	Recording = 1
Television set	Off = 0	On = 1
Relay	Not energized = 0	Energized = 1
Transistor	Cutoff = 0	Saturated = 1
Voltage	0 V = 0	5 V = 1

For the AND gate truth table of Fig. 7-5, there are only two possibilities for each terminal: 0 and 1. The first row in the truth table shows that if both inputs (A and B) are at 0, the output is at 0. The second and third rows show that if one input is 0 and the other is 1, the output is 0. The last row shows that if both inputs are at 1, the output is 1. Note that the last column (L) shows that there is a 0 in the output at all times, except when both inputs are at 1.

The example circuit in Fig. 7-5 shows two switches in series with a lamp and a battery. The switches can be called the *inputs*, and the lamp the *output*. If you let 0 = Switch OFF, and 1 = Switch ON, then all of the possible combinations for OFF and ON are shown in the first two columns of the truth table. If you let 0 = Lamp OFF and 1 = Lamp ON, then the third column of the truth table shows that the lamp is always OFF, unless both switches are closed. Note that every possible condition of the switches and the lamp are shown in the truth table. The trick is to represent the conditions of the switches and lamp with 0s and 1s.

As shown in Fig. 7-5, the multiplication sign is used to represent AND. When you see $A \times B$, or $A \bullet B$, or AB, you should read it as A and B, never A times B in logic systems.

The OR gate

Having gone into detail with the AND gate of Fig. 7-5, makes it easier to discuss the remaining basic gates. The four important features of the OR gate are shown in Fig. 7-6.

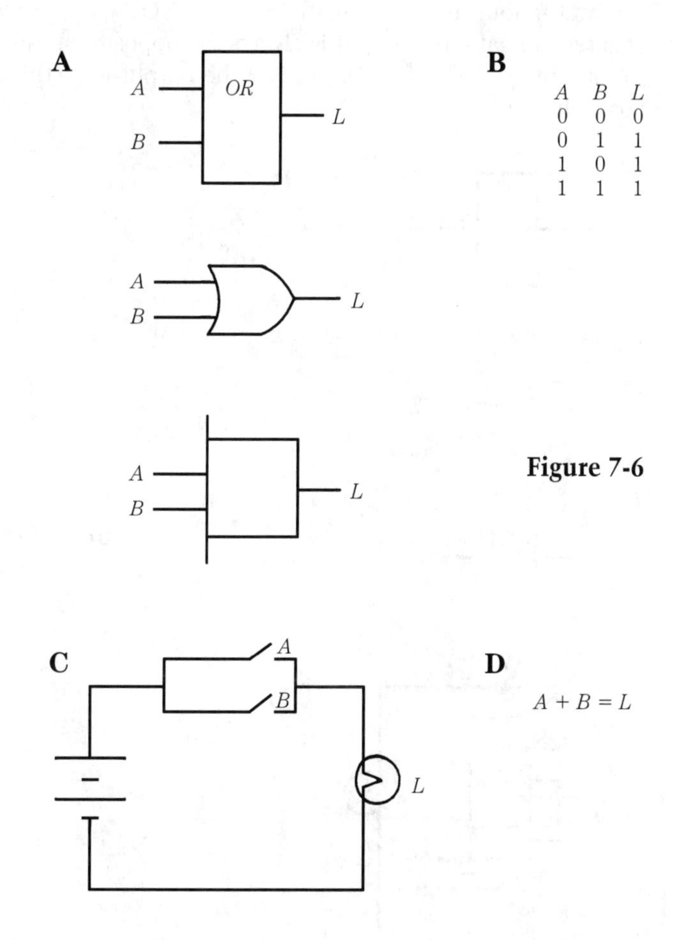

Figure 7-6

Memorize the symbols. It would be a good idea to practice drawing the AND and OR symbols so that you will get them fixed in your mind.

The truth table and the equivalent circuit show that there is an output (L) whenever there is a logic 1 at either or both inputs. This is sometimes called inclusive OR because it includes the possibility of both inputs being present. The following example will make this more clear.

Suppose an instruction sheet says: "The signal amplitude can be increased by adjusting R3 or R4." Does that mean you can adjust both R3 and R4 to get more amplitude, or does it mean that you can adjust either R3 or R4 but not both? If it means you can adjust R3 or R4 or both, then it is an inclusive OR situation. If it means to adjust one or the other, but not both, then it is an EXCLUSIVE OR. More will be said about the EXCLUSIVE OR condition later in this section.

Note that the lamp will be ON in the circuit of Fig. 7-6 if either A or B is in the 1 condition or if they are both in the 1 condition. Read $A + B$ as A OR B in logic systems.

The NOT gate

Figure 7-7 shows the four important features of the NOT gate, often called an *inverter*. With this type of gate, the output is always the opposite of the input, so if the input is 1 the output is 0, and if the input is 0 the output is 1 (the truth table shows the two possibilities).

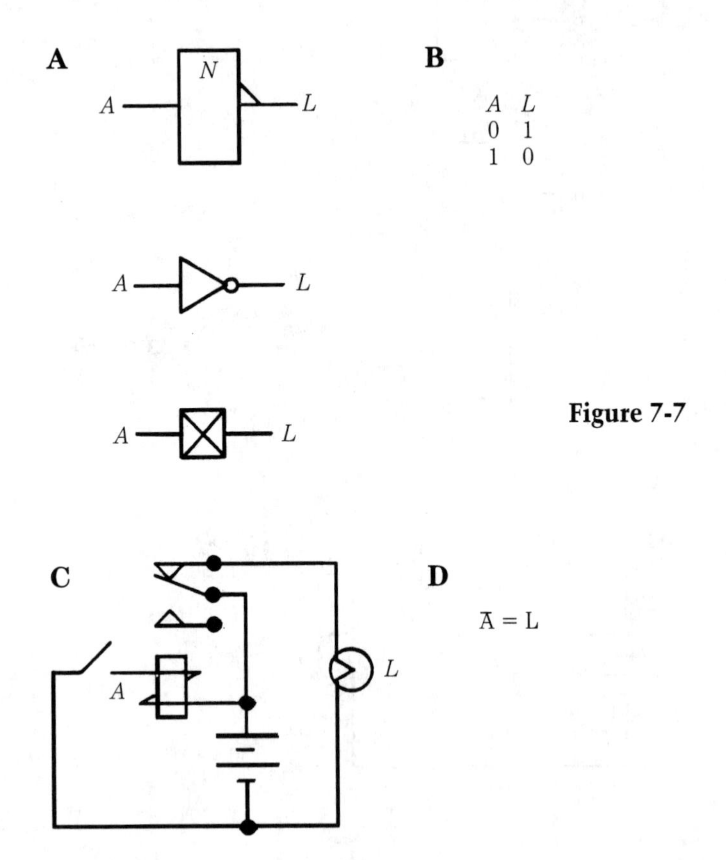

Figure 7-7

In the relay circuit of Fig. 7-7, the lamp is ON whenever the circuit is NOT energized, that is, whenever switch A is open. When switch A is closed, the relay is energized and the lamp is OFF.

Anytime there is a bar over a letter, such as $\overline{A}$, it means that the expression is negated. In other words, $\overline{AB}$ means NOT *A* AND *B*.

Two symbols for inverters are shown in Fig. 7-7. The *active high* inverter requires a logic 1 input to get a logic 0 output. The *active low* inverter requires a logic 0 to get a logic 1 output.

In the case of the active high device, a logic 0 is needed at the output for some reason dictated by a circuit requirement. For the active low device, a logic 1 is needed in the output circuit.

The NAND gate

A NAND gate is a *N*egated *AND* gate. You can see from the symbols that a NOT and an AND can be combined to get a NAND (Fig. 7-8).

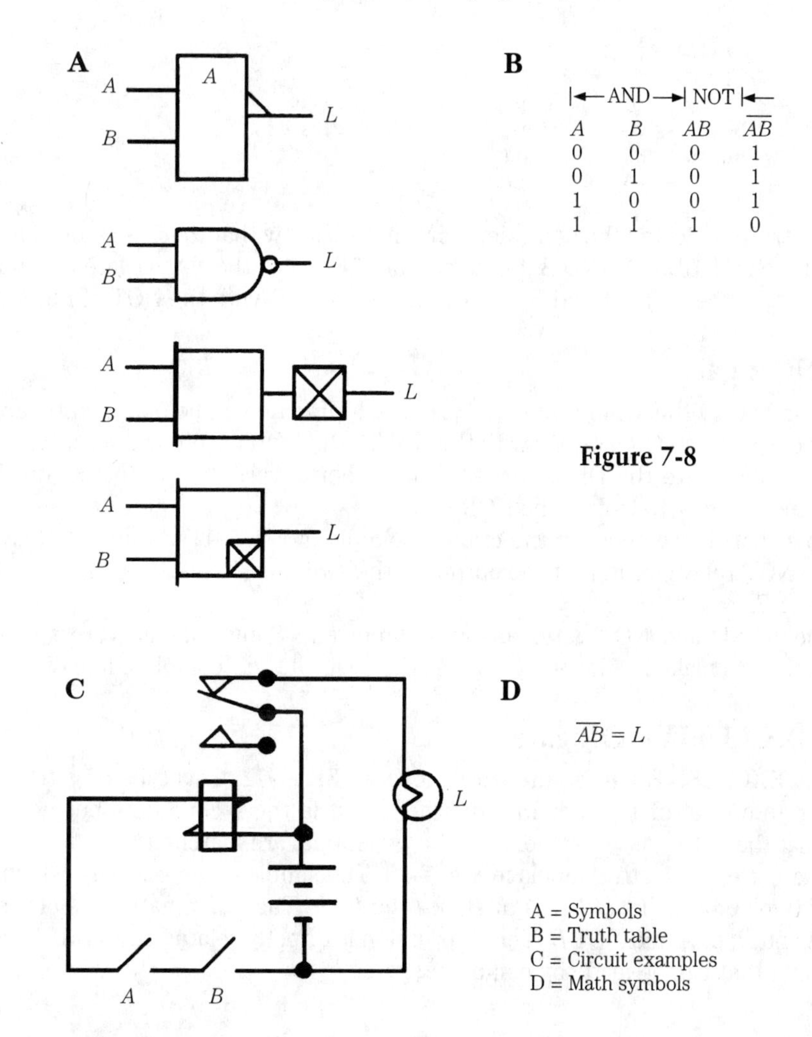

A	*B*	*AB*	$\overline{AB}$
0	0	0	1
0	1	0	1
1	0	0	1
1	1	1	0

Figure 7-8

The truth table is obtained by inverting every term in the L column of an AND truth table. This is very important because it shows that the output of two gates in combination can be obtained from their truth tables. The third column shows the output of the AND gate (AB), and the fourth column shows the output after the AND is inverted. Figure 7-9 shows this in another way.

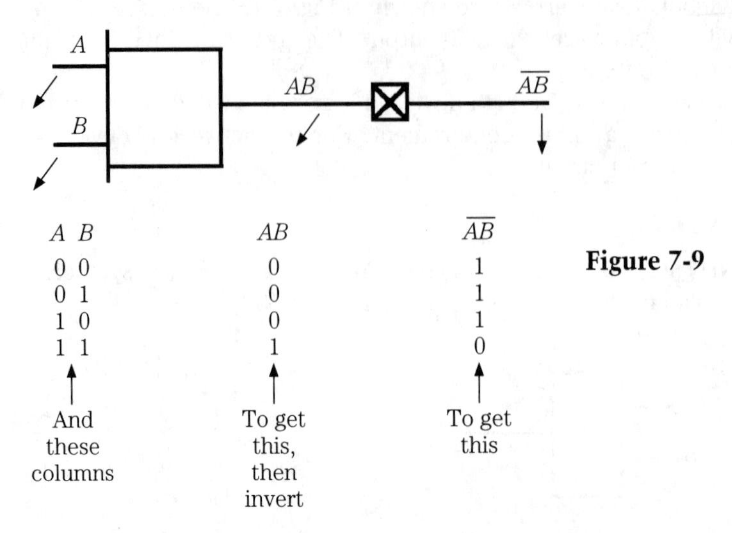

A B	AB	$\overline{AB}$
0 0	0	1
0 1	0	1
1 0	0	1
1 1	1	0

Figure 7-9

The truth table and the equivalent circuit both show that there is an output only when there is NOT an A AND B. In the circuit of Fig. 7-8, the NAND gate is obtained by combining the AND (A and B in series as in Fig. 7-5) with the NOT of Fig. 7-7.

The NOR gate

As shown by the symbols in Fig. 7-10, the NOR gate can be obtained by inverting the output of an OR gate. As with the NAND, the NOR output can be obtained by a truth table. Write the OR truth table first, then invert the output. Figure 7-11 shows another way to look at the NOR gate.

Note that the switches in the circuit example of Fig. 7-11 are in an OR layout, and the NOT relay circuit is at the output of the switches. Read the symbol $\overline{A + B}$ as NOT *A* or *B*.

The NAND and NOR gates are very important. Some manufacturers of integrated circuits make NAND or NOR gates the basis of their complete line of gates.

The EXCLUSIVE OR gate

The EXCLUSIVE OR features are shown in Fig. 7-12. This type of gate means that for an output of 1 you can have either input in the 1 condition, but not both. Note that the output is zero whenever the two inputs are matched.

There are two math symbols in Fig. 7-12. The simplest one is $A + B$, which you should think of as being *either A or B, but not both*. The other math symbol is read: NOT *A* and *B*, or *A* and NOT *B*. This expression is directly related to an EXCLUSIVE OR circuit that can be made by combining gates.

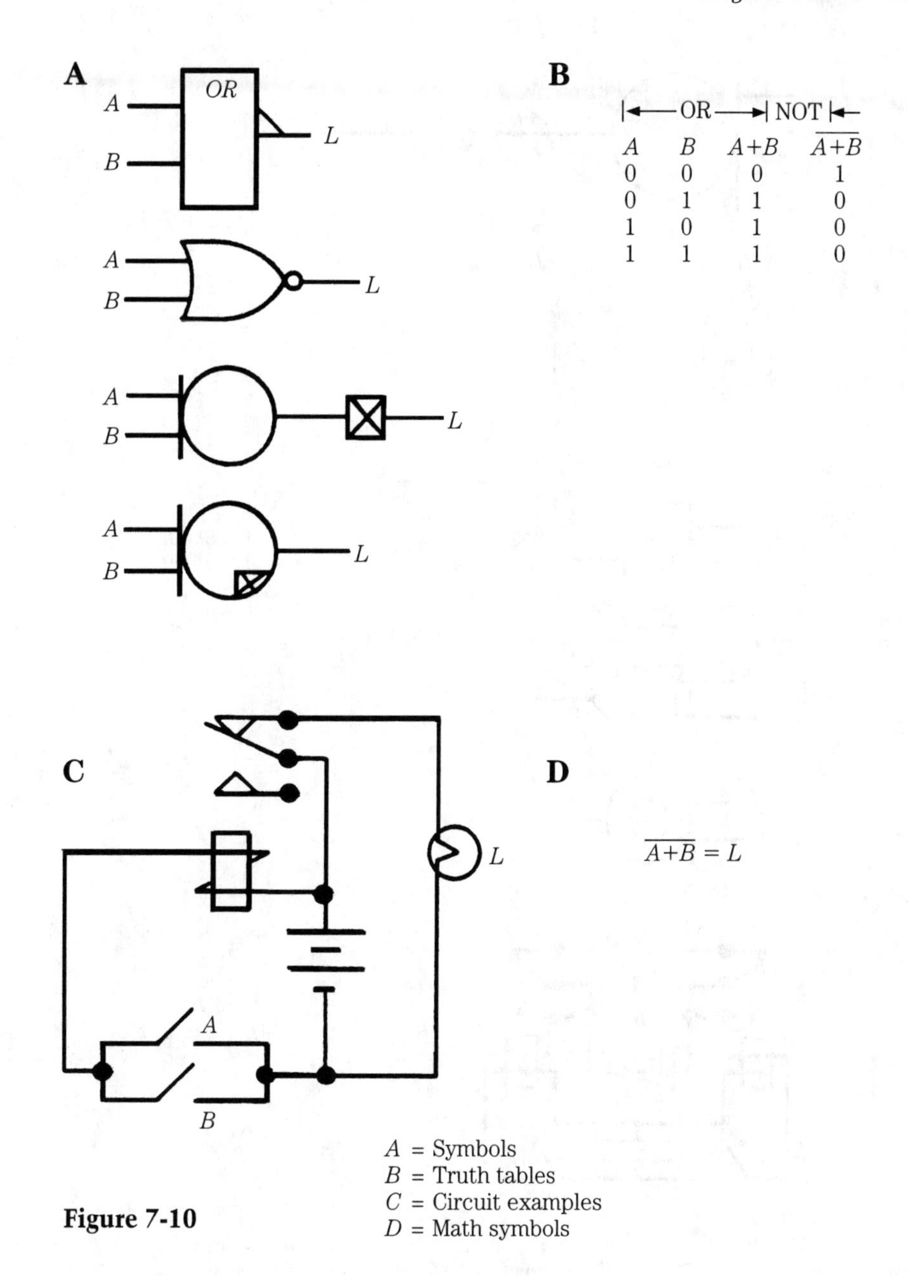

Figure 7-10

A = Symbols
B = Truth tables
C = Circuit examples
D = Math symbols

Logic families

Before you've done much work in logic systems, you will run across the names of the various logic families. Examples are TTL, ECL, and CMOS. Those names are actually descriptions of how the gates are made. For example, CMOS means that Complementary *MOS*fets are used to make gates, and ECL means that the gates use *E*mitter-*C*oupled amplifiers as *L*ogic gates.

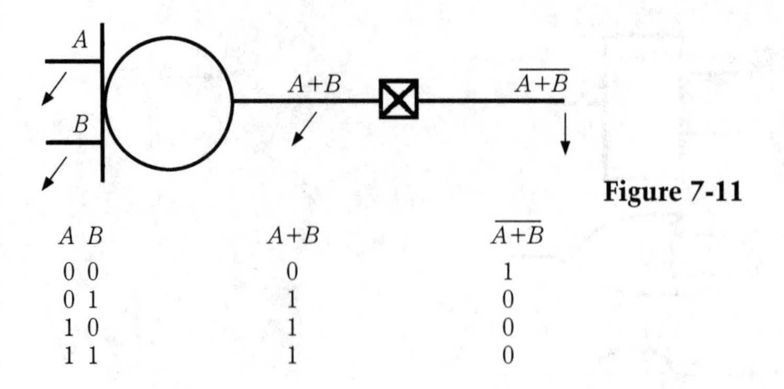

A B	A+B	$\overline{A+B}$
0 0	0	1
0 1	1	0
1 0	1	0
1 1	1	0

Figure 7-11

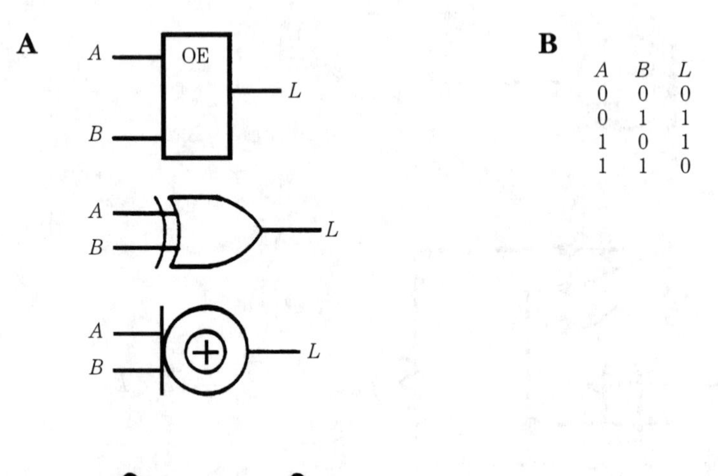

A	B	L
0	0	0
0	1	1
1	0	1
1	1	0

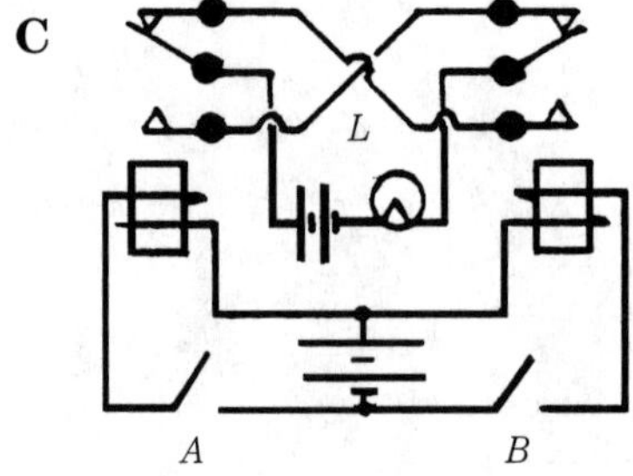

D

$$A \oplus B = L$$
$$\bar{A}B + A\bar{B} = L$$

Figure 7-12

A = Symbols
B = Truth tables
C = Circuit examples
D = Math symbols

It isn't necessary, or desirable, to spend a lot of time learning the electronic circuitry used to make the gates in each family. A two-input NAND gate does the same thing regardless of which family it comes from.

One reason for having the different families is that different manufacturers have their own preferred, patented way of making gates. As a technician, there are certain identifying characteristics of families that you should know. They are listed below.

Transistor-Transistor Logic (TTL)

This integrated circuit (IC) family of gates is the most popular, and has the largest number of circuits available. A regulated +5 V power supply is *required* for operating these circuits! The +5-V supply *usually*, but not *always*, goes to pin 14, and the common lead for the supply *usually*, but not *always*, goes to pin 7. The identifying numbers for the ICs with the basic gates are 54 or 74. For example, 5401 and 7401 are numbers for the IC shown in Fig. 7-13.

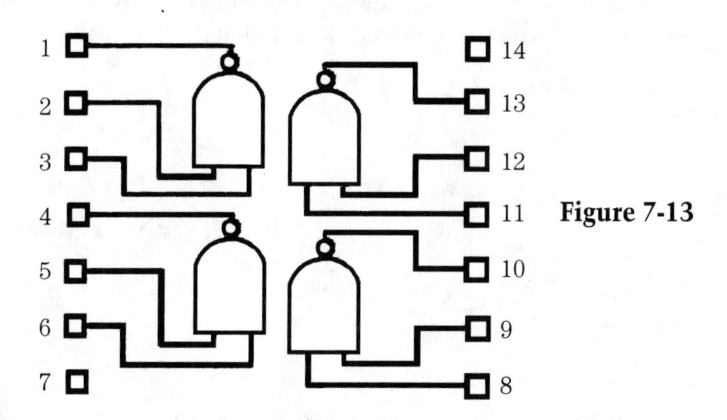

Figure 7-13

CMOS

As mentioned before, these circuits use MOSFETs. Therefore, static charges can destroy older types. Most of the newer CMOS gates are buffered and have zener diodes at their terminals, which prevents destruction by static charges. The older gates, however, are not protected.

CMOS logic ICs can be operated with any voltage from +5 to +15 V, and the supply doesn't have to be regulated, one of their most important advantages over TTL. Their disadvantages are the possible destruction by static charges and that they are often slower in operation. The CMOS ICs often have numbers in the 4000s. Pins 7 and 14 are often used as power-supply connections for 14-pin DIPs. For 16-pin DIPs, pins 8 and 16 are often used for the supply connections.

You might hear that a CMOS IC can be substituted for a TTL IC because the CMOS will operate on +5 V. Don't you believe it! Even if the pinouts are identical, they can't be interchanged because of the difference in *propagation delay*. This is simply a measure of how long it takes for the signal to go through the gate (measured in nanoseconds). Compared to TTL gates, it takes longer to get a signal through a CMOS gate when it is operating at 5 V. If it is interconnected with TTL gates, the signals just won't show up on time at some point down the line. To repeat, CMOS gates won't work in a TTL circuit.

A special series of CMOS chips have exactly the same pinouts as the equivalent TTL. They have a C in the middle of their number. As an example, a 74C76 is a CMOS integrated circuit that has the same pinout as a TTL 7476. (Both are dual flip-flops.)

Manufacturers have booklets giving the pinouts for their logic families. CMOS is an RCA specialty, TTL is a Texas Instruments specialty. Other companies, like Motorola, make them under their own names (like McMos) and numbering systems (1400 Series).

Unlike tubes and transistors, where the identifying numbers, like 6L6 and 2N2905, are the same device regardless of which company makes them, logic ICs do not always have identical numbers.

An ENABLE gate is shown in Fig. 7-14. The switch marked "ENABLE" is usually an electronic device that permits switching with a voltage. As an example, the transistor in the circuit enables the AND whenever the base is positive. Regardless of how the enable is switched ON, the result is that the input signal appears at the output.

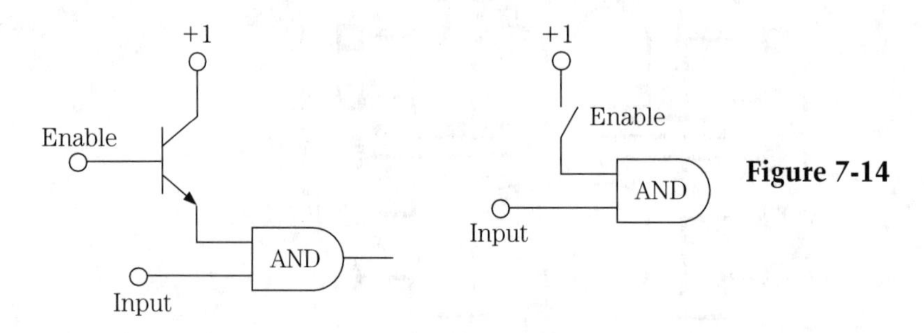

Figure 7-14

When the input signal is positive there are two logic 1 signals into the AND so that the output is logic 1. When the enable is at 0 V, there is a logic 0 into one of the AND terminals and the output of the NAND is 0 V.

Figure 7-15 shows two examples of tri-state devices. The ENABLE lead is used to switch the output between two different states:

ENABLE at Logic 0 = Output Terminal Open
ENABLE at Logic 1 = Output Normal for the device

These tri-state devices are used to disconnect a logic gate or circuit from a common line. An example is shown in Fig. 7-16. Two circuits share the same output line.

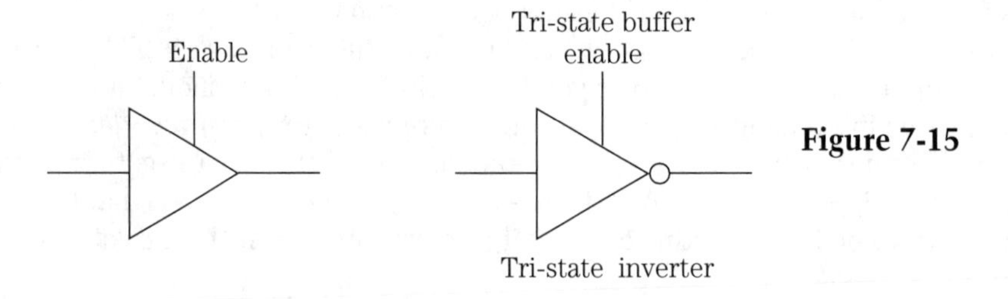

Figure 7-15

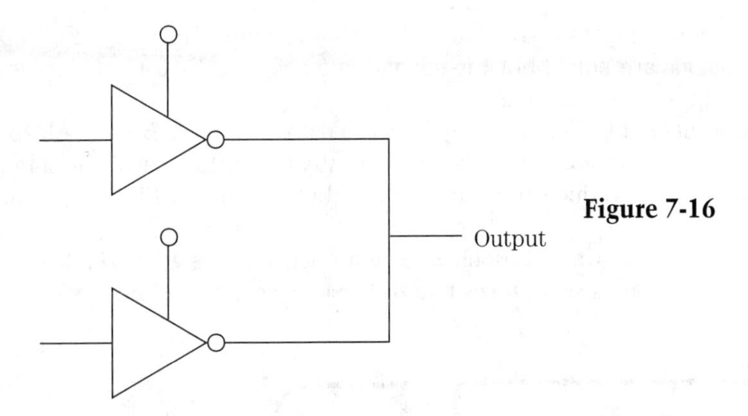

Figure 7-16

The ENABLES determine which of the circuits can access that line. The ENABLES are connected to a component (not shown) called a *flip-flop*. It allows an input to ENABLE A or B, but not both.

The circuit in Fig. 7-16 is often used in microprocessors. The output is one of 8 or 16 lines. That combination of lines is called a *bus*.

Two types of circuits are made with gates. One is called *static* and it operates without a *clock signal*. Clock signals are square wave or pulse voltages. The second type of circuit is called *dynamic*. It uses a clock signal to put the circuit operations through individual steps.

Circuits made of individual gates

Let's start with the individual-gate circuits that are combined into circuits. Figure 7-17 shows an example.

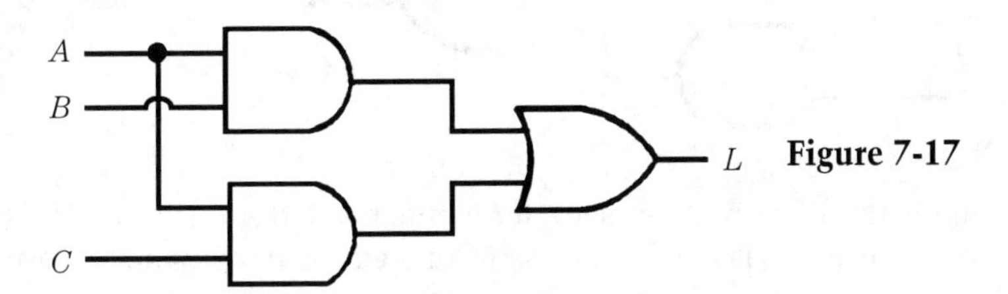

Figure 7-17

Writing output equations for combined logic gates

One way to find the output of the complete circuit is start by writing the outputs of each gate. Figure 7-17 shows how this is done. The output of the AND gate marked "X" is *A* AND *B* (*AB*). The output of the AND gate marked "Y" is *A* AND *C* (*AC*). So, the two inputs to the OR gate (marked "Z") are *AB* and *AC*. These signals are combined in the OR gate to get *A* AND *B* OR *A* AND *C*. This is written with symbols as $AB + AC$.

The circuit in Fig. 7-17 can be used to demonstrate how *Boolean algebra* can be used in logic. This is only an introduction to Boolean algebra, and more will be said on the subject in this chapter.

With ordinary algebra, you can factor $AB + AC$ into $A(B + C)$ (Fig. 7-18). The two expressions are equal, but the second one suggests that a simpler circuit, like the one in Fig. 7-18, is possible.

The outputs of Fig. 7-18 and Fig. 7-19 are $AB + AC$ and $A(B + C)$. Algebra shows that these two expressions are just different ways of writing the same thing. Therefore, both circuits produce the same output, but the one in Fig. 7-18 requires one less gate.

This example shows how Boolean algebra helps in design work. It is also useful to the technician because it shows him or her how certain circuits are obtained.

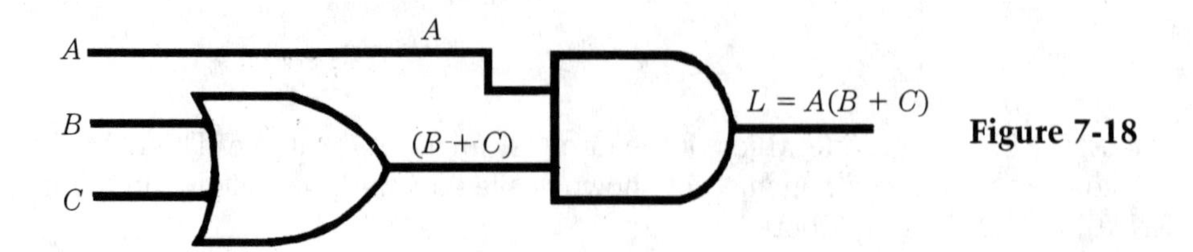

Figure 7-18

Truth tables for combined logic gates

You can make a truth table for a circuit by using the procedure that was just used (Fig. 7-19). The first row of the truth table is for $A = 0$ and $B = 0$. Those values are shown without parentheses. The numbers without parentheses throughout the circuit are based upon the two zero inputs.

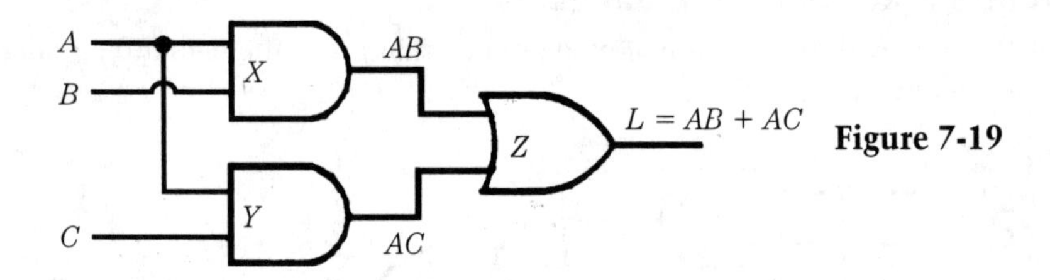

Figure 7-19

The numbers with parentheses are for inputs marked $A = (0)$ and $B = (1)$. The output with those inputs is (1). Use the same procedure to write the last two rows of the truth table.

An interesting circuit is obtained by adding an inverter to the combination of gates shown in Fig. 7-19. The resulting truth table becomes:

A	*B*	*L*
0	0	1
0	1	0
1	0	0
1	1	1

This is a truth table for a logic comparator. The circuit has an output whenever the two inputs are identical, and has been called an *EXCLUSIVE NOR*.

Finding the output signals of combined gate circuits

You can determine the output signal of a logic circuit in the same way as you determine the output logic level. You simply draw the output signal for each gate, one at a time, until you have drawn the output signal for the complete circuit. Remember that the signals are either high or low, representing 1 and 0 logic values.

As an example, when the two signals into the EXCLUSIVE OR gate of Fig. 7-20 are both 0 or both 1, the output is low (output = 0). If one of the input signals is 1 and the other is 0, the output signal will have a high output of 1. The output signal (L) is shown in Fig. 7-20.

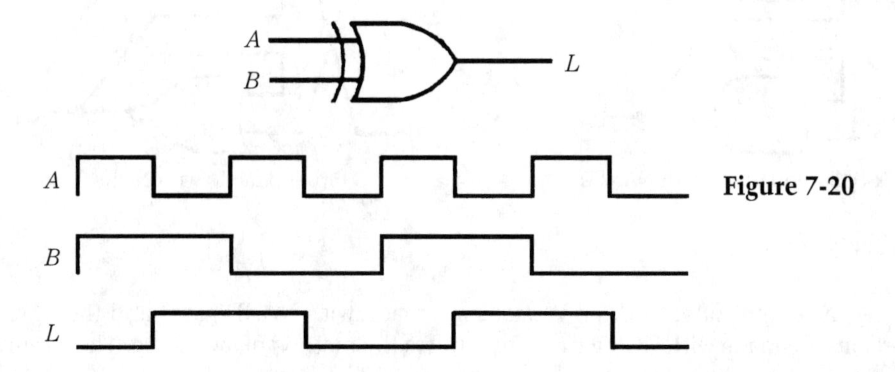

Figure 7-20

It helps to line up the signals vertically so that you can draw the output signals of each gate, and of the circuit.

Aligning the signals makes it possible to get the right time spaces. This is shown again in Fig. 7-21.

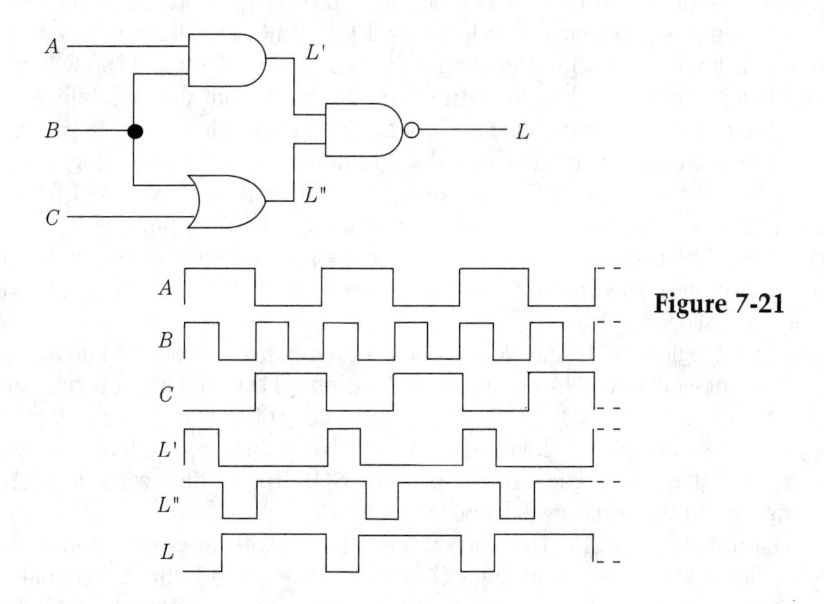

Figure 7-21

The R-S flip-flop

Figure 7-22 shows two versions of a very popular circuit that is made by combining basic gates. This circuit is known by the names *R-S flip-flop*, *S-R flip-flop*, *latch*, and *bounceless switch*. Although it is discussed as a circuit made with basic gates, you can also buy versions of it on a single chip.

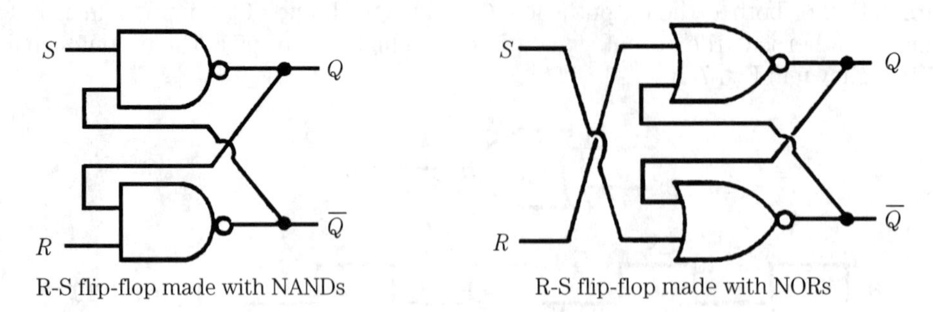

R-S flip-flop made with NANDs

R-S flip-flop made with NORs

Figure 7-22

As a general rule, the TTL versions are made with NAND gates and the CMOS versions are made with NOR gates. This affects the method of switching. The circuits will be compared in this section. Refer to Fig. 7-23.

The letters S and R at the input terminals stand for *set* and *reset*, but the letters Q and $\overline{Q}$ at the output terminals have no special meaning. In fact, as shown in Fig. 7-23, you will sometimes see 1 and 0 instead of Q and $\overline{Q}$ (Don't confuse those circuit identifiers with logic levels 1 and 0).

For any flip-flop, there are only two possible states: *high* and *low*. A high condition occurs when the Q terminal is at logic level 1 and the $\overline{Q}$ terminal is at logic level 0. A low condition occurs when the Q terminal is at logic level 0 and the $\overline{Q}$ terminal is at logic level 1. Figure 7-23 shows the two conditions that are possible, and includes a table of names that can be used instead of high and low.

Before continuing this review, you should know that the *normal* logic level of both input terminals is logic 1 for the NAND R-S flip-flop. It is switched from one condition to another by switching either S or R to logic 0. The input can be momentarily switched then returned to logic 1, or it can be permanently switched. In either case, the condition of the flip-flop can only be changed if the proper input terminal is switched to logic 0.

The normal logic level of the NOR R-S flip-flop input terminals is logic 0. An important difference between NAND and NOR flip-flops is that the input terminals are normally at logic 1 for NAND R-S flip-flops and logic 0 for NOR R-S flip-flops. if a square wave is used to switch the flip-flop, the NAND type will switch on the trailing edge when the square wave goes from 1 to 0. The NOR R-S flip-flop will switch on the leading edge when the square wave goes from 0 to 1.

The statements regarding the normal input logic levels are in reference to the simple circuits of Fig. 7-22. By adding NOT gates to the S and R input terminals, the normal input conditions will be reversed. This is likely to be done for the NAND type

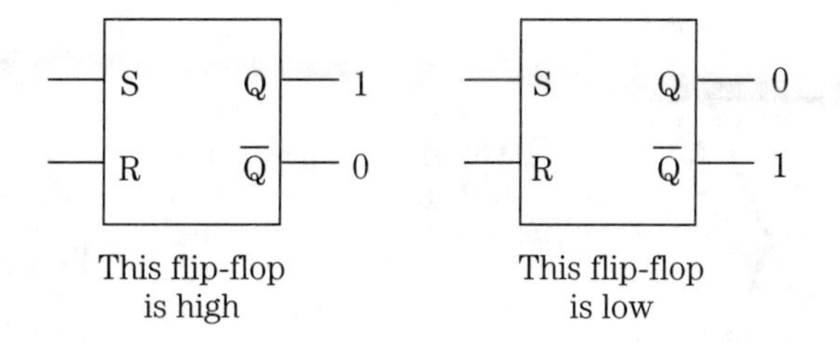

Condition	Name of condition
Q = Logic 1 $\overline{Q}$ = Logic 0	The flip-flop is said to be on, or high, or in a logic 1 condition
Q = Logic 0 $\overline{Q}$ = Logic 1	The flip-flop is said to be off, or low, or in a logic 0 condition

Figure 7-23

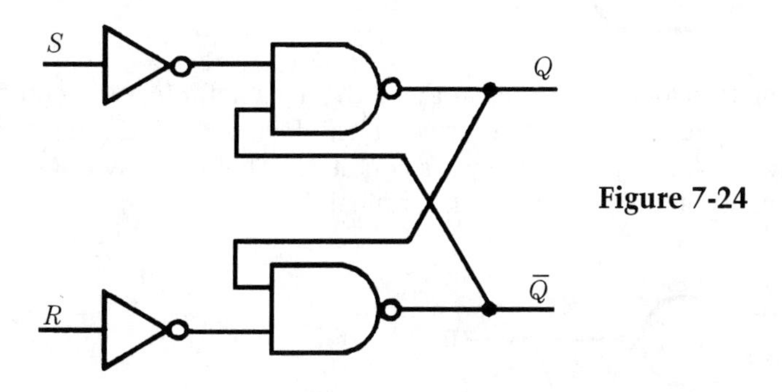

Figure 7-24

so that *positive logic* can be used (see Fig. 7-24). With positive logic, the flip-flops are switched with logic 1, and they are in their normal condition with logic 0.

Now let's review the signals for a NAND S-R flip-flop and see how it is switched. Figure 7-25 shows the NAND flip-flop with its normal logic levels. The numbers are the normal values when the flip-flop is in the *high* condition. If you switch terminal S to 0, the output of the upper NAND will still be 1 because it will have two 0s at the input. Therefore, there will be no change in the input and output logic levels of either gate when S is switched to 0, provided the flip-flop is in the high condition.

Returning the S terminal to its normal logic 1 value does not affect the condition of the flip-flop. To summarize, the S terminal was switched from 1 to 0 and back to 1 on a high NAND flip-flop, but there was no change in the flip-flop condition. Now let's review what happens when you switch lead R to logic 0. See Fig. 7-26.

If you switch the R terminal to logic 0, as shown in parentheses in Fig. 7-26, the output of the lower NAND gate changes to 1. This, in turn, changes the condition of the upper NAND because it will now have two logic 1 inputs and a logic 0 output.

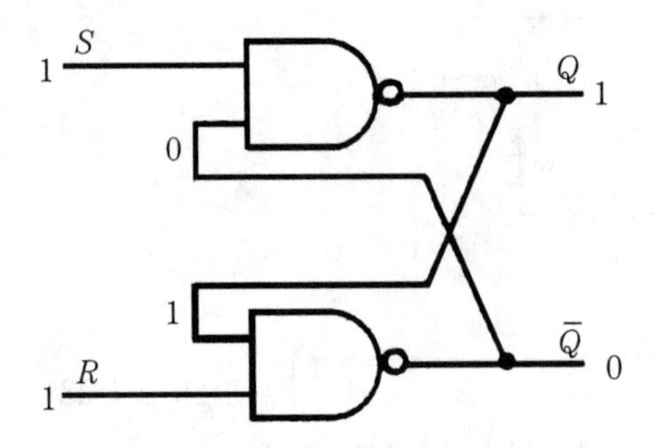

This flip-flop has its input terminals at the standard logic level and the flip-flop is high. Switching S will not change its condition.

Figure 7-25

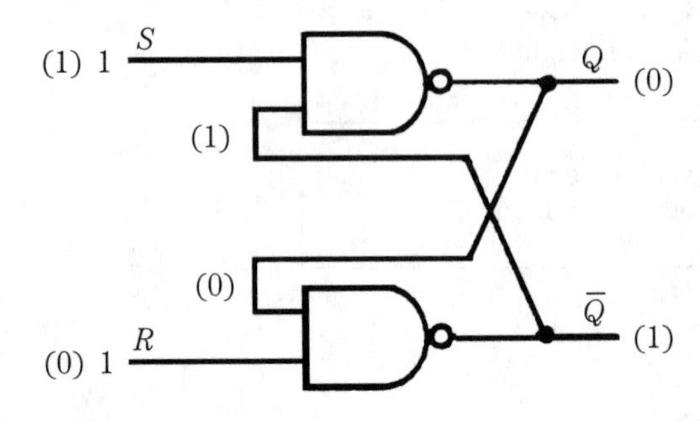

The flip-flop is switched by changing terminal *R* to logic 0. Numbers not in parenthesis show condition after *R* is switched to 0.

Figure 7-26

Returning R to logic level 1, as in Fig. 7-27, does not affect the condition of the NAND flip-flop. Furthermore, if you switch lead R to 0 when the flip-flop is low, as shown in Fig. 7-27, it will not change its condition. The only way to get the flip-flop back to the high condition is to switch terminal S to 0.

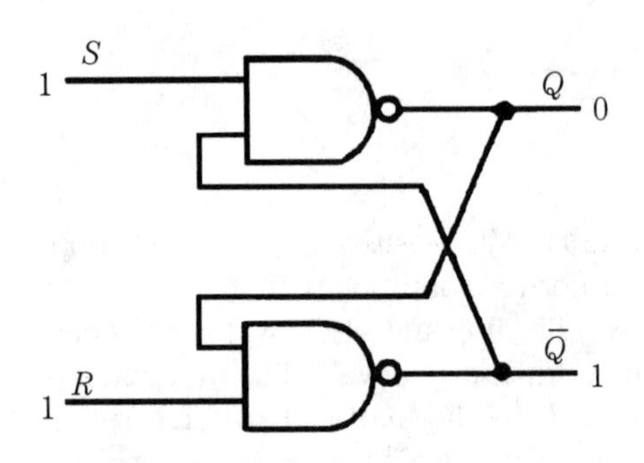

The input terminals are returned to their standard logic level. The output is not changed.

Figure 7-27

Figure 7-28 shows a summary of input logic levels that will change the condition of a NAND flip-flop, and the changes in input logic levels that have no effect on the condition of the flip-flop. Study this illustration very carefully until you are sure that you can answer questions about the action of an R-S flip-flop.

The conditions for the NAND and NOR R-S flip-flops can be described with truth tables, as shown in Fig. 7-29.

Note that the normal high condition of the NAND type occurs when the inputs are at logic 1, and the normal high for the NOR type occurs when the inputs are at logic 0. These are starting points for the truth tables. From there, they are switched

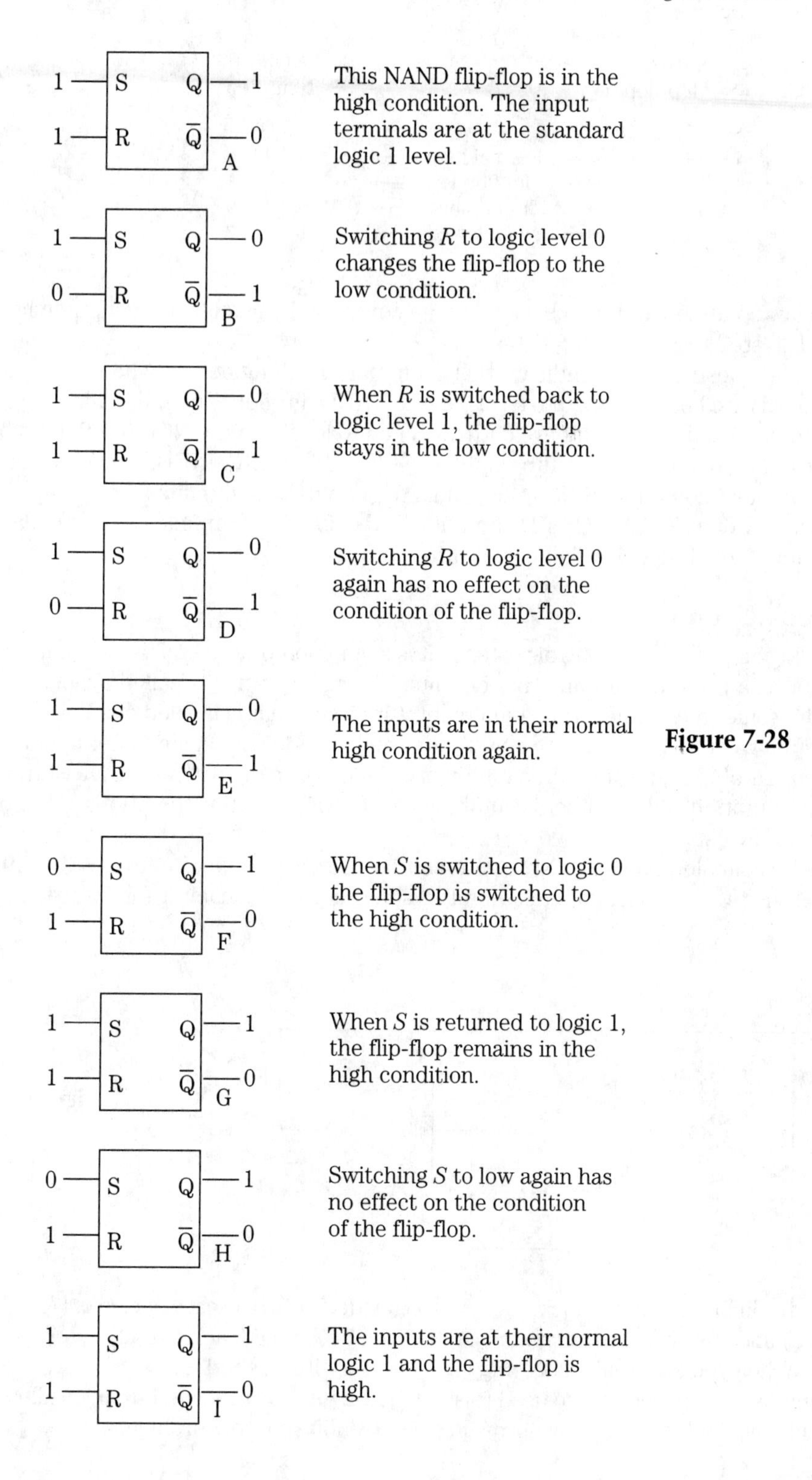
A
This NAND flip-flop is in the high condition. The input terminals are at the standard logic 1 level.
B
Switching R to logic level 0 changes the flip-flop to the low condition.
C
When R is switched back to logic level 1, the flip-flop stays in the low condition.
D
Switching R to logic level 0 again has no effect on the condition of the flip-flop.
E
The inputs are in their normal high condition again.
F
When S is switched to logic 0 the flip-flop is switched to the high condition.
G
When S is returned to logic 1, the flip-flop remains in the high condition.
H
Switching S to low again has no effect on the condition of the flip-flop.
I
The inputs are at their normal logic 1 and the flip-flop is high.

Figure 7-28

NAND S-R flip-flop					NOR S-R flip-flop			
S	R	Q	$\bar{Q}$		S	R	Q	$\bar{Q}$
1	1	1	0	← Normal high →	0	0	1	0
1	0	0	1	← Flip-flop low →	0	1	0	1
0	1	1	0	← Flip-flop high →	1	0	1	0
0	0	?	?	← Not allowed →	1	1	?	?

Figure 7-29

to a low condition and then back to a high condition by changing the input logic levels delivered to S and R.

The question marks indicate that an important thing occurs when the S and R terminals are both switched to level 0 in the NAND flip-flop. The same thing happens when the S and R terminals are both switched to level 1 in the NOR flip-flop. In both cases, the condition of the flip-flop CANNOT BE DETERMINED, which is called a *not allowed* condition. S-R flip-flops can only be used in circuits where there is no possibility of a NOT ALLOWED condition. If this is a disadvantage in a particular application, the designer will choose a different type of flip-flop.

Clock circuits

In computers and control systems, it is a common practice to use a square-wave signal as a reference for all other circuits. The square wave is called a clock signal, and it sometimes comes from a multivibrator circuit that is labeled *clock*.

You will remember that the NOT gate can be made with an amplifier operated at either cutoff or saturation. In a multivibrator circuit made with two transistors, one is in saturation and the other is cutoff. On each half cycle, the operations of the transistors reverse.

It seems logical that a NOT gate can be used as a multivibrator (Fig. 7-30). A tipoff that this is a form of multivibrator is in the use of capacitor for feedback.

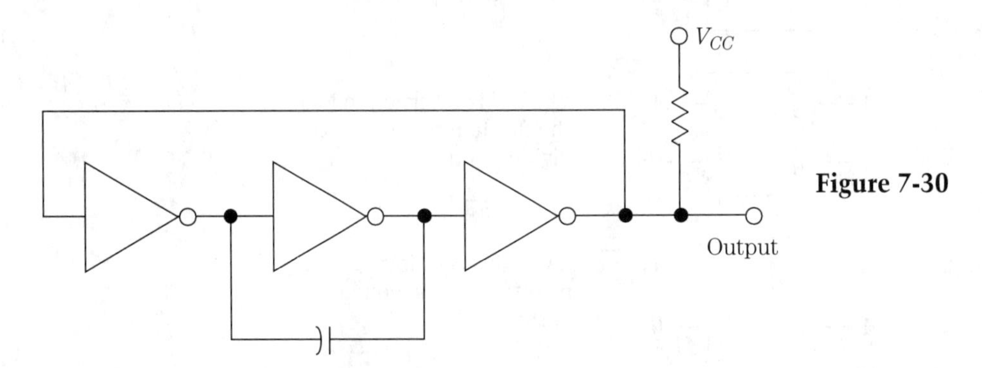

Figure 7-30

If a highly accurate clock frequency is needed, as in the case of clocks (for telling time) and stop watches, a crystal can be used in the feedback circuit.

When you are troubleshooting logic circuitry it is a good idea to check the clock signal. A logic probe can be used to show that a pulse is present, but an oscilloscope can be needed to study waveforms in some troubleshooting problems.

Combined logic circuits on IC chips

There are a great number of circuits made of gates and packaged in integrated circuit form. It would not be possible to list them all in a single chapter.

J-K flip-flops

The R-S flip-flop has a "not-allowed" condition that could make it useless in some applications. The J-K flip-flop has no such disallowed input. It is made by taking basic R-S flip-flops and combining them with some other basic gates. The overall result is a more elaborate, but also more useful circuit.

Figure 7-31 shows the IC package for a typical J-K flip-flop. Normally, two or more such flip-flops are located in a single integrated package. If you replace the S and R with J and K in the flip-flops of Fig. 7-21, the operation would be the same. However, there are additional terminals, as shown in Fig. 7-31, that also affect the operation. There are J-K flip-flops in all logic families. The discussion here is for a J-K flip-flop in the TTL family.

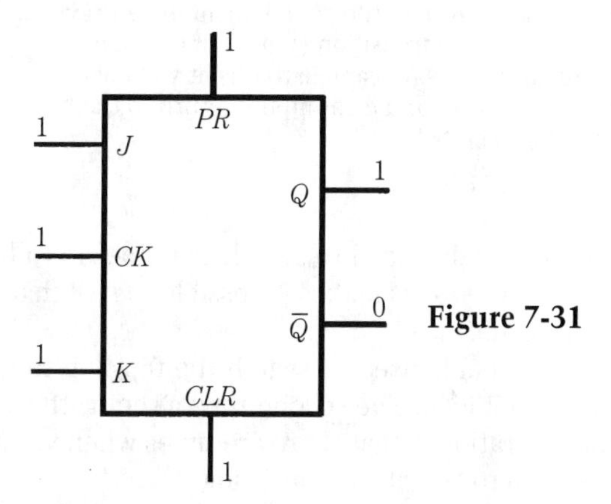

Figure 7-31

The CK terminal is used for a clock input signal. The J-K flip-flop cannot change by switching J or K to logic 0 unless the CK terminal is switched to 0 at the same time. Figure 7-31 shows this. When K is switched to 0, but CK is held at 1, the flip-flop condition does not change. However, when the CK terminal is switched to 0 and K is switched to 0, then the condition of the flip-flop changes to low. This arrangement assures that the flip-flop is not accidentally switched with a noise signal or with a glitch.

The 5th and 6th rows of the truth table in Fig. 7-32 shows that the condition of the flip-flop changes when the CK terminal goes to 0. This is indicated by the symbol **. The SET and PRESET terminals are held at logic 1 for this operation.

PR Preset	CLR clear	Clock	J	K	Outputs Q	$\overline{Q}$
L	H	X	X	X	H	L
H	L	X	X	X	L	H
L	L	X	X	X	H*	H*
H	H	**	L	L	Q_0	$\overline{Q}_0$
H	H	**	H	L	H	L
H	H	**	L	H	L	H
H	H	**	H	H	Toggle	
H	H	H	X	X	Q_0	$\overline{Q}_0$

Figure 7-32

H = High level (steady state)
L = Low level (steady state)
X = Irrelevant
** = Transition from high to low level
Q_0 = The level of Q before the indicated input conditions were established
toggle = Each output changes to the complement of its previous level on each active transition (pulse) of the clock
* = This configuration is nonstable; that is, it will not persist when preset and clear inputs return to their inactive (high) state

The PR (preset) terminal is used to switch the flip-flop to high, as is shown in the first row of the truth table. This makes it possible to switch all flip-flops in a system to high at the same time.

The CLR (clear) terminal is used to switch the flip-flop to a low condition, regardless of what condition it is in. The second row of the truth table shows this. The reason for having this operation is that there are times when you want all of the flip-flops in a system to switch to low at the same time.

The fourth row shows that when both inputs are switched to low the flip-flop does not change its condition. Remember that this is a not allowed condition for the R-S flip-flop.

The toggle condition, which is the 7th row of the truth table, shows that if all input terminals except the clock input are held at logic 1, then the flip-flop will change condition every time the CLK terminal is switched to 0. The CLK terminal is then called a *toggle signal.*

Schmitt triggers

Figure 7-33 shows the logic symbol for a Schmitt trigger. You probably studied the tube or transistor equivalent at some time in your work as a technician.

Basically, the Schmitt trigger is used to produce a square-wave output for almost any input waveform. A sine-wave input will produce a very good square-wave output. The TTL 74132 integrated circuit contains four Schmitt triggers.

Figure 7-33

Examples of systems

Because you have already had technical training before you started using this book, you must be familiar with the problems of troubleshooting. Regardless of which field your training has been in, you know that troubleshooting is a logical procedure. We like to think of it as a common-sense procedure.

Two very basic systems and a circuit in another system will be covered. The discussion is mainly about troubleshooting procedures that you might use to find a problem in systems. Don't be misled by the fact that these systems contain only a few parts. The procedures we discuss here are the same as for systems that have many integrated circuits mounted on printed circuit boards, or "mother boards."

We will start this discussion with the logic probe of Fig. 7-34. The reason we call this a system is because it has a number of different basic gates that contribute to the complete operation of the system. In other words, as far as we are concerned, a system is something that operates by itself and is not dependent upon other circuits for its operation. In general, a system usually contains a number of different basic circuits.

How the probe works

First, look at the circuit and decide what the various components are used for. This is a probe for checking for logic 1 or 0, or for a pulse in a TTL circuit.

Resistor R1 and diode X1 form a protective circuit to prevent high negative voltages from being delivered to the input of an AND gate. Remember that TTL gates are

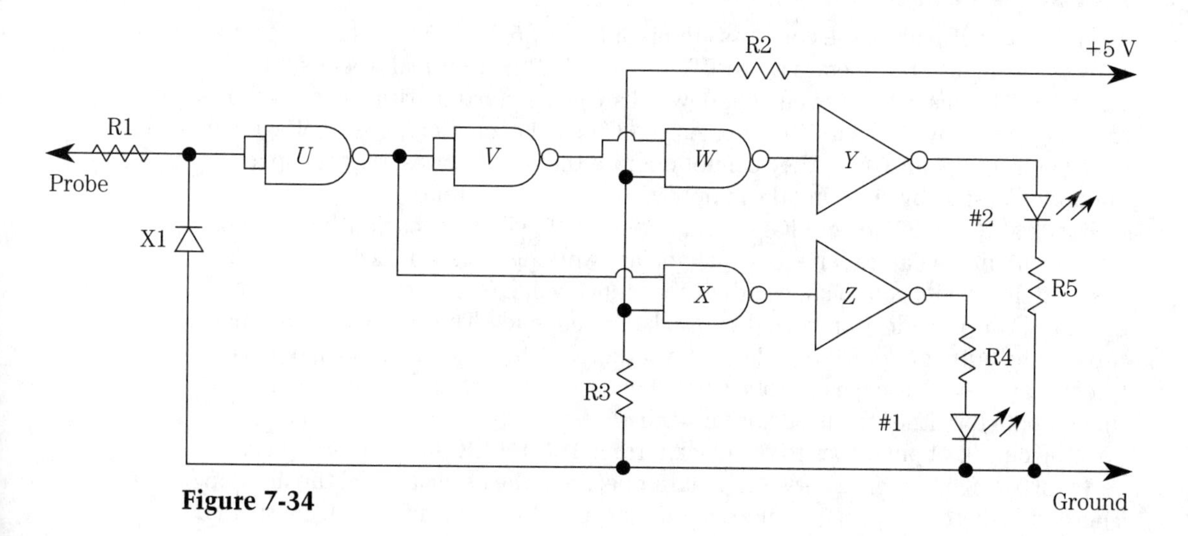

Figure 7-34

designed for positive logic and a high negative voltage can produce an internal breakdown. This would occur, for example, if you should accidentally probe into some point where there is 8 to 10 V of sine-wave or square-wave ac voltage.

NAND gates U and V are both connected as NOT gates. NAND gates W and X have two inputs. One input, applied to both W and X, is obtained from voltage divided R2 and R3. In most NAND logic circuits, a logic 1 corresponds to a +3 V or greater, so you want an output when the probed level is 3 V. That's why you need the voltage divider—to lower the operating level. The overall result is that NAND's W and X have a logic 1 (3 V) delivered to one of the terminals at all times.

The second input terminal on gate X is delivered to the output of NOT gate U, and the second input lead on gate W is connected to the output of NOT gate V.

The outputs of the two NAND gates are delivered to inverters Y and Z, and each inverter feeds an LED. Notice that the two LEDs have series-limiting resistors R4 and R5 to limit the amount of current through those diodes. This is very important: You should never connect a logic gate directly to an LED. The low forward resistance will overload the gate output circuitry and will likely destroy it.

When the probe is connected to a point in the circuit where the logic value is 1, a logic 1 is delivered to the input of NOT U. The output of that gate is 0, which is inverted by V again at the input of NOT V. Now there are two 1s delivered to W, and a 0 delivered to NOT Y. The output of NOT Y is logic 1, so LED #2 is ON when the probe is touching a logic 1.

If the probe is touching a logic 0 point, there is an output of logic 1 from gate U. Therefore, there are two logic 1s into gate X, and the output of that gate is logic 0, which is inverted to logic 1 by gate Z. Therefore, LED #1 is on when the probe is touching a logic 0.

You can easily determine that LED #1 is off when the probe is touching a logic 1, and LED #2 is off when the probe is touching a logic 0.

Electronic coin toss

Now we will analyze the coin toss circuit of Fig. 7-35. Notice that three C cells at 4.5 V are being used to operate these TTL circuits, but their normal power supply input is 5 V. This means that even with new cells you are starting with a lower voltage than suggested by the manufacturer. As the C cells become old, their voltage will drop below the point where they cannot operate the NAND gates, so it is especially important to start by checking the supply voltage in this system.

Look at IC1-a, IC1-b, and IC1-c, and notice the feedback capacitor between pin 4 and pin 3. Immediately you know that you are working with a multivibrator circuit. The switch permits you to choose the capacitance value of C_1, or C_1 in parallel with C_2, or C_3. This provides control of the oscillator frequency. There should be a square wave at the output of pin 6 regardless of whether the "toss" switch is open or closed. Checking this with a scope or probe would be a good place to start if you are not getting outputs for either the heads or tails LEDs.

The output of pin 6 goes to the clock terminal of the J-K flip-flop through an inverter, IC1-d. This inverter serves as a buffer between the oscillator and the flip-flop. There is something special to note about the flip-flop compared with most TTL types. It toggles when J and K are grounded, so it is a CMOS flip-flop.

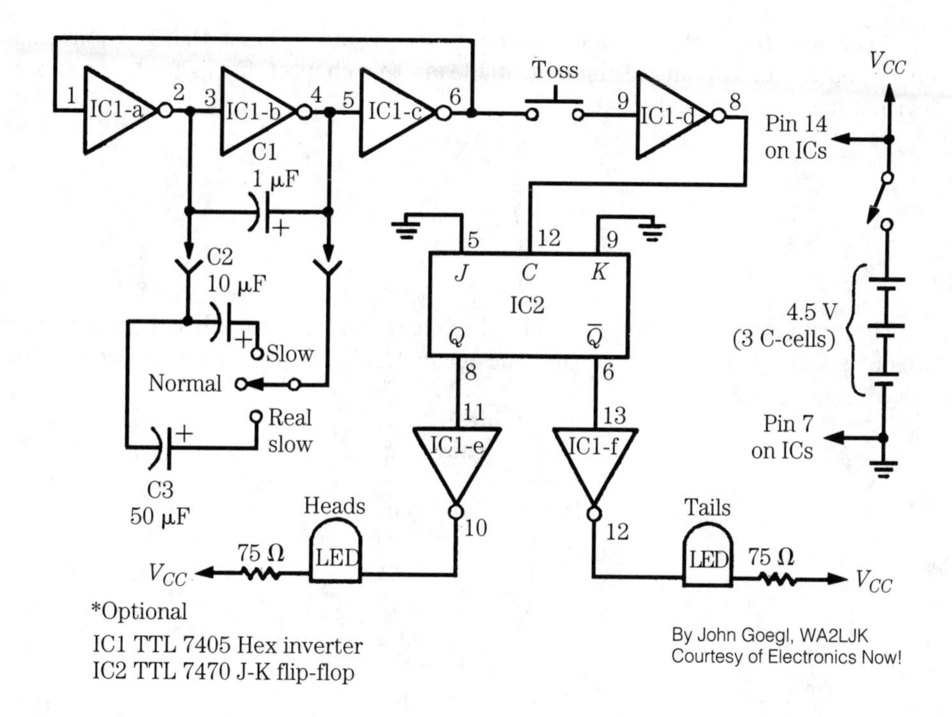

Figure 7-35

The square wave at the clock terminal causes the flip-flop to go back and forth, giving a logic 1 alternately at Q and $\overline{Q}$. The output of these terminals goes through buffers (IC1-e and IC1-f) to the LEDs and the current-limiting resistors.

When you push the toss button, the LEDs are alternately lit at such a high rate that you cannot tell when the heads and when the tails light is on. When you lift the toss switch, whatever position the J-K flip-flop is in (high or low) will determine whether you get a heads light or a tails light.

Counting

Although it is convenient for humans to count by tens, it is not so easy for an electronic device. In order to count by tens electronically, it is necessary to have a device that has ten stable states. Although such devices exist, they are more expensive and more complicated than two-state devices.

Here are some examples of the many devices with two stable states:

Devices	*States*	
Switches	On	Off
Diodes	Conducting	Cutoff
Transistors	Saturated	Cutoff
Relay	Energized	Not energized
Lamp	Glowing	Off

Devices with two stable states can be used to count by twos, that is, to make a binary count. Refer to any of the truth tables in this chapter. If it is for a two-input device it has the following choices:

A	B
0	0
0	1
1	0
1	1

The numbers are the first numbers of a binary count. Notice that there are only two digits: 0 and 1. Table 7-2 compares the decimal and binary counts up to decimal 16.

Table 7-2
Binary counting

Decimal	Binary
0	00000
1	00001
2	00010
3	00011
4	00100
5	00101
6	00110
7	00111
8	01000
9	01001
10	01010
11	01011
12	01100
13	01101
14	01110
15	01111
16	10000

As shown in Table 7-2, every time all of the digits have been used a new column is started. In the tens system, the digits are 0 through 9. In the binary count the digits are 0 and 1.

Two other popular counts are count by 8s (octal system) and by 16s (hexadecimal). Table 7-3 gives examples of counts in those systems. As with the counts in Table 7-2, a new column is always started when all of the symbols in a system have been used.

The zeros in front of the decimal numbers and binary numbers (and all numbering systems) do not affect their values. So, 0010 is the same as 101. The zeros in front of the numbers are only used for setting places. In other words, 00101 is a number within a system that counts from 00000 to 11111.

Table 7-3
Octal and hexadecimal counting

Decimal	Octal	Hexadecimal
0	00	00
1	01	01
2	02	02
3	03	03
4	04	04
5	05	05
6	06	06
7	07	07
8	10	08
9	11	09
10	12	0A
11	13	0B
12	14	0C
13	15	0D
14	16	0E
15	17	0F
16	20	10

The radix of a numbering system tells how many symbols there are in that system. The decimal system has a radix of 10 and the binary system has a radix of 2.

It is possible to have more than one radix for a given number. As an example, the number 10000 is the seventeenth count in the binary system and the tenthousandth count in the decimal system.

When there is confusion between numbers in different counting systems the radix is usually given as a subscript. In the above example, 10000_{10} is a decimal number and 10000_2 is a binary number.

As part of your preparation for taking certification and license tests, you should know how to count in the systems shown in Tables 7-2 and 7-3.

Additional comments on Boolean algebra

A Boolean equation is a shorthand method of identifying digital circuits. Each equation explains the relationship between input and output logic levels. It is a way of remembering what the gate does. However, in practice, the Boolean equations serve a much greater purpose.

Designers use Boolean algebra to determine such things as the minimum number of switches, or logic gates, or relays needed to accomplish a given circuit function. The designers accomplish that by writing Boolean equations for the circuit and manipulating those equations in the same way that algebra equations are manipulated with Xs and Ys.

Technicians are not usually required to design circuits, so an extensive mathematical treatise on Boolean algebra is not helpful if the ultimate goal is to learn how to pass most license and certification tests. However, it is a good idea for technicians to know the equations and understand their basic meaning.

The Boolean basic equations are listed in Table 7-4. These equations use A and B to show relationships. That is an arbitrary identification. Another book might use different letters or numbers to identify those same relationships.

Table 7-4
Rules of Boolean algebra

$A \times 1 = A$	(X means AND)
$A \times 0 = 0$	
$A \times A = A$	
$A + 1 = 1$	(+ means OR)
$A + 0 = A$	
$A + A = A$	(Inclusive OR)
$(A + B)\ (A + C) = A + BC$	
$A \times \overline{A} = 0$	($\overline{A}$ means NOT A)
$A + \overline{A} = 1$	
$A + \overline{A}B = A + B$	
$\overline{A}B + A\overline{B} = L$	(Exclusive OR)
$AB + \overline{A}\,\overline{B} = L$	(Logic comparator)
$\overline{A} \times \overline{B} = L$	NAND
$\overline{A} + \overline{B} = L$	NOR
$\overline{\overline{A}} = \overline{A}$	
$\overline{\overline{A}} = A$	
$\overline{AB} = \overline{A} + \overline{B}$	DeMorgan's Theorem
$\overline{A + B} = \overline{A}\,\overline{B}$	(Break the bar and change the sign)

Each of the Boolean equations can be demonstrated by using the basic logic gates that have been discussed so far. This is a good practice for learning to recognize some very important basic relationships in logic systems.

Some technicians have found the Boolean equations and equivalent circuits helpful for learning digital circuits. For, remembering DeMorgan's Laws, technicians use the line "break the bar and change the sign."

To become proficient in the use of Boolean algebra you must know the rules of algebra. That is not a requirement of the FCC license tests. It might be a requirement of some certification tests. You are urged, again, to write to the certifying agency about their requirements. You will find them very helpful for answering questions.

There is another subject in digital theory that is not required in FCC tests, but it might be encountered in certification tests. It is the conversion of numbers from one radix to another.

Again, write to the certification agencies and get their latest information on that subject.

Divide by two

Figure 7-36 shows a CMOS flip-flop in a divide-by-two connection. The flip-flop is connected to toggle (change output states) on the rising edge of each clock pulse. From the clock and $\overline{Q}$ output, you can see that the toggled flip-flop divides the clock frequency by 2.

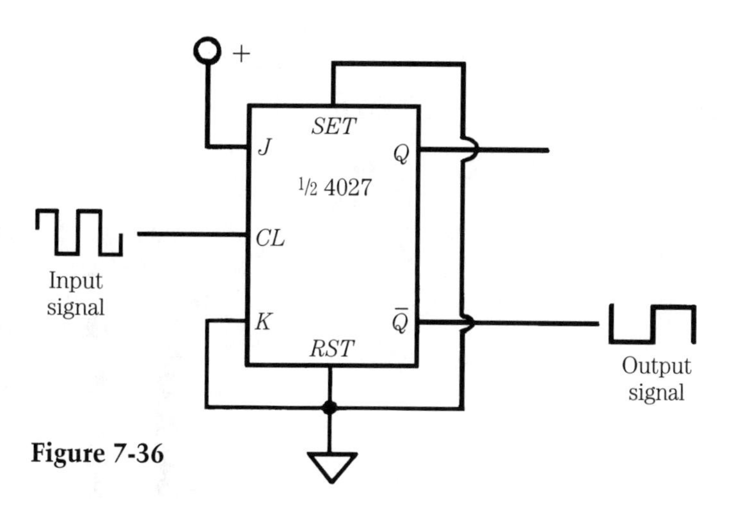

Figure 7-36

The output of a counter usually starts with the output of Q used as the least significant digit. Therefore, two flip-flops are needed in a divide-by-two circuit. This is shown in Fig. 7-37.

The J, K, S, and R connections are not shown in Fig. 7-37. It is the way you will usually see this type of circuit.

An overall view of counting systems

Figure 7-37 shows the basic components that are used in making a counter. Study this illustration carefully. In some cases the individual components are combined into an integrated circuit, but they are treated separately here.

The system in Fig. 7-37 displays two decimal digits. You can easily extend this basic design for as many digits as you need. Designers of systems that use digital counters are faced with a trade-off between cost and accuracy. Adding more digits means a higher cost, and a higher cost can mean fewer sales of a piece of equipment. On the other hand, the greater the number of digits, the greater the accuracy.

We have said this before, but it is important enough to repeat: it is a good idea to build some of the circuits described in this chapter. Doing so will help you get some good "hands-on" experience to go along with your study. The pin numbers and circuit connections for the circuits are given on the drawings.

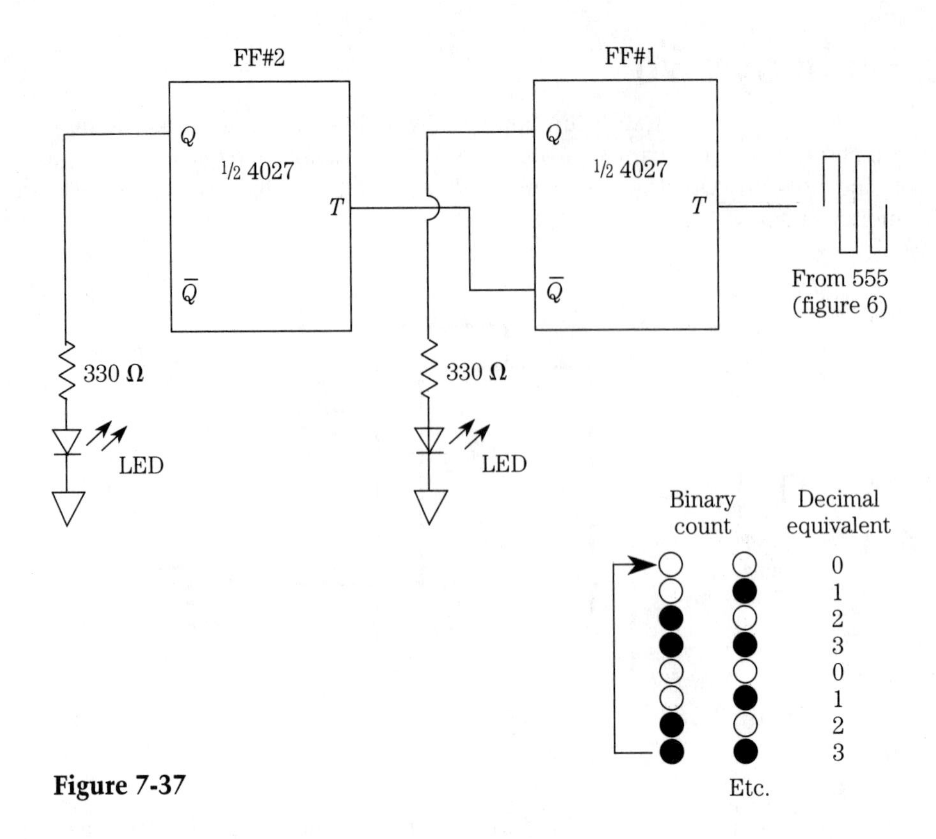

Figure 7-37

Here is a brief summary of the different sections in the counting system of Fig. 7-37:

Clock

This part of the system causes the count to go from digit to digit. Each square wave causes the binary readout to increase by one digit.

Units counter

This part of the system controls the units part of the readout. Note that there are four flip-flops in this section. Four flip-flops are needed to count 0 through 9. The units counter gives the least significant bit. For example, in binary 10, the zero is the least significant bit (LSB). The LSB is the bit on the right-hand side of a number.

Example What is the LSB of decimal number 7241?
Answer 1; the most significant bit (MSB) is 7.

Tens counter

This part of the system controls the tens part of the readout, and is identical in construction to the units counter. However, the input signal comes from the units counter rather than from the clock. Each time the units counter goes from 9 back to 0, a pulse is supplied to the tens counter to increase the tens digit by one count.

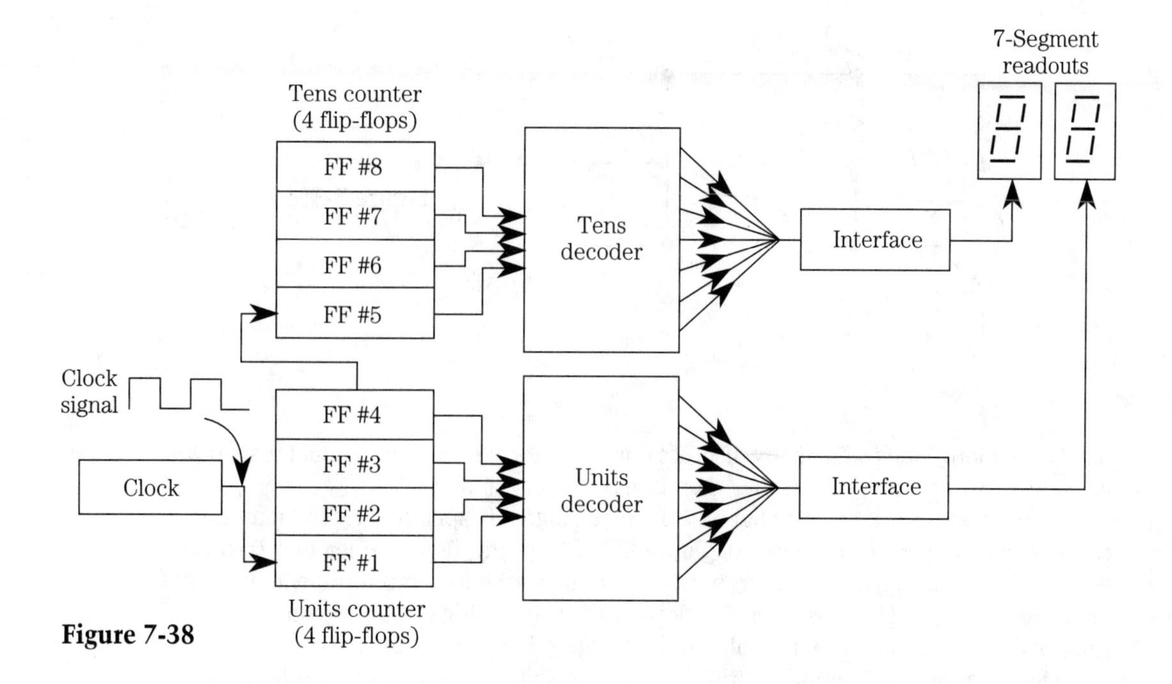

Figure 7-38

Example The units digit changes from 9 to 0, and the tens counter changes from 0 to 1 (Fig. 7-38).

Units decoder

The flip-flops in the units counter actually count by binary numbers. The binary count must be converted to a decimal count in order to get a decimal readout. That is the reason for the units decoder.

Tens decoder

Converts the tens counter binary output to a tens decimal readout.

Interface

In many cases, the units and tens decoders cannot supply enough voltage or power to operate the digital readout. The interface circuit acts like an amplifier in such cases.

The readout might require a different voltage polarity than the output of the decoder. This is another reason for an interface.

Because the clock is such an important part of the digital system, it is necessary to be able to recognize the various types of clock circuits. Important types are given in this section.

Discrete

Clock signals can be square waves, rectangular waves, or short duration pulses. Multivibrators that use *discrete* components, such as the one in Fig. 7-39, can be

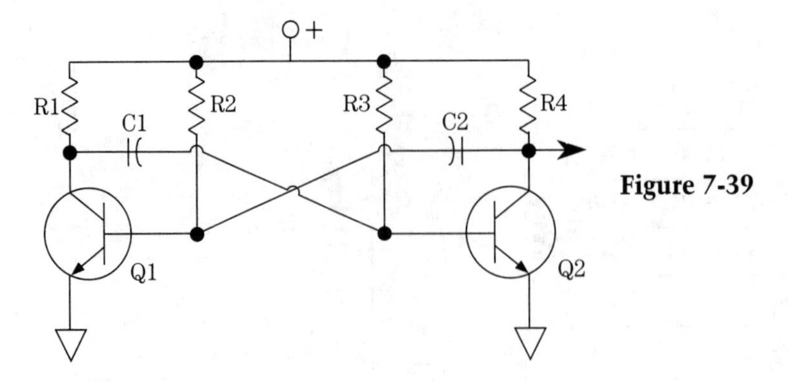

Figure 7-39

used for making clock signals. By discrete components, we mean individual resistors, capacitors, transistors, etc., rather than integrated circuits.

In many applications, the shape of the clock signal is specified by the manufacturer. For example, in order to toggle a 4027 CMOS flip-flop, the leading (rising) edge of the clock signal must go from the minimum to the maximum value in five microseconds or less. For most TTL flip-flops (FF), the trailing edge of the clock pulse toggles the FF. In that case, the fall (trailing edge) shape is specified.

The operation of the multivibrator can be explained in terms of switching. One transistor, for example Q1, conducts and switches Q2 off. Capacitor C1 delivers the switch-off signal to the base of Q2. Transistor Q1 conducts until C1 discharges. Then, Q2 conducts and switches Q1 off through C2. Transistor Q2 conducts until C2 discharges, then Q1 conducts and again turns Q2 off.

Resistors R1 and R4 are load resistors for Q1 and Q2. Resistors R2 and R3 provide forward bias for Q1 and Q2.

Schmitt trigger

A sine wave can be converted to a square wave by using a *Schmitt trigger*. The square wave or pulse can be used as a clock signal (Fig. 7-40).

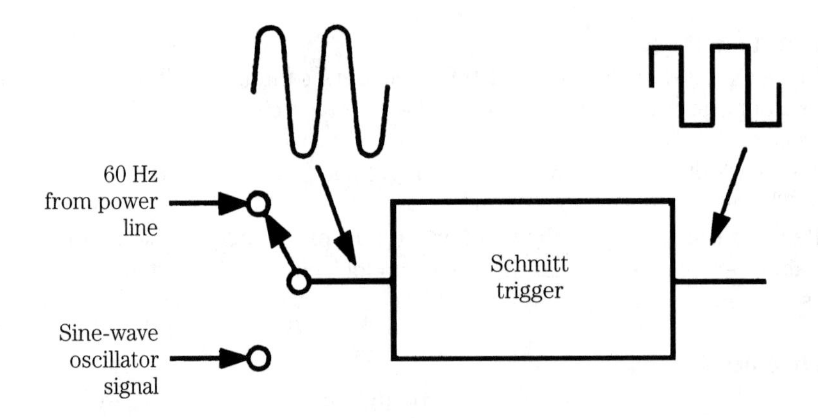

Figure 7-40

A Schmitt trigger has only two possible output levels: *high* and *low*. When the input voltage is below a certain level, the output is low. When the output voltage rises above that certain level, the output switches immediately to high. The switching level depends upon the design of the Schmitt trigger.

The symbols shown in Fig. 7-40 are used in integrated circuit logic systems. The NAND symbol is often used with TTL systems, and the NOR symbol is often used with CMOS circuitry.

As shown in Fig. 7-40, two types of input signals can be used in Schmitt trigger clocks for counting systems. Both inputs are sine waves; any sine-wave oscillator can be used. If it is important to have a very accurate frequency, then it is customary to use a custom-controlled oscillator or the 60-Hz power line frequency.

555 type

The 555 timer of Fig. 7-41 is a very popular IC and is a convenient way to get a low-frequency clock signal. The output waveform of the stable circuit in Fig. 7-41 is either a square wave or rectangular wave, depending upon the values chosen for R1 and R2. The values shown in the circuit produce a pulse with a convenient low frequency for experiment 21 counting circuits described in this monograph. Choose a value of capacitance between 0.1 to 0.3 μF to give a visible pulse output as observed with the LED. To slow the pulse down, add capacitors in parallel with C in Fig. 7-41.

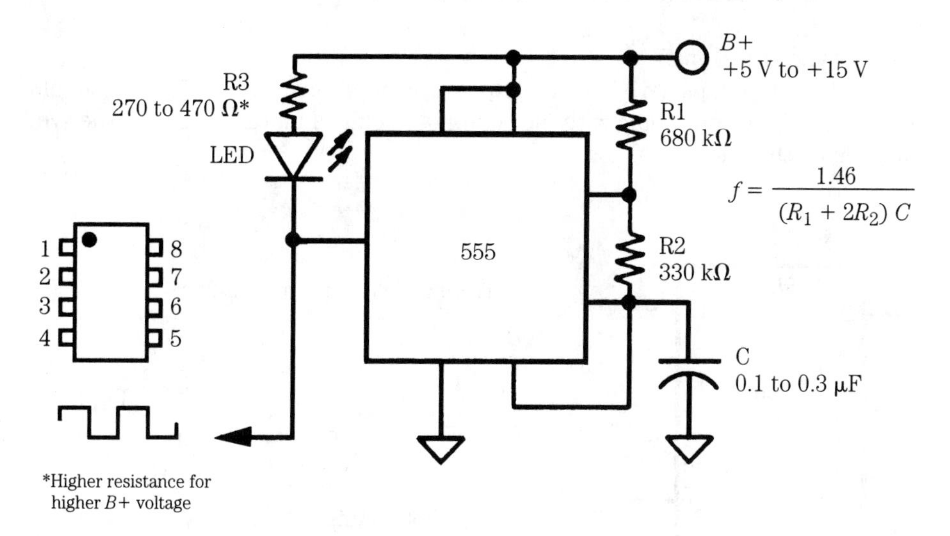

Figure 7-41

Clocks made with logic gates

Figure 7-42 shows an example of a clock made with integrated logic circuits. The gates are used for switching and the RC circuits determine the oscillator frequency. Adjustments for frequency and duty cycle are included.

Duty cycle = On time/Total time for 1 cycle

There are a number of circuits where it is desirable to have a crystal-controlled clock circuit. Examples are digital clocks and digital stop watches. The precise fre-

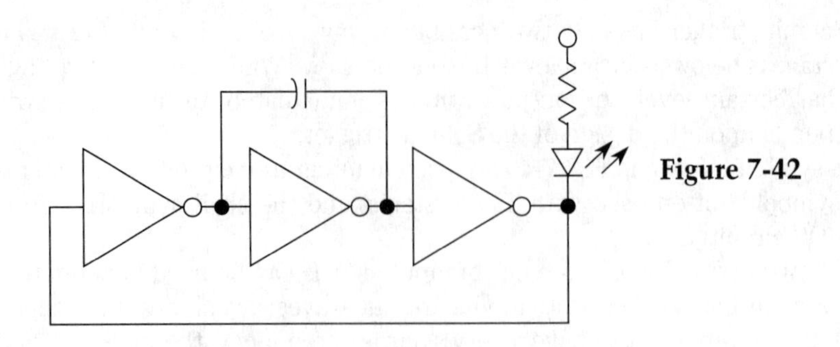

Figure 7-42

quency control is necessary to keep exact time intervals in the display that shows one-second steps. The CMOS crystal-controlled circuit in Fig. 7-43 is the simplest form of clock with an exact frequency.

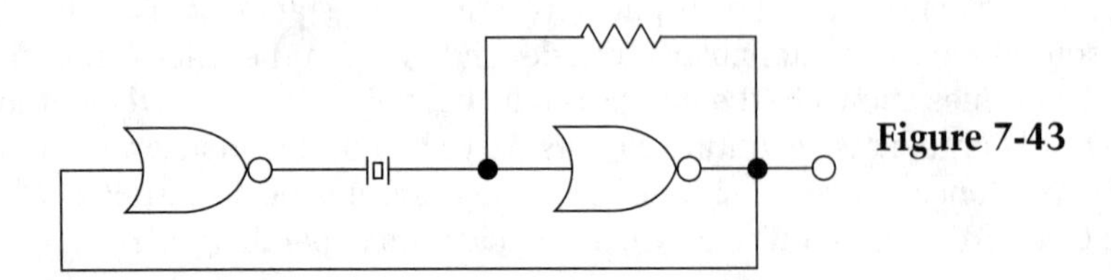

Figure 7-43

D-type flip-flops

R-S and J-K flip-flops have been reviewed. One additional type of flip-flop should be reviewed before continuing with the counting circuits. Figure 7-44 shows the symbol and the truth table.

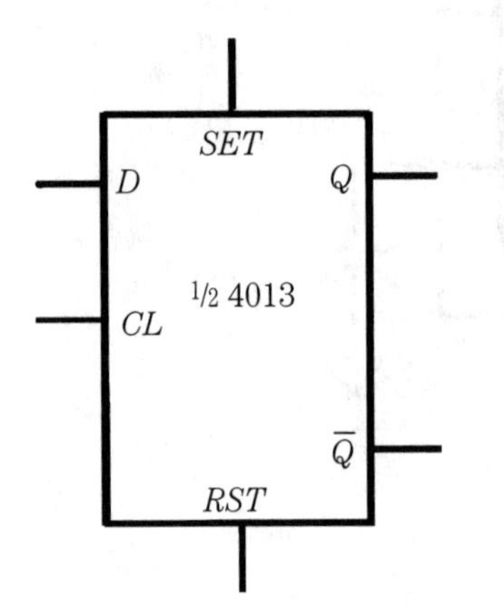

Truth table for 9D-type flip-flop

D	*CL*	*Q*	$\overline{Q}$	*SET*	*RST*
1	↑	1	0	0	0
0	↑	0	1	0	0
X	X	1	0	1	0
X	X	0	1	0	1

The symbols in this truth table have the same meanings as those in Fig. 7-17.

Figure 7-44

The divide-by-four circuit

If a second flip-flop is added, as shown in Fig. 7-45, it will divide the output of flip-flop #1 by two again. In other words, it will divide the clock frequency by four. The LEDs connected in Fig. 7-45 will actually display a binary count.

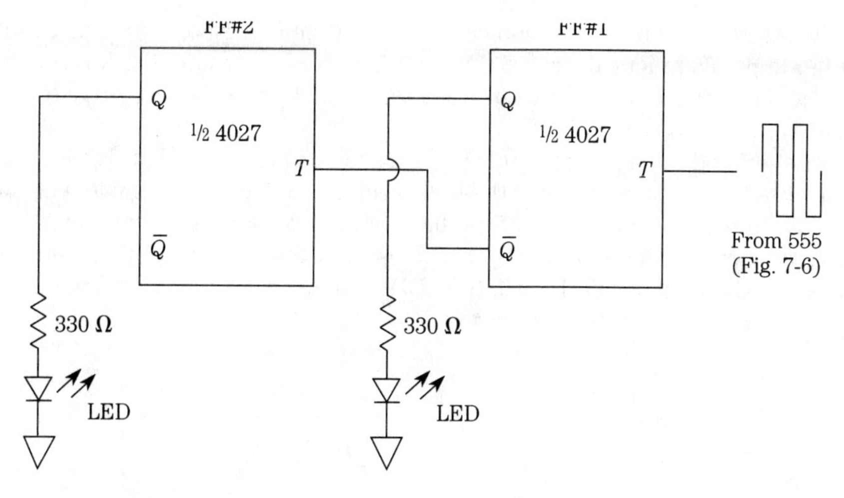

Figure 7-45

If another flip-flop is added to the binary counter of Fig. 7-45, the count will go to 8. Four toggled flip-flops will count to 16.

Control of flip-flops

In the circuit of Fig. 7-46, the K terminal of the first flip-flop (FF #1) can be switched from logic 1 to logic 0. When K is switched to logic 0 the counting stops, even though the clock is still running.

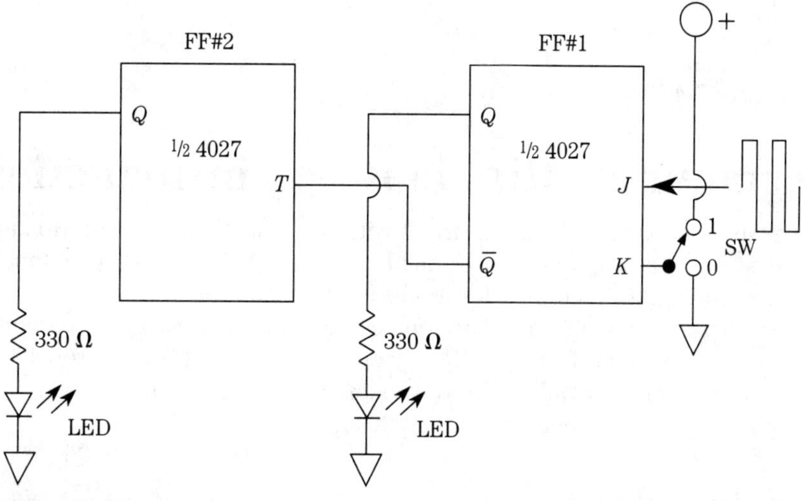

Figure 7-46

The reason the count stops is because the first flip-flop cannot toggle unless both J and K are at a logic 1 level. Because the second flip-flop gets its signal from FF #1, it cannot change because FF #1 cannot change.

Therefore, to stop a count you can switch either J, or K, or both J and K of the first flip-flop. In a practical circuit, the manufacturer can ground the K terminals on all of the flip-flops in a counter to stop the clock. That prevents accidental triggering because of noise signals.

Another method of stopping the count is shown in Fig. 7-47. In this case, you reset the flip-flop by switching all of the RST terminals to logic 1. Whenever you switch the RST terminals to logic 1, all of the flip-flops go to a low condition and all of the LEDs stop glowing. In other words, the LEDs show binary number 0. As long as the RST terminals are at a logic 1 level, the LEDs will display a binary number 0.

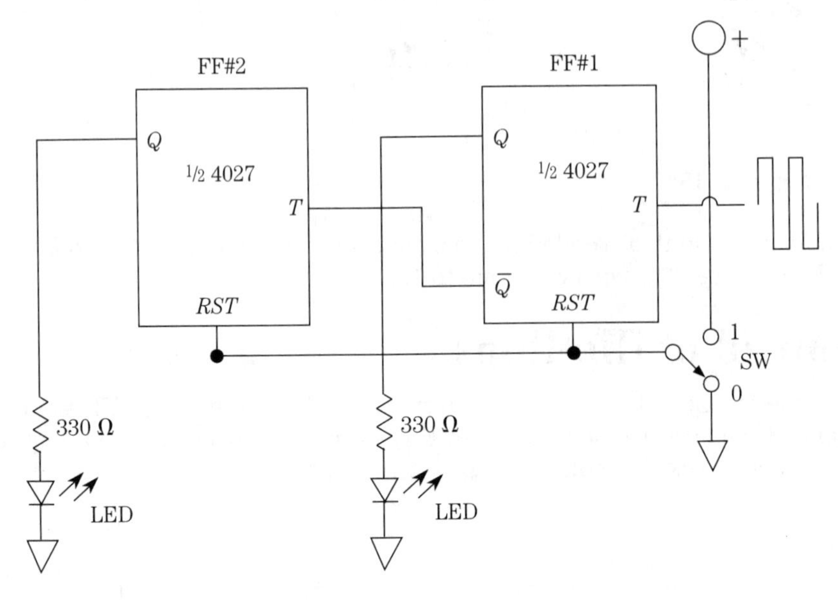

Figure 7-47

Programming flip-flops—an introduction

It is an easy matter to make a circuit with flip-flops that will count to any number and stop. Also, a circuit can be made to count to any desired number, then restart the count at 0. Counting down is as easy as counting up.

Both the units counter and the tens counter in Fig. 7-47 are programmed to count 0 through 9, then start over. Each time the units counter goes from 9 to 0, the tens counter receives a pulse that advances it one count. The result is that the counter goes from 00 to 99 then starts at 00 again.

The decoder

The third section in the basic counting system of Fig. 7-47 is comprised of two *decoders*. The output of the flip-flop counter is a binary-coded decimal value. Generally, the code is as follows:

Output of units counter (*Binary code*)				Decimal *Value*	Output of units counter (*Binary code*)				Decimal *Value*
#4	#3	#2	#1		#4	#3	#2	#1	
(flip-flop number)					(flip-flop number)				
0	0	0	0	0	0	1	0	1	5
0	0	0	1	1	0	1	1	0	6
0	0	1	0	2	0	1	1	1	7
0	0	1	1	3	1	0	0	0	8
0	1	0	0	4	1	0	0	1	9

It is more convenient for us to read decimal numbers, so the binary code is converted into a decimal readout. This can be accomplished in two different ways (Fig. 7-48).

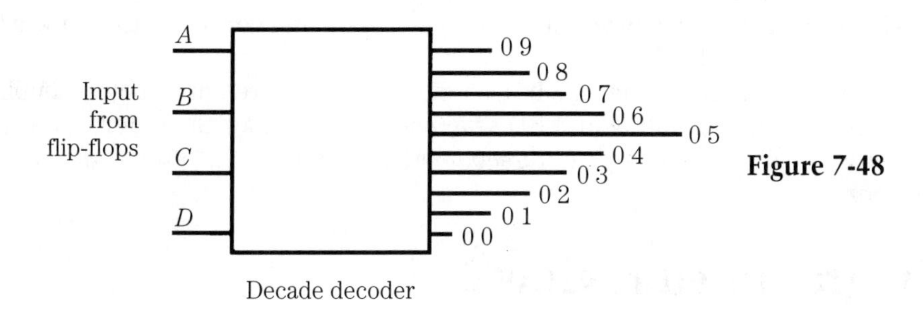

Figure 7-48

One type of decoder has four signal input terminals for A, B, C, and D. The connections to the decoder are as follows:

FF #1 goes to A
FF #2 goes to B
FF #3 goes to C
FF #4 goes to D

The decoder has 10 individual output terminals representing decimal numbers 0 through 9. As an example, if the input signal code shows binary 0111 (ABCD) there will be a logic 1 at decimal terminal #7.

A second kind of decoder changes the four binary code signals to seven different output combinations for a seven-segment display. Figure 7-49 shows how the seven-segment display shows decimal numbers 0 through 9. This type of display is very popular for clocks, stop watches, and digital tuner displays.

Seven-segment displays can use incandescent lamps, light-emitting diodes (LEDs), or liquid-crystal displays (LCDs). A *driver* might be required to deliver enough power to operate the display if incandescent lamps are used because most decoders cannot supply the high currents required by the lamp filaments.

Light-emitting diodes have a light output that is visible in the dark. Liquid-crystal displays usually operate with reflected light, but require very low power for their operation. Backlighting can be supplied to liquid-crystal displays to show in the dark.

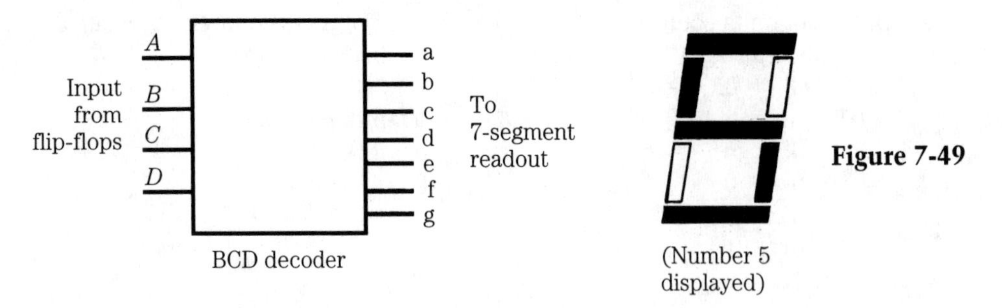

Figure 7-49

The interface

Some displays require a higher power or higher voltage than can be delivered by the decoders. Manufacturers have made interface ICs available that can take the relatively low-voltage, low-power output of the decoder and convert it to whatever is required for the display.

A CMOS or TTL decoder might also incorporate the required power amplification needed for operating an LED seven-segment display. Such devices would properly be called *seven-segment decoder/driver*, but that is often shortened to *decoder*.

Programmed review

Start with Block 1. Pick the answer you believe is correct. Go to the next block and check your answer. All answers are in italics. There is only one choice for each block. There is some material in this section that was not covered in the chapter.

Block 1

What is the international VHF digital selective calling channel?

A. 2182 kHz. C. 156.525 MHz.
B. 156.35 MHz. D. 500 kHz.

Block 2

The correct answer is C.

Here is your next question: Which VHF channel is used only for digital selective calling?

A. Channel 70. C. Channel 22A.
B. Channel 16. D. Channel 6.

Block 3

The correct answer is A.

Here is your next question: What advantage does a logic probe have over a voltmeter for monitoring logic states in a circuit?

A. A logic probe has fewer leads to connect to a circuit than a voltmeter.
B. A logic probe can be used to test analog and digital circuits.
C. A logic probe can be powered by commercial ac lines.
D. A logic probe is smaller and shows a simplified readout.

Block 4

The correct answer is D.

Here is your next question: What piece of test equipment can be used to directly indicate high and low logic states?

A. A galvanometer. C. A logic probe.
B. An electroscope. D. A Wheatstone bridge.

Block 5

The correct answer is C.

Here is your next question: What is a logic probe used to indicate?

A. A short-circuit fault in a digital-logic circuit.
B. An open-circuit failure in a digital-logic circuit.
C. A high-impedance ground loop.
D. High and low logic states in a digital-logic circuit.

Block 6

The correct answer is D.

Here is your next question: What piece of test equipment besides an oscilloscope can be used to indicate pulse conditions in a digital-logic circuit?

A. A logic probe. C. An electroscope.
B. A galvanometer. D. A Wheatstone bridge.

Block 7

The correct answer is A.

Here is your next question: What do the initials TTL stand for?

A. Resistor-transistor logic. C. Diode-transistor logic.
B. Transistor-transistor logic. D. Emitter-coupled logic.

Block 8

The correct answer is B.

Here is your next question: What is the recommended power supply voltage for TTL series integrated circuits?

A. 12.00 V. C. 5.00 V.
B. 50.00 V. D. 13.60 V.

Block 9

The correct answer is C.

Here is your next question: What logic state do the inputs of a TTL device assume if they are left open?

A. A high logic state.
B. A low logic state.
C. The device becomes randomized and will not provide consistent high or low logic states.
D. Open inputs on a TTL device are ignored.

Block 10

The correct answer is A.

Here is your next question: What level of input voltage is high in a TTL device operating with a 5 V power supply?

A. 2.0 to 5.5 V. C. 1.0 to 1.5 V.
B. 1.5 to 3.0 V. D. –5.0 to –2.0 V.

Block 11

The correct answer is A.

Here is your next question: What level of input voltage is low in a TTL device operating with a 5 V power supply?

A. –0.20 to –5.5 V. C. –0.6 to 0.8 V.
B. 2.0 to 5.5 V. D. –0.8 to 0.4 V.

Block 12

The correct answer is C.

Here is your next question: Why do circuits containing TTL devices have several bypass capacitors per printed circuit board?

A. To prevent RFI to receivers.
B. To keep the switching noise within the circuit, thus eliminating RFI.
C. To filter out switching harmonics.
D. To prevent switching transients from appearing on the supply line.

Block 13

The correct answer is D.

Here is your next question: What is a CMOS IC?

A. A chip with only P-channel transistors.
B. A chip with P-channel and N-channel transistors.
C. A chip with only N-channel transistors.
D. A chip with only bipolar transistors.

Block 14

The correct answer is B.

Here is your next question: What is one major advantage of CMOS over other devices?

A. Small size. C. Low cost.
B. Low current consumption. D. Ease of circuit design.

Block 15

The correct answer is B.

Here is your next question: Why do CMOS digital integrated circuits have high immunity to noise on the input signal or power supply?

A. Larger bypass capacitors are used in CMOS circuit design.
B. The input switching threshold is about two times the power supply voltage.
C. The input switching threshold is about one-half the power supply voltage.
D. Input signals are stronger.

Block 16

The correct answer is C.

Here is your next question: What is a flip-flop circuit?

A. A binary sequential logic element with one stable state.
B. A binary sequential logic element with eight stable states.
C. A binary sequential logic element with four stable states.
D. A binary sequential logic element with two stable states.

Block 17

The correct answer is D.

Here is your next question: How many bits of information can be stored in a single flip-flop circuit?

A. 1. C. 3.
B. 2. D. 4.

Block 18

The correct answer is A.

Here is your next question: What is a bistable multivibrator circuit?

A. An AND gate. C. A flip-flop.
B. An OR gate. D. A clock.

Block 19

The correct answer is C.

Here is your next question: How many output changes are obtained for every two trigger pulses applied to the input of a bistable T flip-flop circuit?

A. No output level changes. C. Two output level changes.
B. One output level change. D. Four output level changes.

Block 20

The correct answer is C.

Here is your next question: The frequency of an ac signal can be divided electronically by what type of digital circuit?

A. A free-running multivibrator. C. A bistable multivibrator.
B. An OR gate. D. An astable multivibrator.

Block 21

The correct answer is C.

Here is your next question: What type of digital IC is also known as a latch?

A. A decade counter. C. A flip-flop.
B. An OR gate. D. An op amp.

Block 22

The correct answer is C.

Here is your next question: How many flip-flops are required to divide a signal frequency by 4?

A. 1. C. 4.
B. 2. D. 8.

Block 23

The correct answer is B.

Here is your next question: What is an astable multivibrator?

A. A circuit that alternates between two stable states.
B. A circuit that alternates between a stable state and an unstable state.
C. A circuit set to block either a 0 pulse or a 1 pulse and pass the other.
D. A circuit that alternates between two unstable states.

Block 24

The correct answer is D.

Here is your next question: What is a monostable multivibrator?

A. A circuit that can be switched momentarily to the opposite binary state and then returns after a set time to its original state.

B. A "clock" circuit that produces a continuous square wave oscillating between 1 and 0.
C. A circuit designed to store one bit of data in either the 0 or the 1 configuration.
D. A circuit that maintains a constant output voltage, regardless of variations in the input voltage.

Block 25

The correct answer is A.

Here is your next question: What is an AND gate?

A. A circuit that produces a logic 1 at its output only if all inputs are logic 1.
B. A circuit that produces a logic 0 at its output only if all inputs are logic 1.
C. A circuit that produces a logic 1 at its output if only one input is a logic 1.
D. A circuit that produces a logic 1 at its output if all inputs are logic 0.

Block 26

The correct answer is A.

Here is your next question: What is a NAND gate?

A. A circuit that produces a logic 0 at its output only when all inputs are logic 0.
B. A circuit that produces a logic 1 at its output only when all inputs are logic 1.
C. A circuit that produces a logic 0 at its output if some but not all of its inputs are logic 1.
D. A circuit that produces a logic 0 at its output only when all inputs are logic 1.

Block 27

The correct answer is D.

Here is your next question: What is an OR gate?

A. A circuit that produces a logic 1 at its output if any input is logic 1.
B. A circuit that produces a logic 0 at its output if any input is logic 1.
C. A circuit that produces a logic 0 at its output if all inputs are logic 1.
D. A circuit that produces a logic 1 at its output if all inputs are logic 0.

Block 28

The correct answer is A.

Here is your next question: What is a NOR gate?

A. A circuit that produces a logic 0 at its output only if all input are logic 0.
B. A circuit that produces a logic 1 at its output only if all inputs are logic 1.
C. A circuit that produces a logic 0 at its output if any or all inputs are logic 1 .
D. A circuit that produces a logic 1 at its output if some but not all of its inputs are logic 1.

Block 29

The correct answer is C.

Here is your next question: What is a NOT gate?

A. A circuit that produces a logic 0 at its output when the input is logic 1 and vice versa.
B. A circuit that does not allow data transmission when its input is high.
C. A circuit that allows data transmission only when its input is high.
D. A circuit that produces a logic 1 at its output when the input is logic 1 and vice versa.

Block 30

The correct answer is A.

Here is your next question: What is a truth table?

A. A table of logic symbols that indicate the high logic states of an op amp.
B. A diagram showing logic states when the digital device's output is true.
C. A list of input combinations and their corresponding outputs that characterizes a digital device's function.
D. A table of logic symbols that indicates the low logic states of an op amp.

Block 31

The correct answer is C.

Here is your next question: In a positive-logic circuit, what level is used to represent a logic 1?

A. A low level.
B. A positive transition level.
C. A negative transition level.
D. A high level.

Block 32

The correct answer is D.

Here is your next question: In a positive-logic circuit, what level is used to represent a logic 0?

A. A low level.
B. A positive transition level.
C. A negative transition level.
D. A high level.

Block 33

The correct answer is A.

Here is your next question: In a negative-logic circuit, what level is used to represent a logic 1?

A. A low level.
B. A positive transition level.
C. A negative transition level.
D. A high level.

Block 34

The correct answer is A.

Here is your next question: In a negative-logic circuit, what level is used to represent a logic 0?

A. A low level. C. A negative transition level.
B. A positive transition level. D. A high level.

Block 35

The correct answer is D.

Here is your next question: How many states does a decade counter digital IC have?

A. 6. C. 15.
B. 10. D. 20.

Block 36

The correct answer is B.

Here is your next question: What is the function of a decade counter digital IC?

A. Decode a decimal number for display on a seven-segment LED display.
B. Produce one output pulse for every 10 input pulses.
C. Produce 10 output pulses for every input pulse.
D. Add two decimal numbers.

Block 37

The correct answer is B.

Quiz

1. Which of the following number could not be an octal number?

   A. 11011. B. 84116.

2. Which of the following logic families must be operated with a +5-V power supply?

   A. Emitter-coupled logic (ECL).
   B. Transistor-transistor logic (TTL).
   C. Complementary MOS (CMOS).
   D. Diode-transistor logic (DTL).

3. Given: The following truth table -

A	B	L
0	0	0
0	1	1
1	0	1
1	1	0

   It is a truth table for ________________.

4. Give the binary number that follows this binary number:

   11001.

5. Which of the following J-K flip-flops normally toggle on the leading edge of the clock pulse?

   A. CMOS. B. TTL.

6. What is the next hexadecimal number in the following count:

   FC.

7. A logic circuit that is not clocked is called a ______ circuit and one that is clocked is called a ________ circuit.

8. What determines whether a manufacturer specifies the rise time or decay time of a clock signal?

9. Why are counters sometimes shown with the clock signal on the right side of the schematic?

10. Why are resistors always connected in series with LEDs?

Quiz answers

1. B. (There is no symbol 8 in octal numbers. The range of symbols is 0 through 7.)

2. B.

3. EXCLUSIVE OR.

4. 11010.

5. A. (This is just a general rule. There are CMOS flip-flops that toggle on the trailing edge.)

6. FD.

7. Static - dynamic.

8. Whether their circuit is rise time or decay time.

9. So the Least Significant Bit (LSB) is on the right and the Most Significant Bit is on the left.

10. To limit the current through the LED.

8
CHAPTER

SBE and other types of certification

Some organizations offer certification in specialized areas of electronics. We believe this book will offer some assistance in passing their tests. At the same time, please understand that we have concentrated on reviewing material for tests that cover a broad range of subjects. The more specialized subjects are outside the range of material covered in this book.

A few organizations offered to supply us with information when we started the books, but later changed their minds. In another case, the material was supposed to be supplied, but did not arrive in time.

This chapter describes the certifications offered, the objectives of the organizations that offer them, and the places you can write for more information.

Some of the organizations, such as SBE, offer extensive training materials that give excellent preparation for passing their tests.

Also listed in this chapter are organizations that offer benefits to technicians holding licenses and/or certifications (for example, W5YI and ARRL). We urge technicians to explore their member benefits, especially their technical publications.

Society of Broadcast Engineers, Inc.

The Society of Broadcast Engineers Certification Program is a service of the SBE, contributing to the advancement of broadcast engineering for the benefit of the entire broadcast industry. Surveys of technicians and engineers employed in the broadcast industry have shown that those who are certified by the SBE earn salaries approximately 15% greater than those without that certification.

Membership in the SBE is not a requirement of the Certification Program. Nonmembers who become certified may receive one year of free membership in the SBE by filling out the membership application.

Complete information on certification is available from the SBE. The address of the SBE is:

Society of Broadcast Engineers, Inc.
Attention Certification Secretary
8445 Keystone Crossing
Suite 140
Indianapolis, IN 46240
(317) 253-1640

Some information on the SBE Program of Certification is provided here, but complete details should be obtained from the society.

There are four grades of SBE certification:

- Broadcast Technologist
- Broadcast Engineer
- Senior Broadcast Engineer
- Professional Broadcast Engineer

The instruction in this book is intended mainly to assist you in preparing for the Broadcast Technologist certification.

Study guides are available from the SBE on preparing for an SBE examination. The study guides are available on computer disk. Computer disks are either 5¼" or 3½", IBM or Macintosh.

The Broadcast Technologist study guide contains instructions and practice questions and answers which are similar to the actual examination.

The Broadcast Technologist certification can be obtained in one of four ways:

- By achieving a passing grade in the proficiency examination. There is no experience requirement to be eligible for the examination.
- By holders of an FCC lifetime General Class license with either two years of continuous satisfactory service in broadcast or related engineering prior to the date of application, or a total of three out of the last five years of satisfactory service in broadcasting or related engineering. (The service record must show no record of discharge for unsatisfactory service or unlawful activity.)
- By holders of a valid FCC Amateur Extra Class license who meet the above service requirement.
- By holders of a valid license equivalent to one of the FCC licenses listed above who meet the service requirement. (*Note:* This provision is to recognize SBE members who do not operate under the rules and regulation of the FCC.)

The more advanced grades of SBE certification, those beyond Broadcast Technologist, require passing an additional written exam plus five or more years of broadcast engineering experience. There are other courses and study guides available which are intended to help you prepare for the higher-level certification exams. Information on those can be obtained from the SBE.

A list of subjects follows that covers the range of subjects you might expect to be covered in the Broadcast Technologist test. The list gives a good idea of what you should know or study before taking the test. Actually, all communications technicians should be knowledgeable in nearly all of the subjects listed. The SBE Broadcast Technologist Study Guide covers the range of subjects in a question-and-answer format, with correct answers identified and many of the answers explained in detail.

Taking the SBE exam

SBE exams are given at scheduled times at any one of many local SBE chapters throughout the United States. Exams are also given at regional and national SBE conventions. Meetings of SBE chapters are held approximately monthly and nonmembers are welcome to attend the technical sessions.

Request further information from the SBE national headquarters on the exams, and where and when local chapters meet.

Topics covered in the SBE Broadcast Technologist Study Guide

FCC Includes questions for both radio and TV relating to:

- Rules and Regulations (FCC part 73)
- Frequency coordination
- EBS (Emergency Broadcast Service)
- Towers (FAA Rules and Regulations)

General Includes questions relating to:

- Basic electronics
- ac theory and wiring practices
- Troubleshooting techniques
- Test equipment
- Digital theory and techniques

Audio Includes questions relating to:

- Radio consoles
- Microphones
- TV audio consoles
- Time code
- Stereo audio
- Digital/dc recording & playback
- Audio tape players/recorders
- Audio head positioning/wear patterns
- Audio levels, standards, and practices
- Wiring practices
- Intercom systems
- Audio test equipment
- Computer editing

Video Includes questions relating to:

- Cameras
- Telecines
- Videotape recorders/players (all formats)
- Routing systems
- Video switchers
- Disk based recorders/players
- Character generators
- Electronic special effects
- Waveform analysis
- Video levels, standards, and practices
- Cable and wiring practices
- Terminal equipment
- Sync generators
- Waveform monitors
- Video test equipment

RF Includes questions for both radio and TV relating to:
- Antenna systems
- STL systems
- Receivers
- Test equipment

AM/FM Includes questions relating to:
- AM/FM transmitters
- Antenna systems
- STL systems
- Receivers
- Test equipment

TV RF Includes questions relating to:
- TV transmitters
- Antenna systems
- STL systems
- Receivers
- Test equipment

Satellite (Sat) Includes questions relating to:
- SNG
- Earth stations
- Fixed uplink transmitters
- SCPC radio
- Data
- Subcarriers
- Multiplex pilots
- Test equipment

Microwave (Micro.) Includes questions relating to:
- ENG
- STLs
- Inter-city systems
- Path specifications
- Data
- Subcarriers
- Test equipment

Safety Includes questions relating to:
- High voltage precautions
- Radio frequency radiation exposure guidelines
- Tower climbing safety
- PCB's
- Test equipment

ARRL

American Radio Relay League (ARRL)
225 Main St.
Newington, CT 06111
(203) 666-1541

The ARRL is the largest organization of radio amateurs in the United States. It serves as the official voice of amateur radio in dealing with government agencies.

The ARRL provides assistance and testing to those who wish to become an amateur radio operator or to upgrade a present license. ARRL publishes a monthly magazine QST for members, and has available a wide variety of literature and student material and services for the radio amateur or prospective amateur. There are many local chapters who willingly provide information and assistance to those interested.

The ARRL cooperates with the VEC (volunteer exam coordinator) service in preparing and administering amateur radio licensing exams. The VEC serves as the interface between you, studying to prepare for a test, and the FCC (Federal Communications Commission) in Gettysburg, PA, which issues amateur radio licenses.

NARTE

NARTE (National Association of Radio and Telecommunications Engineers)
PO Box 678
Medway, MA 02053
(508) 533-8333

The NARTE certification program for technicians consists of four levels of examinations. Each is a written test consisting of 100 written multiple-choice questions. The examination levels are Class IV, Class III, Class II, and Class I, with the Class IV being the entry level exam and Class I being the highest level technician exam. Their certifications include broadcasting and telecommunications.

NARTE also offers examinations for three classes of engineers. Further information on engineer level exams, and on other NARTE examinations or ways of becoming certified by NARTE, may be obtained from NARTE.

The practice exams in this book may be similar to many parts of the actual exams given by NARTE. But, anyone studying this book preparing for a NARTE certification exam should be aware that there may be some topics not covered, and that there may be other qualifications required. Be sure to contact NARTE for further information. Actual examinations are administered by authorized proctors at local colleges and universities located throughout the country.

At the present time, NARTE does not sell or offer practice examinations, but NARTE does recommend specific lists of study references for each level of examination.

SDA

Satellite Dealers Association (SDA)
607 North Jackson Street
Greencastle, IN 46135
(317) 653-8262

The SDA provides a CSI or Certified Installer Program for certification for satellite installer technicians. This is a journeyman certification that follows the ETA Associate certification.

The study guide called Improving TV Reception is available from SDA.

SCTE

Society of Cable TV Engineers (SCTE)
669 Exton Commons
Exton, PA 19341
(215) 363-5898

SCTE provides certification for cable TV technicians and engineers. The certification programs are called Broadband Communications Technician (BCT) and Broadband Communications Engineers (BCE).

These programs are based on professionalism, experience, and technical knowledge. These qualifications require completion of a written examination. Full information is available from SCTE.

Marine Radio Certificates

Marine Radio Certification
The National Marine Electronics Association (NMEA)
67 Boston Post Road
Waterford, CT 06385
(205) 473-1793

There are three grades of Certified Marine Electronics Technicians (CMETs): basic CMET, advanced ACMET, and senior SCMET. There is a CMET Handbook. All applicants for CMET must pass written examination among other things, including being sponsored by an NMEA member. For full information, write or call NMEA.

NICET

National Institute for Certification in Engineering Technologies (NICET)
1420 King Street
Alexandria, VA 22314
(703) 684-2035

NICET certification requires a considerable amount of dedication, study, and concentration. Most of their programs are not related to electronics, but to many other specialties, such as Industrial Plant Process Piping, Transportation/Highway Maintenance, Fire Protection/Fire Alarms, etc. Some of those that are related to electronics include Industrial Instrumentation and Electrical/Electronics Engineering Technology.

Further information can be obtained by contacting NICET.

NABER

NABER (National Association of Business and Educational Radio)
1501 Duke St.
Alexandria, VA 22314
(800) 759-0300

The NABER Certification Exam is a written test consisting of 150 multiple choice questions. The examination covers four primary areas: two-way radio and systems, troubleshooting, FCC Rules Part 90, soldering and hand-tool use, and equipment installation.

W5YI

W5YI Group, Inc.
PO Box 565101
Dallas, TX 753566
(817) 461-6443

The W5YI Group provides many services to the field of amateur radio. The W5YI Report is published twice a month and contains various news items on amateur radio and emerging electronics. It is the nations oldest ham radio newsletter. The W5YI Group also has available amateur radio license preparation materials and many items that assist the amateur operator. Write or call W5YI for further information.

9
CHAPTER

Element 1—Laws and operating practices

If you are planning to take an FCC test, consider this to be required material. Also, reread the material in chapter 6 before taking any FCC test.

The questions in this chapter are also given in the book, *The TAB Source Book for Communications Licensing and Certification Examinations*, however the questions and answers have been scrambled. That gives you a new experience in answering questions on this subject. You should review this material periodically.

1. 3:00 PM Central Standard Time is:

   A. 100 UTC. B. 2100 UTC. C. 1800 UTC. D. 0300 UTC.

2. 10 statute miles per hour equals how many knots?

   A. 11.5. B. 8.7. C. 5. D. 3.

3. Which of the following is an acceptable method of solder removal from holes in a printed board?

   A. Compressed air. C. Soldering iron and a suction device.
   B. Toothpick. D. Power drill.

4. 100 statute miles equals how many nautical miles?

   A. 87. B. 108. C. 173. D. 13.

5. 6:00 PM PST is equal to what time in UTC?

   A. 0200. B. 1800. C. 2300. D. 1300.

6. An auto alarm signal consists of two sine-wave audio tones transmitted alternately at what frequencies?

   A. 121.5 and 243 MHz. C. 1300 and 220 kHz.
   B. 500 and 100 kHz. D. None of the above.

7. What is not a good soldering practice in electronic circuits?

   A. Use adequate heat.
   B. Clean parts sufficiently.
   C. Prevent corrosion by never using flux.
   D. Be certain parts do not move while solder is cooling.

8. One statute mile equals how many nautical miles?

   A. 3.8. B. 1.5. C. 0.87. D. 0.7.

9. When soldering electronic circuits be sure to:

   A. use sufficient heat. C. heat wires until sweating begins.
   B. use maximum heat. D. use minimum solder.

10. 2.3 statute miles equals how many nautical miles?

    A. 2. B. 1.5. C. 1.73. D. 1.

11. What is the purpose of flux?

    A. Removes oxides from surfaces to be joined.
    B. Prevents oxidation during soldering.
    C. Acid cleans printed circuit connections.
    D. Both A and B.

12. Which of these will be useful for insulation at UHF?

    A. Rubber. B. Mica. C. Wax-impregnated paper. D. Lead.

13. The condition of a storage battery is determined with a:

    A. hygrometer. B. manometer. C. FET. D. hydrometer.

14. The ILS localizer measures what deviation of an aircraft?

    A. Horizontal. C. Ground speed.
    B. Vertical. D. Distance between aircraft.

15. What is the approximate transmission time for a facsimile picture transmitted by a radio station?

    A. Approximately 6 minutes per frame at 240 lines per minute.
    B. Approximately 3.3 minutes per frame at 240 lines per minute.
    C. Approximately 6 seconds per frame at 240 lines per minute.
    D. 1.60 seconds per frame at 240 lines per minute.

16. What is the term for the transmission of printed pictures by radio for the purpose of a permanent display?

    A. Television. B. Facsimile. C. Xerography. D. ACSSB.

17. In facsimile, how are variations in picture brightness and darkness converted into voltage variations?

A. With an LED.
B. With a Hall-effect transistor.
C. With a photodetector.
D. With an optoisolator.

18. What is an ascending pass for a satellite?

A. A pass from west to east.
B. A pass from east to west.
C. A pass from south to north.
D. A pass from north to south.

19. What is a descending pass for a satellite?

A. A pass from north to south.
B. A pass from west to east.
C. A pass from east to west.
D. A pass from south to north.

20. Waveguides are not utilized at VHF or UHF frequencies because:

A. characteristic impedance.
B. resistance to high-frequency waves.
C. large dimensions of waveguides are not practical.
D. grounding problems.

21. 2300 UTC time is:

A. 6 AM EST. B. 10 AM EST. C. 3 PM PST. D. 2 PM CST.

22. What is the period of a satellite?

A. An orbital arc that extends from 60 degrees.
B. The point on an orbit where satellite height is minimum.
C. The amount of time it takes for a satellite to complete one orbit.
D. The time it takes a satellite to travel from perigee to apogee.

23. What is a linear transponder?

A. A repeater that passes only linear or binary signals.
B. A device that receives and retransmits signals of any mode in a certain passband.
C. An amplifier for SSB transmissions.
D. A device used to change an FM emission to an AM emission.

24. What are the two basic types of linear transponders used in satellites?

A. Inverting and noninverting.
B. Geostationary and elliptical.
C. Phase 2 and Phase 3.
D. Amplitude modulated and frequency modulated.

25. Why does the down-link frequency appear to vary by several kHz during a low-earth-orbit satellite pass?

A. The distance between the satellite and ground station is changing, causing the Kepler effect.
B. The distance between the satellite and ground station is changing, causing the Bernoulli effect.

C. The distance between the satellite and ground station is changing, causing the Boyle's law effect.
D. The distance between the satellite and ground station is changing, causing the Doppler effect.

26. Why does the received signal from a satellite stabilized by a computer-pulsed electromagnet exhibit a fairly rapid pulsed fading effect?

A. Because the satellite is rotating.
B. Because of the ionospheric absorption.
C. Because of the satellite's low orbital altitude.
D. Because of the Doppler effect.

27. What is facsimile?

A. The transmission of characters by radioteletype that form a picture when printed.
B. The transmission of still pictures by slow-scan television.
C. The transmission of video by television.
D. The transmission of printed pictures for permanent display on paper.

28. What is the modern standard scan rate for a facsimile picture transmitted by a radio station?

A. The modern standard is 240 lines per minute.
B. The modern standard is 50 lines per minute.
C. The modern standard is 150 lines per second.
D. The modern standard is 60 lines per second.

29. In the International Phonetic Alphabet, the letter E, M, and S are represented by the words:

A. Echo, Michigan, Sonar.
B. Equator, Mike, Sonar.
C. Echo, Mike, Sierra.
D. Element, Mister, Scooter.

30. What is the international radiotelephone distress call?

A. "SOS, SOS, SOS; this is;" followed by the callsign of the station (repeated 3 times).
B. "Mayday, mayday, mayday; this is;" followed by the callsign (or name, if no callsign assigned) of the mobile station in distress, spoken three times.
C. For radiotelephone use, any words or message which will attract attention may be used.
D. The alternating two-tone signal produced by the radiotelephone alarm signal generator.

31. What has most priority:

A. URGENT. B. DISTRESS. C. SAFETY. D. SECURITY.

32. When and how can class-A and B EPIRBs be tested?

A. Within the first 5 minutes of the hour; tests not to exceed 3 audible sweeps or one second, whichever is longer.
B. Within first 3 minutes of hour; tests not to exceed 30 seconds.
C. Within first 1 minute of hour, test not to exceed 1 minute.
D. At any time ship is at sea.

33. When is the Silent Period on 2182 kHz, when only emergency communications may occur?

A. One minute at the beginning of every hour and half hour.
B. At all times.
C. No designated period; silence is maintained only when a distress call is received.
D. Three minutes at the beginning of every hour and half hour.

34. What is the frequency range of UHF?

A. 0.3 to 3 kHz. B. 0.3 to 3 MHz. C. 30 to 300 kHz. D. 300 to 3000 MHz.

35. A room temperature of +30.0° Celsius is equivalent to how many degrees Fahrenheit?

A. 104. B. 83. C. 95. D. 86.

36. Who is required to make entries on a required service or maintenance log?

A. The licensed operator or a person whom he or she designates.
B. The operator responsible for the station operation or maintenance.
C. Any commercial radio operator holding at least a Restricted Radiotelephone Operator Permit.
D. The technician who actually makes the adjustments to the equipment.

37. What is a requirement of every commercial operator on duty and in charge of a transmitting system?

A. A copy of the Proof-of-Passing Certificate (PPC) must be on display at the transmitter location.
B. The original license or a photocopy must be posted or in the operator's personal possession and available for inspection.
C. The FCC Form 756 certifying the operator's qualifications must be readily available at the transmitting system site.
D. A copy of the operator's license must be supplied to the radio station's supervisor as evidence of technical qualification.

38. What is distress traffic?

A. In radiotelegraphy, SOS sent as a single character; in radiotelephony, the speaking of the word, "Mayday."
B. Health and welfare messages concerning the immediate protection of property and safety of human life.
C. Internationally recognized communications relating to emergency situations.
D. All messages relative to the immediate assistance required by a ship, aircraft or other vehicle in imminent danger.

39. What is a maritime mobile repeater station?

    A. A fixed land station used to extend the communications range of ship and coast stations.
    B. An automatic on-board radio station that facilitates the transmissions of safety communications aboard ship.
    C. A mobile radio station that links two or more public coast stations.
    D. A one-way low-power communications system used in the maneuvering of vessels.

40. What is an urgency transmission?

    A. A radio distress transmission that impacts the protection of on-board personnel.
    B. Health and welfare traffic that impacts the protection of on-board personnel.
    C. A communications alert that important personal messages must be transmitted.
    D. A communications transmission concerning the safety of a ship, aircraft or other vehicle, or of some person on board or within sight.

41. Atmospheric noise or static is not a great problem:

    A. at frequencies below 20 MHz. C. at frequencies above 1 MHz.
    B. at frequencies below 5 MHz. D. at frequencies above 30 MHz.

42. Frequencies which have substantially straight-line propagation characteristics similar to that of light waves are:

    A. frequencies below 500 kHz.
    B. frequencies between 500 kHz and 1,000 kHz.
    C. frequencies between 1,000 kHz and 3,000 kHz.
    D. frequencies above 50,000 kHz.

43. What is the standard video level between the sync tip and the whitest white at TV camera outputs and modulator inputs?

    A. 1 V peak-to-peak. B. 120 IEEE units. C. 12 Vdc. D. 5 V RMS.

44. What is the standard video level, in percent PEV, for black?

    A. 0%. B. 12.5%. C. 70%. D. 100%.

45. What is the standard video level, in percent PEV, for white?

    A. 0%. B. 12.5%. C. 70%. D. 100%.

46. What is the standard video level, in percent PEV, for blanking?

    A. 0%. B. 12.5%. C. 75%. D. 100%.

47. What is the Global Maritime Distress and Safety System (GMDSS)?

    A. An automated ship-to-shore distress alerting system using satellite and advanced terrestrial communications systems.
    B. An emergency radio service employing analog and manual safety apparatus.
    C. An association of radio officers trained in emergency procedures.
    D. The international organization charged with the safety of ocean-going vessels.

48. What authority does the Marine Radio Operator Permit confer?

    A. Grants authority to operate commercial broadcast stations and repair associated equipment.
    B. Allows the radio operator to maintain equipment in the Business Radio Service.
    C. Confers authority to operate licensed radio stations in the Aviation, Marine, and International Fixed Public Radio Services.
    D. The nontransferable right to install, operate, and maintain any type-accepted radio transmitter.

49. Which of the following persons are ineligible to be issued a commercial radio operator license?

    A. Individuals who are unable to send and receive correctly by telephone spoken messages in English.
    B. Handicapped person with uncorrected disabilities which affect their ability to perform all duties required of commercial radio operators.
    C. Foreign maritime radio operators unless they are certified by the International Maritime Organization (IMO).
    D. U. S. Military radio operators who are still on active duty.

50. If you are listening to an FM radio station at 100.6 MHz on a car radio when an airplane in the vicinity is transmitting 121.2 MHz and your car radio receives interference the possible problem could be:

    A. improper shielding in receive. C. image frequency.
    B. poor Q receiver. D. intermodulation or coupling.

51. One nautical mile is equal to how many statute miles?

    A. 1.5. B. 8.3. C. 1.73. D. 1.15.

52. 1.73 nautical miles equals how many statute miles?

    A. 2. B. 1.5. C. 1.73. D. 1.

53. Solder is:

    A. 50% lead, 50% tin. C. 60% lead, 40% tin.
    B. 40% lead, 60% tin. D. 70% lead, 30% tin.

54. What ferrite device can be used instead of a duplexer to isolate a microwave transmitter and receiver when both are connected to the same antenna?

    A. Isolator. B. Circulator. C. Magnetron. D. Simplex.

55. What type of antenna can be used to minimize the effects of spin modulation and Faraday rotation?

    A. A nonpolarized antenna. C. An isotropic antenna.
    B. A circularly polarized antenna. D. A log-periodic dipole array.

56. What is blanking in a video signal?

    A. Synchronization of the horizontal and vertical sync pulses.

B. Turning off the scanning beam while it is traveling from right to left and from bottom to top.
C. Turning off the scanning beam at the conclusion of a transmission.
D. Transmitting a black-and-white test pattern.

57. Which VHF channel is used only for digital selective calling?

A. Channel 70. B. Channel 16. C. Channel 22A. D. Channel 6.

58. VHF ship station transmitters must have the capability of reducing carrier power to:

A. 1 W. B. 10 W. C. 25 W. D. 50 W.

59. The system of substituting words for corresponding letters is called:

A. international code system. C. mnemonic system.
B. phonetic system. D. 10 codes.

60. How long should station logs be retained when there are no entries relating to distress or disaster situations?

A. For a period of three years from the date of entry, unless notified by the FCC.
B. Until authorized by the Commission in writing to destroy them.
C. Indefinitely, or until destruction is specifically authorized by the U. S. Coast Guard.
D. For a period of one year from the date of entry.

61. The auto alarm device for generating signals shall be:

A. tested monthly using a dummy load.
B. tested every three months using a dummy load.
C. tested weekly using a dummy load.
D. none of the above.

62. Licensed radiotelephone operators are not required on board ships for:

A. voluntarily equipped ship stations on domestic voyages operating on VHF channels.
B. ship radar, provided the equipment is nontunable, pulse-type Magnetron, and can be operated by means of exclusively external controls.
C. installation of a VHF transmitter in a ship station, where the work is performed by or under the immediate supervision of the licensee of the ship station.
D. any of the above.

63. A 25-MHz amplitude-modulated transmitter's actual carrier frequency is 25.00025 MHz without modulation and is 24.99950 MHz when modulated. What statement is true?

A. If the allowed frequency tolerance is 0.001% this is an illegal transmission.
B. If the allowed frequency tolerance is 0.002% this is an illegal transmission.
C. Modulation should not change carrier frequency.
D. If the authorized frequency tolerance is 0.005% for the 25-MHz band this transmitter is operating legally.

64. What safety signal call word is spoke three times, followed by the station call letters spoken three times, to announce a storm warning, danger to navigation, or special aid to navigation?

    A. PAN. B. MAYDAY. C. SECURITY. D. SAFETY.

65. When your transmission is ended and you expect no response, say:

    A. BREAK. B. OVER. C. ROGER. D. CLEAR.

66. When attempting to contact other vessels on Channel 16:

    A. limit calling to 30 seconds.
    B. if no answer is received, wait 2 minutes before calling vessel again.
    C. channel 16 is used for emergencies only.
    D. both A and B.

67. When a message has been received and will be complied with say:

    A. MAYDAY. B. OVER. C. ROGER. D. WILCO.

68. The FCC may suspend an operator license upon proof that the operator:

    A. has assisted another to obtain a license by fraudulent means.
    B. has willfully damaged transmitter equipment.
    C. has transmitted obscene language.
    D. any of the above.

69. What channel must compulsorily equipped vessels monitor at all times in the open sea?

    A. Channel 8, 156.4 MHz. C. Channel 22A, 157.1 MHz.
    B. Channel 16, 156.8 MHz. D. Channel 6, 156.3 MHz.

70. When testing is conducted on 2182 kHz or 156.8 MHz testing should not continue for more than __________ in any 5 minute period.

    A. 10 seconds. B. 1 minute. C. 2 minutes. D. none of the above.

71. Tests of survival craft radio equipment, except EPIRBs and two-way radiotelephone equipment, must be conducted:

    A. at weekly intervals while the ship is at sea.
    B. within 24 hours prior to departure when a test has not been conducted within a week of departure.
    C. both A and B.
    D. when required by the Commission.

72. Each cargo ship of the United States that is equipped with a radiotelephone station for compliance with Part II of Title III of the Communications Act shall while being navigated outside of a harbor or port keep a continuous watch on:

    A. 2182 kHz.
    B. 156.8 MHz.

C. both A and B.
D. cargo ships are exempt from radio watch regulations.

73. When may you test a radiotelephone transmitter on the air?
    A. Between midnight and 6:00 AM local time.
    B. Only when authorized by the Commission.
    C. At any time as necessary to ensure proper operation.
    D. After reducing transmitter power to 1 W.

74. What is the required daytime range of a radiotelephone station aboard a 900-ton ocean-going cargo vessel?
    A. 25 miles. B. 50 miles. C. 150 miles. D. 500 miles.

75. What do you do if the transmitter aboard your ship is operating off-frequency, overmodulating, or distorting?
    A. Reduce to low power. C. Reduce audio volume level.
    B. Stop transmitting. D. Make a notation in station operating log.

76. What is the authorized frequency for an on-board ship repeater for use with a mobile transmitter operating at 467.750 MHz?
    A. 457.525 MHz. B. 467.775 MHz. C. 467.800 MHz. D. 467.825 MHz.

77. Survival-craft EPIRBs are tested:
    A. with a manually activated test switch.
    B. with a dummy load having the equivalent impedance of the antenna affixed to the EPIRB.
    C. with radiation reduced to a level not to exceed 25 microvolts per meter.
    D. all of the above.

78. Marine transmitter should be modulated between:
    A. 75% to 100%. B. 70% to 105%. C. 85% to 100%. D. 75% to 120%.

79. What is a good practice when speaking into a microphone in a noisy location?
    A. Overmodulation. C. Increase monitor audio gain.
    B. Change phase in audio circuits. D. Shield microphone with hands.

80. When pausing briefly for station copying message to acknowledge, say:
    A. BREAK. B. OVER. C. WILCO. D. STOP.

81. Overmodulation is often caused by:
    A. turning down audio gain control. C. weather conditions.
    B. station frequency drift. D. shouting into microphone.

82. An operator or maintainer must hold a General Radiotelephone Operator License to:
    A. adjust or repair FCC licensed transmitters in the aviation, maritime, and international fixed public radio services.

B. operate voluntarily equipped ship maritime mobile or aircraft transmitters with more than 1,000 watts of peak envelope power.
C. operate radiotelephone equipment with more than 1,500 watts of peak envelope power on cargo ships over 300 gross tons.
D. all of the above.

83. What is the radiotelephony calling and distress frequency?

A. 500 kHz. B. 500R122JA. C. 2182 kHz. D. 2182R2647.

84. If a ship radio transmitter signal becomes distorted:

A. cease operations. C. use minimum modulation.
B. reduce transmitter power. D. reduce audio amplitude.

85. A reserve power source must be able to power all radio equipment plus an emergency light system for how long?

A. 24 hours. B. 12 hours. C. 8 hours. D. 6 hours.

86. Frequencies used for portable communications on board ship:

A. 9300 to 9500 MHz. C. 2900 to 3100 MHz.
B. 1636 to 1644 MHz. D. 457.525 to 467.825 MHz.

87. In the FCC rules the frequency band from 30 to 300 MHz is also known as:

A. Very High Frequency (VHF). C. Medium Frequency (MF).
B. Ultra High Frequency (UHF). D. High Frequency (HF).

88. What channel must VHF-FM equipped vessels monitor at all times the station is operated?

A. Channel 8; 156.4 MHz. C. Channel 5A; 156.25 MHz.
B. Channel 16; 156.8 MHz. D. Channel 1A; 156.07 MHz.

89. When testing is conducted within the 2170 to 2194 kHz and 156.75 to 156.85 MHz, transmissions should not continue for more than ________ in any 15-minute period.

A. 15 seconds. B. 1 minute. C. 5 minutes. D. No limitation.

90. What emergency radio testing is required for cargo ships?

A. Tests must be conducted weekly while ship is at sea.
B. Full-power carrier tests into dummy load.
C. Specific gravity check in lead-acid batteries, or voltage under load for dry-cell batteries.
D. All of the above.

91. The master or owner of a vessel must apply how many days in advance for an FCC ship inspection?

A. 60 days. B. 30 days. C. 3 days. D. 24 hours.

92. Where do you submit an application for inspection of a ship radio station?
    A. To a Commercial Operator Licensing Examination Manager (COLE Manager).
    B. To the Federal Communications Commission, Washington, DC 20554.
    C. To the Engineer-in-Charge of the FCC District Office nearest the proposed place of inspection.
    D. To the nearest International Maritime Organization (IMO) review facility.

93. What are the antenna requirements of a VHF telephony coast, marine utility, or ship station?
    A. The shore or on-board antenna must be vertically polarized.
    B. The antenna array must be type accepted for 30 to 200 MHz operation by the FCC.
    C. The horizontally polarized antenna must be positioned so as not to cause excessive interference to other stations.
    D. The antenna must be capable of being energized by an output in excess of 100 W.

94. What regulations govern the use and operation of FCC-licensed ship stations operating in international waters?
    A. The regulations of the International Maritime Organization (IMO) and Radio Officers Union.
    B. Part 80 of the FCC Rules plus the International Radio Regulations and agreements to which the United States is a party.
    C. The Maritime Mobile Directives of the International Telecommunication Union.
    D. Those of the FCC's Aviation and Marine Branch, PRB, Washington, DC 20554.

95. Which of the following transmissions are not authorized in the Maritime Service?
    A. Communications from vessels in dry dock undergoing repairs.
    B. Message handling on behalf of third parties for which a charge is rendered.
    C. Needless or superfluous radiocommunications.
    D. Transmissions to test the operating performance of on-board station equipment.

96. When should both the callsign and the name of the ship be mentioned during radiotelephone transmissions?
    A. At all times.
    B. During an emergency.
    C. When transmitting on 2182 kHz.
    D. Within 100 miles of any shore.

97. How often is the auto alarm tested?
    A. During the 5-minute silent period.
    B. Monthly on 121.5 MHz using a dummy load.
    C. Weekly on frequencies other than the 2182-kHz distress frequency using a dummy antenna.
    D. Each day on 2182 kHz using a dummy antenna.

98. One nautical mile is approximately equal to how many statute miles?

    A. 1.61 statute miles. C. 1.15 statute miles.
    B. 1.83 statute miles. D. 1.47 statute miles.

99. Omega operates in what frequency band?

    A. Below 3 kHz. B. 3 to 30 kHz. C. 30 300 kHz. D. 300 to 3000 kHz.

100. Shipboard transmitters using F3E emission (FM voice) may not exceed what carrier power?

    A. 500 W. B. 250 W. C. 100 W. D. 25 W.

101. Loran C operates in what frequency band?

    A. VHF; 30 to 300 MHz. C. MF; 300 to 3000 kHz.
    B. HF; 3 to 30 MHz. D. LF; 30 to 300 kHz.

102. What is a ship earth station?

    A. A maritime mobile-satellite station located at a coast station.
    B. A mobile satellite station located on board a vessel.
    C. A communications system which provides line-of-sight communications between vessels at sea and coast stations.
    D. An automated ship-to-shore distress alerting system.

103. What is the internationally recognized urgency signal?

    A. The letters "TTT" transmitted three times by radiotelegraphy.
    B. Three oral repetitions of the word "safety" sent before the call.
    C. The word "pan" spoken three times before the urgent call.
    D. The pronouncement of the word "mayday."

104. What is a safety transmission?

    A. A radiotelephony warning preceded by the words "pan."
    B. Health and welfare traffic concerning the protection of human life.
    C. A communications transmission which indicates that a station is preparing to transmit an important navigation or weather warning.
    D. A radiotelegraphy alert preceded by the letters "XXX" sent three times.

105. What is a requirement of all marine transmitting apparatus used aboard United States vessels?

    A. Only equipment that has been type accepted by the FCC for Part 80 operations is authorized.
    B. Equipment must be approved by the U. S. Coast Guard for maritime mobile use.
    C. Certification is required by the International Maritime Organization (IMO).
    D. Programming of all maritime channels must be performed by a licensed Marine Radio Operator.

106. A marine public coast station operator may not charge a fee for what type of communication?

   A. Port Authority transmissions. C. Distress.
   B. Storm updates. D. All of the above.

107. Which of the following represent the first three letters of the phonetic alphabet?

   A. Alpha Bravo Charlie. C. Adam Baker Charlie.
   B. Adam Baker Charlie. D. Adam Brown Chuck.

108. Two way communications with both stations operating on the same frequency is:

   A. radiotelephone. B. duplex. C. simplex. D. multiplex.

109. When a ship is sold:

   A. new owner must apply for a new license.
   B. FCC inspection of equipment is required.
   C. old license is valid until it expires.
   D. continue to operate; license automatically transfers with ownership.

110. What is the second in order of priority?

   A. URGENT. B. DISTRESS. C. SAFETY. D. MAYDAY.

111. Portable ship units, hand-helds or walkie-talkies used as an associated ship unit:

   A. must operate with 1 W and be able to transmit on Channel 16.
   B. may communicate only with the mother ship and other portable units and small boats belonging to mother ship.
   C. must not transmit from shore or to other vessels.
   D. all of the above.

112. The HF (high frequency) band is:

   A. 3 to 30 MHz. B. 3 to 30 GHz. C. 30 to 300 MHz. D. 300 to 3000 MHz.

113. What is the procedure for testing a 2182-kHz ship radiotelephone transmitter with full carrier power while out at sea?

   A. Reduce to low power, then transmit the test tone.
   B. Switch transmitter to another frequency before testing.
   C. Simply say: "This is (call letters) testing." If all meters indicate normal values, it is assumed transmitter is operating properly.
   D. It is not permitted to test on the air.

114. If your transmitter is producing spurious harmonics or is operating at a deviation from the technical requirements of the station authorization:

   A. Continue operating until returning to port.
   B. Repair problem within 24 hours.
   C. Cease transmission.
   D. Reduce power immediately.

115. As an alternative to keeping watch on a working frequency in the band 1600 to 4000 kHz, an operator must tune station receiver to monitor 2182 kHz:

    A. at all times.
    B. during distress calls only.
    C. during daytime hours of service.
    D. during the silence periods each hour.

116. To indicate a response is expected, say:

    A. WILCO. B. ROGER. C. OVER. D. BREAK.

117. When all of a transmission has been received, say:

    A. ATTENTION. B. ROGER. C. RECEIVED. D. WILCO.

118. What information must be included in a DISTRESS message?

    A. Name of vessel.
    B. Location.
    C. Type of distress and specifics of help requested.
    D. All of the above.

119. Maritime MF radiotelephone silence periods begin at _______ and _______ minutes past the UTC hour.

    A. :15, :45. B. :00, :30. C. :20, :40. D. :05, :35.

120. Identify a ship station's radiotelephone transmission by:

    A. country of registration.
    B. callsign.
    C. name of the vessel.
    D. both B and C.

121. Maritime emergency radios should be tested:

    A. before each voyage.
    B. weekly while the ship is at sea.
    C. every 24 hours.
    D. both A and B.

122. The urgency signal concerning the safety of a ship, aircraft, or person shall be sent only on the authority of:

    A. master of ship.
    B. person responsible for mobile station.
    C. either A or B above.
    D. an FCC licensed operator.

123. Survival craft emergency transmitter tests may not be made:

    A. for more than 10 seconds.
    B. without using station callsign, followed by the word "test."
    C. within 5 minutes of a previous test.
    D. all of the above.

124. International laws and regulations require a silent period on 2182 kHz:

    A. for three minutes immediately after the hour.
    B. for three minutes immediately after the half-hour.
    C. for the first minute of every quarter-hour.
    D. both A and B above.

125. How should the 2182-kHz auto alarm be tested?

   A. On a different frequency into antenna.
   B. On a different frequency into dummy load.
   C. On 2182 kHz into dummy load.
   D. On 2182 kHz into antenna.

126. Each cargo ship of the United States that is equipped with a radiotelephone station for compliance with the Safety Convention shall, while at sea:

   A. not transmit on 2182 kHz during emergency conditions.
   B. keep the radiotelephone transmitter operating at full 100% carrier power for maximum reception on 2182 kHz.
   C. reduce peak envelope power on 156.8 MHz during emergencies.
   D. keep continuous watch on 2182 kHz using a watch receiver having a loudspeaker and auto alarm distress frequency watch receiver.

127. Who determines when a ship station may transmit routine traffic destined for a coast or government station in the maritime mobile service?

   A. Shipboard radio officers may transmit traffic when it will not interfere with on-going radiocommunications.
   B. The order and time of transmission and permissible type of message traffic is decided by the licensed on-duty operator.
   C. Ship stations must comply with instructions given by the coast or Government station.
   D. The precedence of conventional radiocommunications is determined by FCC and international regulation.

128. Who is responsible for payment of all charges accruing to other facilities for the handling or forwarding of messages?

   A. The licensee of the ship station transmitting the message.
   B. The third party for whom the message traffic was originated.
   C. The master of the ship jointly with the station licensee.
   D. The licensed commercial radio operator transmitting the radiocommunication.

129. Ordinarily, how often would a station using a telephony emission identify?

   A. At least every 10 minutes.
   B. At 15-minute intervals unless public correspondence is in progress.
   C. At the beginning and end of each communication and at 15-minute intervals.
   D. At 20-minute intervals.

130. When does a maritime radar transmitter identify its station?

   A. By radiotelegraphy at the onset and termination of operation.
   B. At 20-minute intervals using an automatic transmitter identification system.
   C. Radar transmitters must not transmit station identification.
   D. By a transmitter identification label (TIL) secured to the transmitter.

131. What is the general obligation of a coast or marine-utility station?

A. To accept and dispatch messages (without charge) that are necessary for the business and operational needs of ships.
B. To acknowledge and receive all calls directed to it by ship or aircraft stations.
C. To transmit lists of callsigns of all fixed and mobile stations for which they have traffic.
D. To broadcast warning and other information for the general benefit of all mariners.

132. Under what license are hand-held transceivers covered when used on board a ship at sea?

A. The ship station license.
B. Under the authority of the licensed operator.
C. Walkie-talkie radios are illegal to use at sea.
D. No license is needed.

133. What should an operator do to prevent interference?

A. Turn off transmitter when not in use.
B. Monitor channel before transmitting.
C. Transmissions should be as brief as possible.
D. Both B and C.

134. Under normal circumstances, what do you do if the transmitter aboard your ship is operating off frequency, overmodulating or distorting?

A. Reduce to low power. C. Reduce audio volume level.
B. Stop transmitting. D. Make a notation in station operating log.

135. The urgency signal has lower priority than:

A. direction finding. B. distress. C. safety. D. security.

136. The primary purpose of bridge-to-bridge communications is:

A. search and rescue emergency calls only.
B. all short-range transmission aboard ship.
C. transmission of Captain's orders from the bridge.
D. navigational communication.

137. What is the international VHF digital selective calling channel?

A. 2182 kHz. B. 156.35 MHz. C. 156.525 MHz. D. 500 kHz.

138. What are the highest priority communications from ships at sea?

A. All critical message traffic authorized by the ship's master.
B. Navigation and meteorological warnings.
C. Distress calls and communications preceded by the international urgency and safety signals.
D. Authorized government communications for which priority right has been claimed.

139. What is the best way for a radio operator to minimize or prevent interference to other stations?

A. By using an omnidirectional antenna pointed away from other stations.
B. Reducing power to a level that will not affect other on-frequency communications.
C. By changing frequency when notified that a radio communication causes interference.
D. Determine that a frequency is not in use by monitoring.

140. Under what circumstances may a ship or aircraft station interfere with a public coast station?

A. Under no circumstances during on-going radio communications.
B. During periods of government priority traffic handling.
C. When it is necessary to transmit a message concerning the safety of navigation or important meteorological warnings.
D. In cases of distress.

141. Portable ship radio transceivers operated as associated ship units:

A. must be operated on the safety and calling frequency 156.8 MHz (Channel 16) or a VHF intership frequency.
B. may not be used from shore without a separate license.
C. must only communicate with the ship station which it is associated or with associated portable ship units.
D. all of the above.

142. Which is a radiotelephony calling and distress frequency?

A. 500 kHz. B. 2182 MHz. C. 156.3 kHz. D. 3113 kHz.

143. What is the priority of communications?

A. Distress, urgency, safety, and radio direction finding.
B. Safety, distress, urgency, and radio direction finding.
C. Distress, safety, radio direction finding, search, and rescue.
D. Radio direction finding, distress, and safety.

144. Cargo ships of 300 to 1600 gross tons should be able to transmit a minimum range of:

A. 75 miles. B. 150 miles. C. 200 miles. D. 300 miles.

145. Radiotelephone stations required to keep logs of their transmission must include:

A. station, date, and time.
B. name of operator on duty.
C. station callsigns with which communication took place.
D. all of the above.

146. Each cargo ship of the United States that is equipped with a radiotelephone station for compliance with Part II of Title III of the Communications Act shall,

while being navigated outside of a harbor or port, keep a continuous and efficient watch on:

A. 2182 kHz.
B. 156.8 MHz.
C. both A and B.
D. monitor all frequencies within the 2000- to 27500-kHz band used for communications.

147. What call should you transmit on Channel 16 if your ship is sinking?

A. SOS three times.
B. Mayday three times.
C. Pan three times.
D. Urgency three times.

148. Who has ultimate control of service at a ship's radio station?

A. The master of the ship.
B. A holder of a First Class Radiotelegraph Certificate with a six-month service endorsement.
C. The radio Officer-in-Charge authorized by the captain of the vessel.
D. An appointed licensed radio operator who agrees to comply with all Radio Regulations in force.

149. What is the power limitation of associated ship stations operating under the authority of a ship station license?

A. The power level authorized to the parent ship station.
B. Associated vessels are prohibited from operating.
C. The minimum power necessary to complete the radio communications.
D. Power is limited to 1 W.

150. How is an associated vessel operating under the authority of another ship station license identified?

A. All vessels are required to have a unique callsign issued by the Federal Communications Commission.
B. With any station callsign self-assigned by the operator of the associated vessel.
C. By the callsign of the station with which it is connected and an appropriate unit designator.
D. Client vessels use the callsign of their parent plus the appropriate ITU regional indicator.

151. On what frequency would a vessel normally call another ship station when using a radiotelephony emission?

A. Only on 2182 kHz in ITU Region 2.
B. On the appropriate calling channel of the ship station at 15 minutes past the hour.
C. On 2182 kHz or 156.800 MHz unless the station knows the called vessel maintains a simultaneous watch on another intership working frequency.
D. On the vessel's unique working radio-channel assigned by the Federal Communications Commission.

152. What is required of a ship station which has established initial contact with another station on 2182 kHz or 156.800 MHz?

   A. The stations must check the radio channel for distress, urgency, and safety calls at least once every 10 minutes.
   B. The stations must change to an authorized working frequency for the transmission of messages.
   C. Radiated power must be minimized so as not to interfere with other stations needing to use the channel.
   D. To expedite safety communications, the vessels must observe radio silence for two out of every 15 minutes.

153. What is the most important practice that a radio operator must learn?

   A. Monitor the channel before transmitting.
   B. Operate with lowest power necessary.
   C. Test a radiotelephone transmitter daily.
   D. Always listen to 121.5 MHz.

154. What is the minimum transmitter power level required by the FCC for a medium frequency transmitter aboard a compulsorily fitted vessel?

   A. At least 100-W single-sideband suppressed carrier power.
   B. At least 60 W PEP.
   C. The power predictably needed to communicate with the nearest public coast station operating on 2182 kHz.
   D. At least 25 W delivered into 50 Ω effective resistance when operated with a primary voltage of 13.6 Vdc.

155. What is a class-A EPIRB?

   A. An alerting device notifying mariners of imminent danger.
   B. A satellite-base maritime distress and safety alerting system.
   C. An automatic, battery-operated emergency position indicating radio beacon that floats free of a sinking ship.
   D. A high-efficiency audio amplifier.

156. What are the radio watch requirements of a voluntary ship?

   A. While licensees are not required to operate the ship radio station, general-purpose watches must be maintained if they do.
   B. Radio watches must be maintained on 500 kHz, 2182 kHz, and 156.800 MHz, but no station logs are required.
   C. Radio watches are optional but logs must be maintained of all medium, high frequency, and VHF radio operation.
   D. Radio watches must be maintained on the 156 to 158 MHz, 1600 to 4000 kHz, and 4000 to 23000 kHz bands.

157. What is the Automated Mutual-Assistance Vessel Rescue system?

A. A voluntary organization or mariners who maintain radio watch on 500 kHz, 2182 kHz, and 156.800 MHz.
B. An international system operated by the Coast Guard providing coordination of search and rescue efforts.
C. A coordinated radio direction finding effort between the Federal Communications Commission and U. S. Coast Guard to assist ships in distress.
D. A satellite-base distress and safety alerting program operated by the U. S. Coast Guard.

158. What is a bridge-to-bridge station?

A. An internal communications system linking the wheel house with the ship's primary radio operating position and other integral ship control points.
B. An inland waterways and coastal radio station serving ship stations operating within the United States.
C. A portable ship station necessary to eliminate frequency application to operate a ship station on board different vessels.
D. A VHF radio station located on a ship's navigational bridge or main control station that is used only for navigational communications.

159. How does a coast station notify a ship that it has a message for the ship?

A. By making a directed transmission on 2182 kHz or 156.800 MHz.
B. The coast station changes to the vessel's known working frequency.
C. By establishing communications using the eight-digit maritime mobile service identification.
D. The coast station may transmit at intervals lists of callsigns in alphabetical order for which they have traffic.

160. Under what circumstances may a coast station using telephony transmit a general call to a group of vessels?

A. Under no circumstances.
B. When announcing or preceding the transmission of distress, urgency, safety, or other important messages.
C. When the vessels are located in international waters beyond 12 miles.
D. When identical traffic is destined for multiple mobile stations within range.

161. What is the proper procedure for making a correction in the station log?

A. The ship's master must be notified, approve and initial all changes to the station log.
B. The mistake may be erased and the correction made and initialed only by the radio operator making the original error.
C. The original person making the entry must strike out the error, initial the correction, and indicate the date of correction.
D. Rewrite the new entry in its entirety directly below the incorrect notation and initial the change.

162. What authorization is required to operate a 350-W PEP maritime voice station on frequencies below 25 MHz aboard a small noncommercial pleasure vessel?

A. Third Class Radiotelegraph Operator's Certificate.
B. General Radiotelephone Operator License.
C. Restricted Radiotelephone Operator Permit.
D. Marine Radio Operator Permit.

163. What is selective calling?

A. A coded transmission directed to a particular ship station.
B. A radiotelephony communication directed at a particular ship station.
C. An electronic device that uses a discriminator circuit to filter out unwanted signals.
D. A telegraphy transmission directed only to another specific radiotelegraph station.

164. In the International Phonetic Alphabet, the letters D, N, and O are represented by the words:

A. Delta, November, Oscar.
B. Denmark, Neptune, Oscar.
C. December, Nebraska, Olive.
D. Delta, Neptune, Olive.

165. What are the technical requirements of a VHF antenna system aboard a vessel?

A. The antenna must provide an amplification factor of at least 2.1 dBi.
B. The antenna must be vertically polarized and nondirectional.
C. The antenna must be capable of radiating a signal a minimum of 150 nautical miles on 156.8 MHz.
D. The antenna must be constructed of corrosion-proof aluminum and capable of proper operation during an emergency.

166. How often must the radiotelephone installation aboard a small passenger boat be inspected?

A. Equipment inspections are required at least once every 12 months.
B. When the vessel is first placed in service and every 2 years thereafter.
C. At least once every 5 years.
D. A minimum of every 3 years, and when the ship is within 75 statute miles of an FCC field office.

167. How far from land may a small passenger vessel operate when equipped only with a VHF radiotelephone installation?

A. No more than 20 nautical miles from the nearest land if within the range of a VHF public coast of U. S. Coast Guard station.
B. No more than 100 nautical miles from the nearest land.
C. No more than 20 nautical miles unless equipped with a reserve power supply.
D. The vessel must remain within the communications range of the nearest coast station at all times.

168. Where must the principal radiotelephone operating position be installed in a ship station?
    A. At the principal radio operating position of the vessel.
    B. In the room or an adjoining room from which the ship is normally steered while at sea.
    C. In the chart room, master's quarters, or wheel house.
    D. At the level of the main wheel house or at least one deck above the ship's main deck.

169. What should a station operator do before making a transmission?
    A. Transmit a general notification that the operator wishes to utilize the channel.
    B. Except for the transmission of distress calls, determine that the frequency is not in use by monitoring the frequency before transmitting.
    C. Check transmitting equipment to be certain it is properly calibrated.
    D. Ask if the frequency is in use.

170. What is the proper procedure for testing a radiotelephone installation?
    A. Transmit the station's callsign, followed by the word "test" on the radio channel being used for the test.
    B. A dummy antenna must be used to ensure the test will not interfere with ongoing communications.
    C. Permission for the voice test must be requested and received from the nearest public coast station.
    D. Short tests must be confined to a single working frequency.

171. What is the minimum radio operator requirement for ships subject to the Great Lakes Radio Agreement?
    A. Third Class Radiotelegraph Operator's Certificate.
    B. General Radiotelephone Operator License.
    C. Marine Radio Operator Permit.
    D. Restricted Radiotelephone Operator Permit.

172. What FCC authorization is required to operate a VHF transmitter on board a vessel voluntarily equipped with radio and sailing on a domestic voyage?
    A. No radio operator license or permit is required.
    B. Marine Radio Operator Permit.
    C. Restricted Radiotelephone Operator Permit.
    D. General Radiotelephone Operator License.

173. On what frequencies does the Communications Act require radio watches by compulsory radiotelephone stations?
    A. Watches are required on 500 kHz and 2182 kHz.
    B. Continuous watch is required on 2182 kHz.
    C. On all frequencies between 405 to 535 kHz, 1605 to 3500 kHz, and 156 to 162 MHz.
    D. Watches are required on 2182 kHz and 156.800 MHz.

174. What is the purpose of the international radiotelephone alarm signal?
    - A. To notify nearby ships of the loss of a person or persons overboard.
    - B. To call attention to the upcoming transmission of an important meteorological warning.
    - C. To alert radio officers monitoring watch frequencies of a forthcoming distress, urgency, or safety message.
    - D. To actuate automatic devices giving an aural alarm to attract the attention of the operator where there is no listening watch on the distress frequency.

175. What is the Communication Act's definition of a "passenger ship?"
    - A. Any ship that is used primarily in commerce for transporting persons to and from harbors or ports.
    - B. A vessel that carries or is licensed or certificated to carry more than 12 passengers.
    - C. Any ship transporting more than six passengers for hire.
    - D. A vessel of any nation that has been inspected and approved as a passenger carrying vessel.

176. What is a distress communication?
    - A. An internationally recognized communication indicating that the sender is threatened by grave and imminent danger, and requests immediate assistance.
    - B. Communications indicating that the calling station has a very urgent message concerning safety.
    - C. Radiocommunications which, if delayed, will adversely affect the safety of life or property.
    - D. An official radiocommunications notification of approaching navigational or meteorological hazards.

177. Who may be granted a ship station license in the maritime service?
    - A. Anyone, including foreign governments.
    - B. Only FCC licensed operators holding a First or Second Class Radiotelegraph Operator's Certificate or the General Radiotelephone Operator License.
    - C. Vessels that have been inspected and approved by the U. S. Coast Guard and Federal Communications Commission.
    - D. The owner or operator of a vessel, or their subsidiaries.

178. Who is responsible for the proper maintenance of station logs?
    - A. The station licensee and the radio operator in charge of the station.
    - B. The station licensee.
    - C. The commercially licensed radio operator in charge of the station.
    - D. The ship's mast and the station licensee.

179. How long should station logs be retained when there are no entries relating to distress or disaster situations?
    - A. Until authorized by the Commission in writing to destroy them.

B. Indefinitely, or until destruction is specifically authorized by the U. S. Coast Guard.
C. For a period of 3 years from the date of entry unless notified by the FCC.
D. For a period of 1 year from the date of entry.

180. Where must ship station logs be kept during a voyage?

A. At the principal radiotelephone operating position.
B. They must be secured in the vessel's strong-box for safekeeping.
C. In the personal custody of the licensed commercial radio operator.
D. All logs are turned over to the ship's master when the radio operator goes off duty.

181. What is the antenna requirement of a radiotelephone installation aboard a passenger vessel?

A. The antenna must be located a minimum of 15 meters from the radiotelegraph antenna.
B. An emergency reserve antenna system must be provided for communications on 156.8 MHz.
C. The antenna must be vertically polarized and as nondirectional and efficient as is practicable for the transmission and reception of ground waves over seawater.
D. All antennas must be tested and the operational results logged at least once during each voyage.

182. If a ship sinks, what device is designed to float free of the mother ship, is turned on automatically, and transmits a distress signal?

A. EPIRB on 121.5 MHz, 243 MHz, or 406.025 MHz.
B. EPIRB on 2182 kHz and 405.025 kHz.
C. Bridge-to-bridge transmitter on 2182 kHz.
D. Auto alarm keyer on any frequency.

183. International laws and regulations require a silent period on 2182 kHz:

A. for 3 minutes immediately after the hour.
B. for 3 minutes immediately after the half-hour.
C. for the first minute of every quarter-hour.
D. both A and B above.

184. How should the 2182 kHz auto alarm be tested?

A. On a different frequency into antenna.
B. On a different frequency into dummy load.
C. On 2182 kHz into antenna.
D. Only under U. S. Coast Guard authorization.

185. What is the average range of VHF marine transmissions?

A. 150 miles. B. 50 miles. C. 20 miles. D. 10 miles.

186. A ship station using VHF bridge-to-bridge Channel 13:

    A. may be identified by callsign and country of origin.
    B. must be identified by callsign and name of vessel.
    C. may be identified by the name of the ship in lieu of callsign.
    D. does not need to identify itself within 100 miles from shore.

187. When using an SSB station on 2182 kHz or VHF-FM on Channel 16:

    A. Preliminary call must not exceed 30 seconds.
    B. If contact is not made, you must wait at least 2 minutes before repeating the call.
    C. Once contact is established, you must switch to a working frequency.
    D. All of the above.

188. By international agreement, which ships must carry radio equipment for the safety of life at sea?

    A. Cargo ships of more than 300 gross tons and vessels carrying more than 12 passengers.
    B. All ships traveling more than 100 miles out to sea.
    C. Cargo ships of more than 100 gross tons and passenger vessels on international deep-sea voyages.
    D. All cargo ships of more than 100 gross tons.

189. Which commercial radio operator license is required to install a VHF transmitter in a voluntarily equipped ship station?

    A. A Marine Radio Operator Permit or higher class of license.
    B. None, if installed by, or under the supervision of, the licensee of the ship station and no modifications are made to any circuits.
    C. A Restricted Radiotelephone Operator Permit or higher class of license.
    D. A General Radiotelephone Operator License.

190. What transmitting equipment is authorized for use by a station in the maritime services?

    A. Transmitters that have been certified by the manufacturer for maritime use.
    B. Unless specifically exempted, only transmitters that are type accepted by the Federal Communications Commission for Part 80 operations.
    C. Equipment that has been inspected and approved by the U. S. Coast Guard.
    D. Transceivers and transmitters that meet all ITU specifications for use in maritime mobile service.

191. When is it legal to transmit high power on Channel 13?

    A. Failure of vessel being called to respond.
    B. In a blind situation, such as rounding a bend in a river.
    C. During an emergency.
    D. All of the above.

192. What must be in operation when no operator is standing watch on a compulsory radio equipped vessel while out at sea?
    A. An auto alarm.
    B. Indicating radio beacon signals.
    C. Distress-Alert signal device.
    D. Radiotelegraph transceiver set to 2182 kHz.

193. When may a bridge-to-bridge transmission be more than 1 W?
    A. When broadcasting a distress message.
    B. When rounding a bend in a river or traveling in a blind spot.
    C. When calling the Coast Guard.
    D. Both A and B.

194. When are EPIRB batteries changed?
    A. After emergency use; after battery life expires.
    B. After emergency use; as per manufacturers instructions marked on outside of transmitter with month and year replacement date.
    C. After emergency use; every 12 months when not used.
    D. Whenever voltage drops to less than 50% of full charge.

195. The radiotelephone distress message consists of:
    A. Mayday spoken three times, callsign and name of vessel in distress.
    B. Particulars of its position, latitude and longitude, and other information which might facilitate rescue, such as length, color and type of vessel, number of persons on board.
    C. Nature of distress and kind of assistance desired.
    D. All of the above.

196. How do the rules define "navigational communications?"
    A. Safety communications pertaining to the maneuvering or directing of vessels' movements.
    B. Important communications concerning the routine of vessels during periods of meteorological crisis.
    C. Telecommunications pertaining to the guidance of maritime vessels in hazardous waters.
    D. Radio signals consisting of weather, sea conditions, notices to mariners, and potential dangers.

197. What type of communications may be exchanged by radioprinter between authorized private coast stations and ships of less than 1600 gross tons?
    A. Public correspondence service may be provided on voyages of more than 24 hours.
    B. All communications, providing they do not exceed 3 minutes after the stations have established contact.

C. Only those communications that concern the business and operational needs of vessels.
D. There are no restrictions.

198. What are the service requirements of all ship stations?

A. Each ship station must receive and acknowledge all communications with any station in the maritime mobile service.
B. Public correspondence must be offered for any person during the hours the radio operator is normally on duty.
C. All ship stations must maintain watch on 500 kHz, 2182 kHz, and 156.800 MHz.
D. Reserve antennas, emergency power sources, and alternate communications installations must be available.

199. When may the operator of a ship station allow an unlicensed person to speak over the transmitter?

A. At no time. Only commercially licensed radio operators may modulate the transmitting apparatus.
B. When the station power does not exceed 200 W peak envelope power.
C. When under the supervision of the licensed operator.
D. During the hours that the radio office is normally off duty.

200. What are the radio operator requirements of a cargo ship with a 1000 W peak-envelope-power radiotelephone station?

A. The operator must hold a General Radiotelephone Operator License or higher class license.
B. The operator must hold a Restricted Radiotelephone Operator Permit or higher class license.
C. The operator must hold a Marine Radio Operator Permit or higher class license.
D. The operator must hold a GMDSS Radio Maintainer's License.

201. What are the radio operator requirements of a small passenger ship carrying more than six passengers equipped with a 1000-watt carrier power radiotelephone station?

A. The operator must hold a General Radiotelephone Operator or higher class license.
B. The operator must hold a Marine Radio Operator Permit or higher class license.
C. The operator must hold a Restricted Radiotelephone Operator Permit or higher class license.
D. The operator must hold a GMDSS Radio Operator's License.

202. Which commercial radio operator license is required to operate a fixed tuned ship radar station with external controls?

A. A radio operator certificate containing a Ship Radar Endorsement.
B. A Marine Radio Operator Permit or higher.

C. Either a First or Second Class Radiotelegraph certificate or a General Radiotelephone Operator License.
D. No radio operator authorization is required.

203. What traffic management service is operated by the U. S. Coast Guard in certain designated water areas to prevent ship collisions, groundings, and environmental harm?

A. Water safety management bureau (WSMB).
B. Vessel traffic service (VTS).
C. Ship movement and safety agency (SMSA).
D. Interdepartmental harbor and port patrol (IHPP).

204. What action must be taken by the owner or operator of a vessel who changes its name?

A. A Request for Ship License Modification (RSLM) must be submitted to the FCC's licensing facility.
B. The Engineer-in-Charge of the nearest FCC field office must be informed.
C. The Federal Communications Commission in Gettysburg, PA, must be notified in writing.
D. Written confirmation must be obtained from the U. S. Coast Guard.

205. When may a shipboard radio operator make a transmission in the maritime services not addressed to a particular station or stations?

A. General CQ calls may only be made when the operator is off duty and another operator is on watch.
B. Only during the transmission of distress, urgency or safety signals or messages, or to test the station.
C. Only when specifically authorized by the master of the ship.
D. When the radio officer is more than 12 miles from shore and the nearest ship or coast station is known.

206. What is the order of priority of radiotelephone communications in the maritime services?

A. Distress calls and signals, followed by communications preceded by urgency and safety signals.
B. Alarm, radio-direction finding, and health and welfare communications.
C. Navigation hazards, meteorological warning, priority traffic.
D. Government precedence, messages concerning safety of life and protection of property, and traffic concerning grave and imminent danger.

207. Which of the following statements is true as to ships subject to the Safety Convention?

A. A cargo ship participates in international commerce by transporting good between harbors.
B. Passenger ships carry 6 or more passengers for hire as opposed to transporting merchandise.

C. A cargo ship is any ship that is not licensed or certificated to carry more than 12 passengers.
D. Cargo ships are FCC inspected on an annual basis while passenger ships undergo U. S. Coast Guard inspections every 6 months.

208. What is a "passenger-carrying vessel" when used in reference to the Great Lake Radio Agreement?
A. A vessel that is licensed or certificated to carry more than 12 passengers.
B. Any ship carrying more than 6 passengers for hire.
C. Any ship, the principal purpose of which is to ferry persons on the Great Lakes and other inland waterways.
D. A ship that is used primarily for transporting persons and goods to and from domestic harbors or ports.

209. How do the FCC's Rules define a power-driven vessel?
A. A ship that is not manually propelled or under sail.
B. Any ship propelled by machinery.
C. A watercraft containing a motor with a power rating of at least 3 HP.
D. A vessel moved by mechanical equipment at a rate of 5 knots or more.

ANSWER SHEET

1. B	2. B	3. C	4. A	5. A
6. D	7. C	8. C	9. C	10. A
11. D	12. B	13. D	14. A	15. B
16. B	17. C	18. C	19. A	20. C
21. B	22. C	23. B	24. A	25. D
26. A	27. D	28. A	29. C	30. B
31. B	32. A	33. D	34. A	35. D
36. B	37. B	38. D	39. A	40. D
41. D	42. D	43. A	44. C	45. B
46. C	47. A	48. C	49. A	50. C
51. D	52. A	53. B	54. B	55. B
56. B	57. A	58. A	59. B	60. D
61. C	62. D	63. D	64. C	65. D
66. D	67. D	68. D	69. B	70. A
71. C	72. C	73. C	74. C	75. B
76. A	77. D	78. A	79. D	80. A
81. D	82. D	83. C	84. A	85. D
86. D	87. A	88. B	89. A	90. D
91. C	92. C	93. A	94. B	95. C
96. B	97. C	98. C	99. B	100. D
101. D	102. B	103. C	104. C	105. A
106. C	107. A	108. C	109. A	110. A
111. D	112. A	113. C	114. C	115. A
116. C	117. B	118. D	119. B	120. B

121. D 122. C 123. D 124. D 125. B
126. D 127. C 128. A 129. C 130. C
131. B 132. A 133. D 134. B 135. B
136. D 137. C 138. C 139. D 140. D
141. D 142. B 143. A 144. B 145. D
146. C 147. B 148. A 149. D 150. C
151. B 152. C 153. B 154. A 155. B
156. C 157. A 158. B 159. D 160. D
161. B 162. C 163. C 164. A 165. A
166. B 167. C 168. A 169. B 170. B
171. A 172. C 173. A 174. D 175. D
176. B 177. A 178. D 179. A 180. C
181. A 182. C 183. A 184. D 185. B
186. C 187. C 188. D 189. A 190. B
191. B 192. D 193. A 194. D 195. B
196. D 197. A 198. C 199. A 200. C
201. C 202. D 203. D 204. B 205. C
206. B 207. A 208. C 209. B 210. B

10

CHAPTER

Associate-level and Journeyman-level CET practice tests

The practice Associate-level tests in this chapter were provided by Electronics Technician Association (ETA). The practice communications test was supplied by International Society of Certified Electronics Technicians (ISCET).

The actual tests given for certification undergo constant revision. Therefore, the tests in this chapter are only for the purpose of testing your knowledge on the wide range of subjects covered in the CET tests.

Before you sit for any CET test, you must write to the certifying organization. Both ETA and ISCET have tests for the Associate and Journeyman Communication certification. Regardless of which type of Journeyman test you want to take, you must pass the Associate-level test! Also refer to the Appendix for more up-to-date information on the latest exams.

Here are the addresses of the two organizations (in alphabetical order):

ETA	ISCET
604 North Jackson Street	708 West Berry Street
Greencastle, IN 46135	Fort Worth, TX 76109

Answers are on pages 367 and 368.

Mathematics

1. In the circuit of Fig. 10-1, the voltage across each component is shown to be 15 volts. The applied voltage (e) from the generator is:

   A. 15 V. B. about 30 V. C. a little over 21 V. D. not quite 11 V.

2. In the circuit of Fig. 10-1, increasing the frequency of the generator will:

   A. increase the circuit current.
   B. decrease the circuit current.
   C. not affect the circuit current.
   D. change the circuit current, but not enough information is given to show how the current will change.

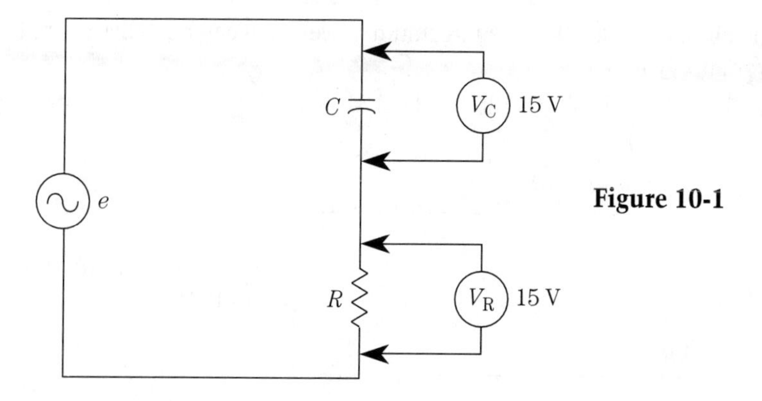

Figure 10-1

3. In the circuit of Fig. 10-1, reducing the capacitance will:
   A. increase the voltage across the resistor.
   B. decrease the voltage across the resistor.
   C. not affect the voltage across the resistor.
   D. decrease the capacitive reactance.

4. The color code of the resistor in the circuit of Fig. 10-2 is:
   A. orange, orange, orange.
   B. orange, white, orange.
   C. yellow, green, orange.
   D. yellow, purple, orange.

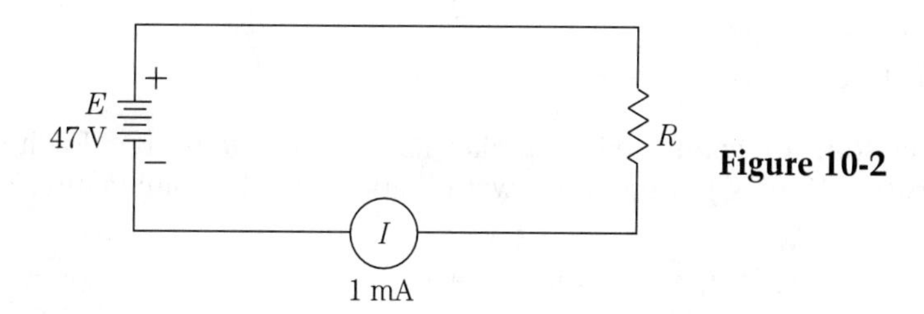

Figure 10-2

5. Regarding the circuit of Fig. 10-3, which of the following statements is NOT correct?
   A. The conductance of R2 is greater than the conductance of R1.
   B. The resistance of the parallel resistor combination is about 8.9 kΩ.
   C. The voltage across the 22-kΩ resistor is greater than the voltage across the 15-kΩ resistor.
   D. There is more current through R2 than there is through R1.

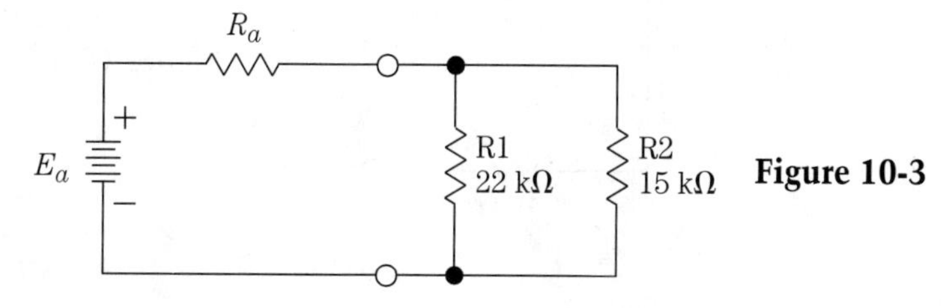

Figure 10-3

6. In the circuit of Fig. 10-4, the maximum power than can be delivered to R_L by the power supply is:

   A. 1.25 W. B. 12.5 W. C. 125 W. D. 8.31 W.

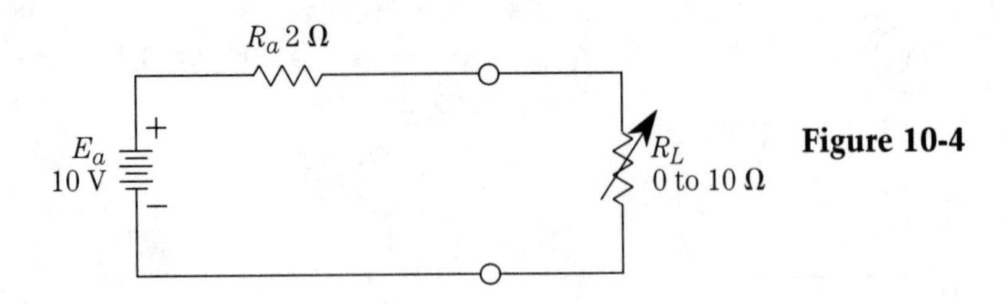

Figure 10-4

7. The efficiency of the circuit in Fig. 10-4 is 50% when the resistance of R_L is:

   A. 1 Ω. B. 2 Ω. C. 4 Ω. D. 10 Ω.

8. The period of a waveform is the time for one complete cycle, and is designated by T. What is the period of a 1000 Hz waveform?

   A. 1000 ms. C. 100 ms.
   B. 100 μs. D. 1000 μs.

9. A resistor is color coded red, red, red, silver. The minimum resistance it can have and still be in tolerance is:

   A. 220 Ω. C. 1980 Ω.
   B. 2200 Ω. D. 2020 Ω.

10. Three resistors are connected in parallel across a voltage source. Each resistor is dissipating 12 W of power. The power dissipated by the combination is:

    A. 36 W. C. 12 W.
    B. 4 W. D. none of these answers is correct.

dc circuits

1. In the circuit of Fig. 10-5, what is the voltage at point (a) with respect to ground?

   A. Cannot be determined from the information given. C. –10 V
   B. 0 V. D. +10 V

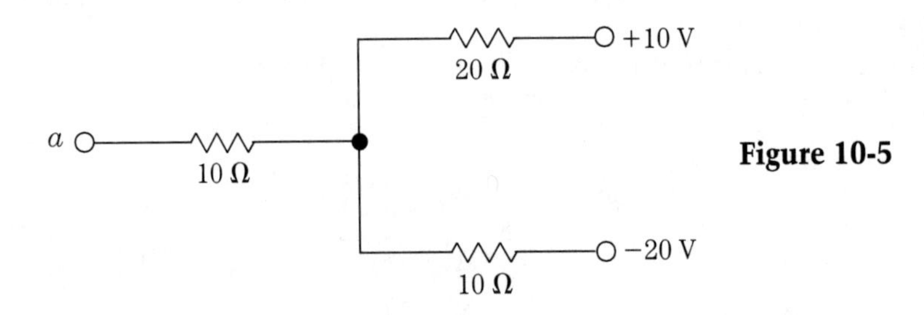

Figure 10-5

2. In the circuit of Fig. 10-6, which capacitor will have the lower value of voltage across it?

   A. C1. B. C2.

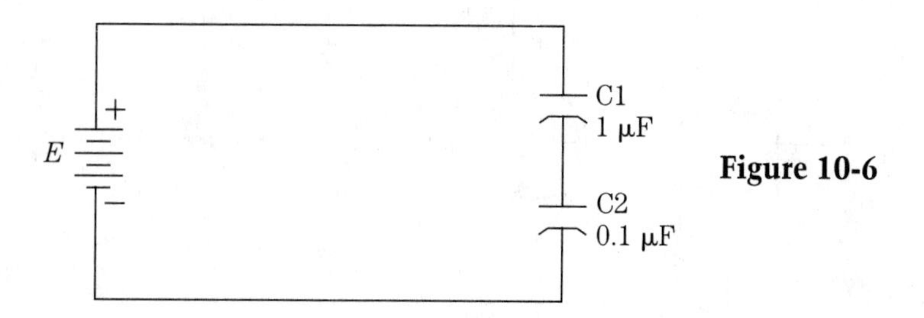

Figure 10-6

3. In the class-A amplifier circuit of Fig. 10-7, the source voltage is 3 V. The drain voltage should be:

   A. 5 V. B. 4 V. C. 3 V. D. 2 V.

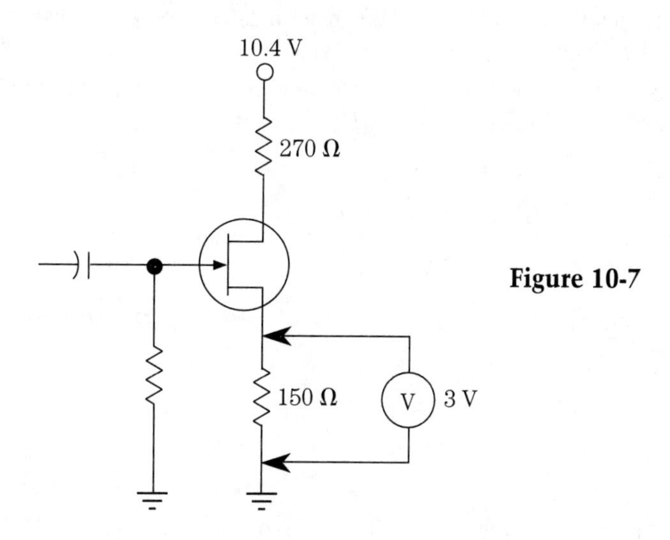

Figure 10-7

4. In the circuit of Fig. 10-8, resistor R1 is open. The collector voltage of the transistor will be:

   A. 0 V.
   B. slightly positive.
   C. the normal amount of negative voltage with respect to ground.
   D. impossible to determine.

5. In the circuit of Fig. 10-9, the capacitor is not charged. The voltage between terminals X and Y is:

   A. 0 V. B. 100 V.

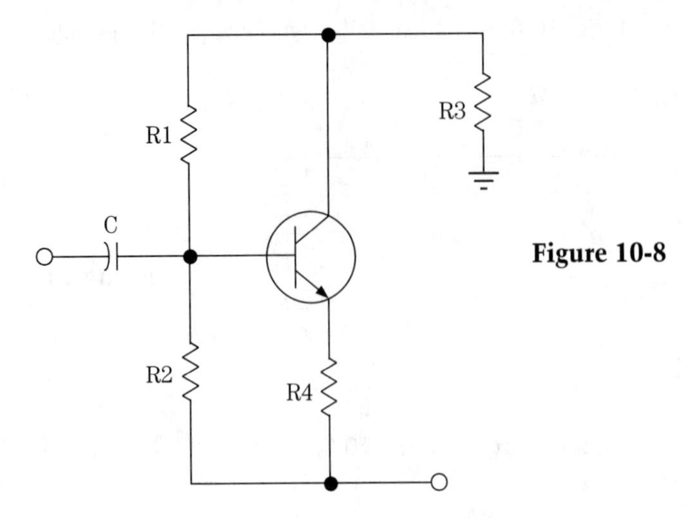

Figure 10-8

6. In the circuit of Fig. 10-9, a 10-MΩ resistor is connected between terminals x and y. The voltage across the resistor will:

   A. decrease to 37 V in 10 seconds. C. decrease to 50 V in 10 seconds.
   B. decrease to 63 V in 10 seconds. D. 0 V after 1 hour.

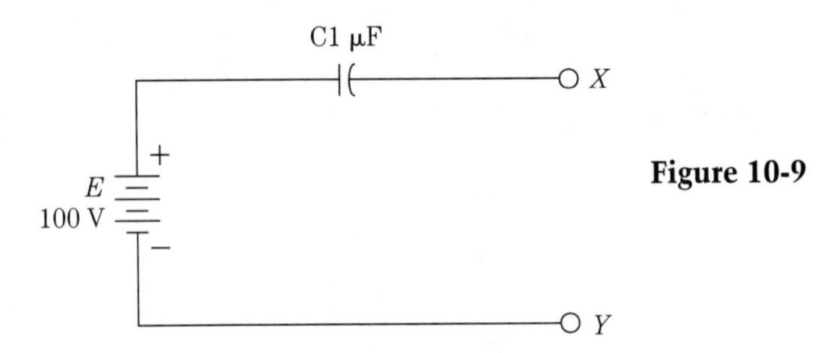

Figure 10-9

7. Refer again to the circuit of Fig. 10-9. Five minutes after the 10-MΩ resistor has been connected across terminals x and y, the voltage across the capacitor will be:

   A. 0 V. B. 100 V.

8. In the circuit of Fig. 10-10A, resistor R1 is open. The voltage between terminals a and b as measured with a high-impedance voltmeter should be:

   A. 50 V. C. 0 V.
   B. 60 V. D. none of those.

9. In the circuit of Fig. 10-10B, the cathode resistor (R_K) is open. Which of the following is correct?

   A. The tube is very likely burned out.
   B. It should be replaced with a resistor with a higher power rating.
   C. The plate voltage should be about 0 V.
   D. The cathode voltage, as measured with a high-impedance meter, should be positive with respect to ground.

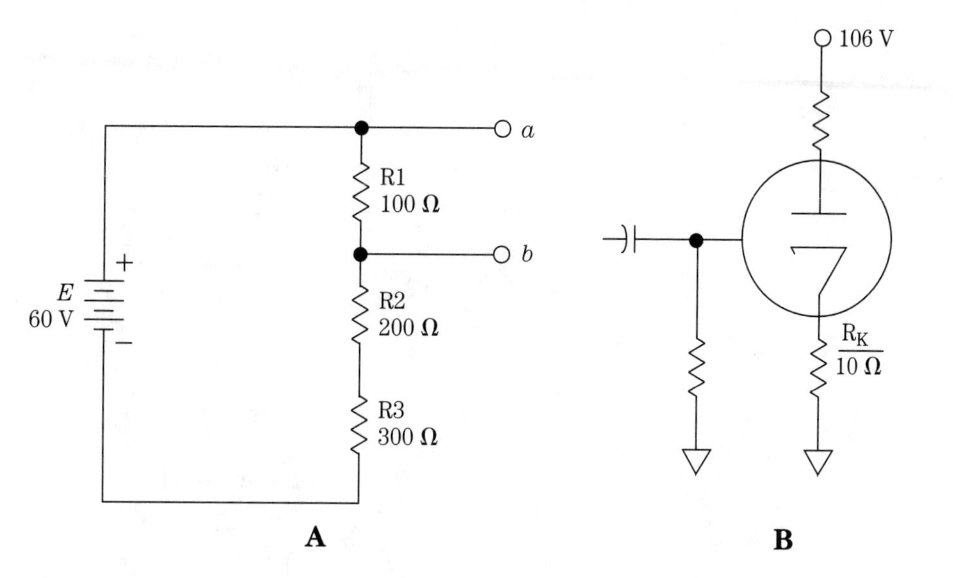

Figure 10-10

10. In a normally operated PNP transistor, the base voltage should be:
    A. positive with respect to the collector. B. negative with respect to the collector.

ac circuits

1. In the circuit of Fig. 10-11, decreasing the generator frequency will:
   A. increase the circuit current. B. decrease the circuit current.

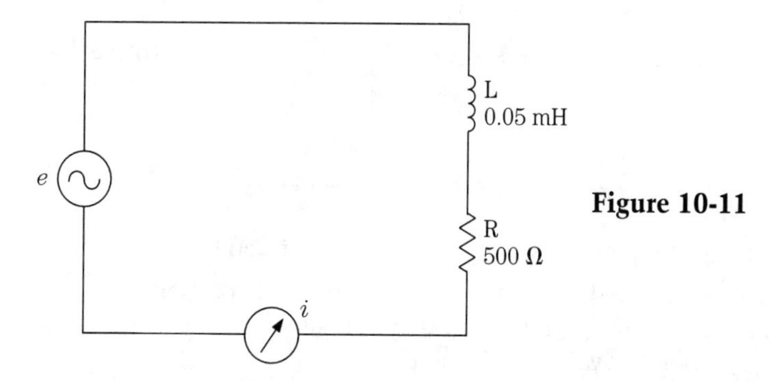

Figure 10-11

2. Figure 10-12 shows the waveform of a voltage across a resistor. A volt-ohm-milliammeter is used to measure this voltage. The voltage reading will be:
   A. about 7 V. C. 5 V.
   B. about 6.36 V. D. none of the above.

3. In the circuit of Fig. 10-13, the product of the RMS 12 voltage (*E*) and RMS current (*I*) will give:
   A. the real power. C. the apparent power.
   B. the reactive power (vars). D. the conjugate impedance.

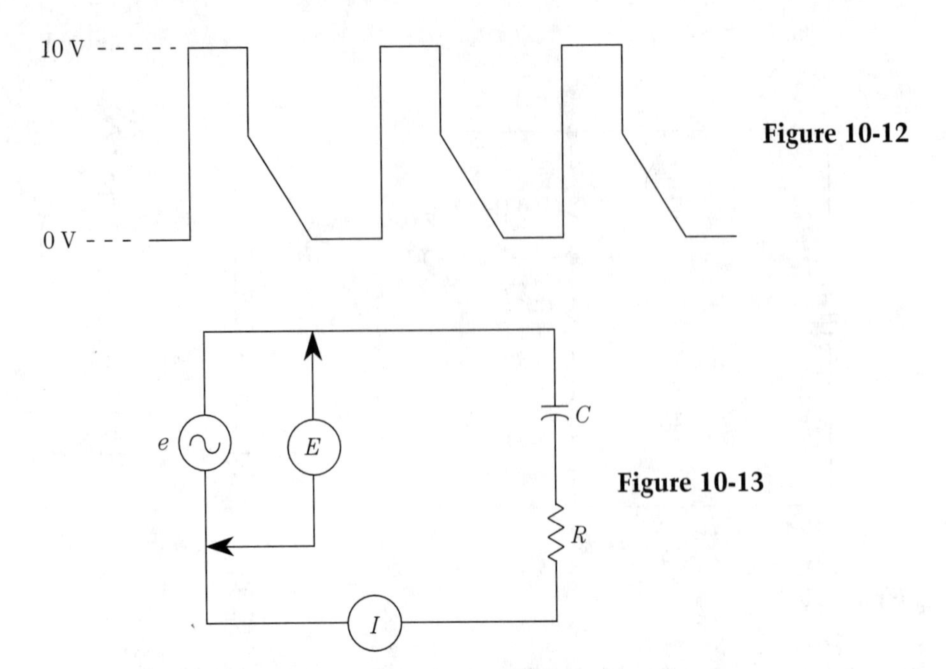

Figure 10-12

Figure 10-13

4. In the power transformer system of Fig. 10-14, the line voltage is too low. The secondary voltage will be increased by moving the contact of switch SW to:

   A. point A. B. point C.

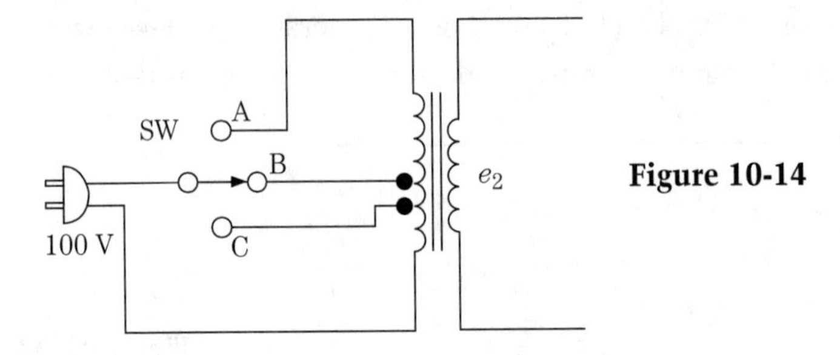

Figure 10-14

5. The resonant frequency (f_r) of a parallel LC circuit:

   A. can be changed with a variable resistor if it is properly connected.
   B. cannot be changed with a variable resistor because the value of f_r depends only upon the values of L and C.

6. The circuit of Fig. 10-15 is an example of:

   A. a high-pass filter. C. a band-pass filter.
   B. a low-pass filter. D. a band-rejection (notch) filter.

7. The capacitive reactance of each capacitor in Fig. 10-16 is 15 Ω. The capacitive reactance seen by the generator is:

   A. 7.5 Ω. B. 15 Ω. C. 30 Ω. D. 225 Ω.

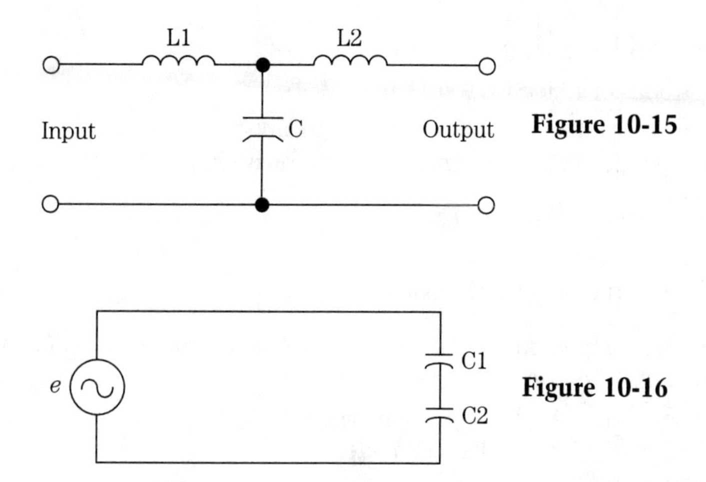

Figure 10-15

Figure 10-16

8. Assume that the circuit of Fig. 10-17 is resonant. Moving the plates of capacitor C closer together will:

   A. raise the resonant frequency. B. lower the resonant frequency.

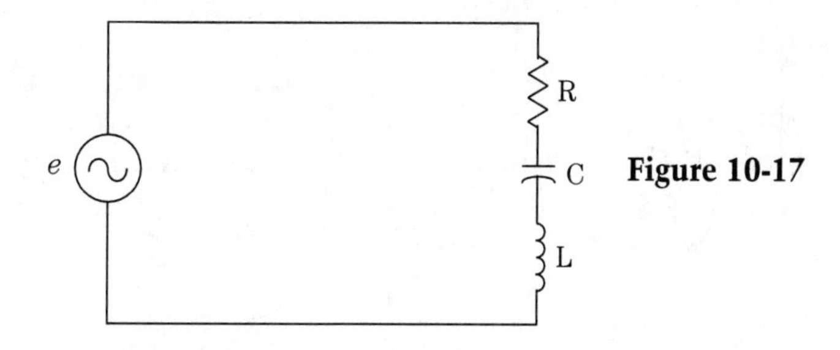

Figure 10-17

9. In the circuit of Fig. 10-18, the frequency of the generator (*e*) can be varied. The effect of switching R into the circuit is to:

   A. increase the bandwidth. B. decrease the bandwidth.

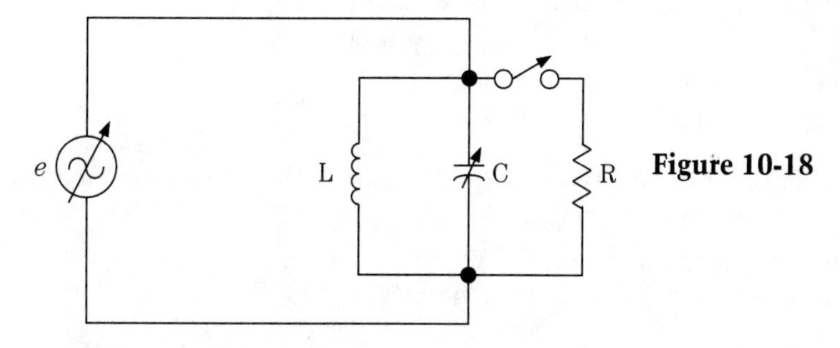

Figure 10-18

10. A ferrite bead in a circuit acts like:

    A. a resistor. C. an inductor.
    B. a capacitor. D. a varactor.

Transistors and semiconductors

1. The electrodes of a MOSFET are called:

   A. emitter, base, drain.
   B. source, gate, collector.
   C. cathode, gate, anode.
   D. none of these answers is correct.

2. Which of the symbols in Fig. 10-19 is for a p-channel MOSFET?

   A. The one marked A.
   B. The one marked B.
   C. The one marked C.
   D. The one marked D.

3. Which of the components of Fig. 10-19 would be most likely used for controlling large amounts of power to a load?

   A. The one marked A.
   B. The one marked B.
   C. The one marked C.
   D. The one marked D.
   E. Tunnel diode.

4. Which of the components of Fig. 10-19 has its electrodes marked base 1, base 2, and emitter?

   A. The one marked A.
   B. The one marked B.
   C. The one marked C.
   D. The one marked D.
   E. Tunnel diode.

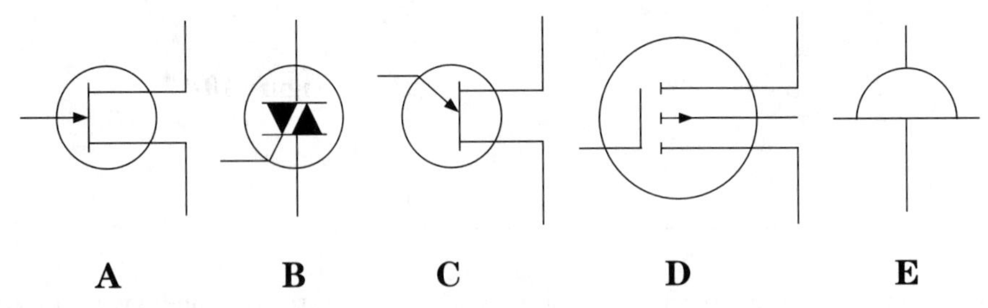

Figure 10-19

5. Figure 10-20 shows a characteristic curve. Which of the components of Fig. 10-19 would have this type of characteristic curve?

   A. The one marked A.
   B. The one marked B.
   C. The one marked C.
   D. The one marked D.
   E. Tunnel diode.

6. Which of the following types of diodes is normally operated with a continuous reverse dc voltage across it?

   A. Small-signal germanium.
   B. Silicon rectifier.
   C. Tunnel.
   D. Zener.

7. Increasing the amount of reverse bias across a varactor diode will:

   A. decrease its capacitance.
   B. increase its capacitance.

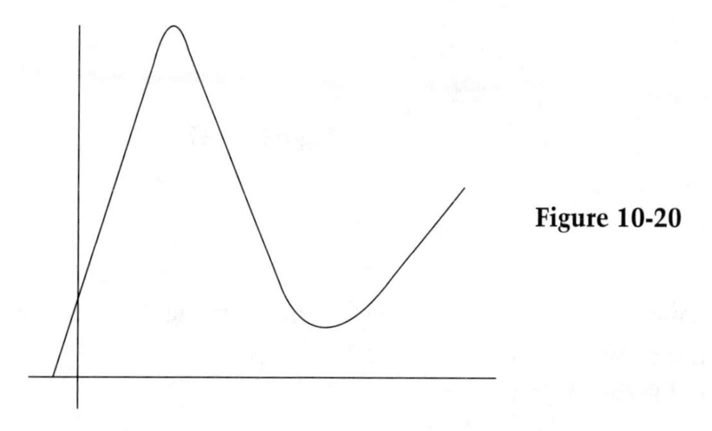

Figure 10-20

C. have no effect on its capacitance.
D. destroy it because a varactor diode must never be operated with a reverse bias voltage.

8. A "+" sign on a semiconductor rectifier diode marks:

   A. the anode lead. B. the cathode lead.

9. In a certain thermistor circuit, the voltage across the thermistor is 3.2 V when the current through it is 16 mA. If the current is increased to 32 mA, the voltage across the thermistor:

   A. will increase to 6.4 V.
   B. will decrease to 1.6 V.
   C. will not change.
   D. cannot be determined from the information given.

10. Which of the following components undergoes a change in resistance when there is a light striking it?

    A. LSR. B. PSR. C. Thermistor. D. LDR.

11. There will be no collector current flow in a PNP transistor when:

    A. its base is positive with respect to its emitter.
    B. its base is negative with respect to its emitter.

12. A certain SCR is conducting through a load. Conduction will be stopped by:

    A. making the gate positive with respect to the cathode.
    B. making the gate negative with respect to the cathode.
    C. making the gate and cathode voltages equal.
    D. none of these answers is correct.

13. Figure 10-21 shows two transistors in the same case. These transistors are:

    A. parallel connected. C. Darlington connected.
    B. series connected. D. Edison connected.

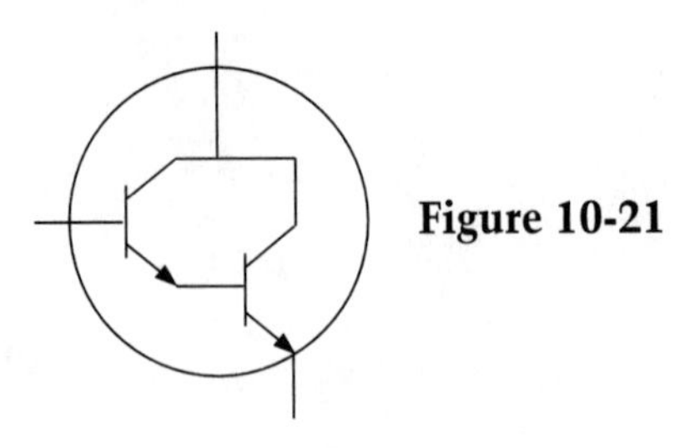

Figure 10-21

14. An advantage of the transistor connection shown in Fig. 10-21 is the:

 A. low power dissipation. C. high beta.
 B. low power-supply requirement. D. easily-matched alpha.

15. Assume that the components in Fig. 10-22 are being used in a class-A amplifier circuit. Which of the following is true?

 A. Both have the correct voltage polarities.
 B. Neither has the correct voltage polarities.
 C. Only the component in A has the correct voltage polarities.
 D. Only the component in B has the correct voltage polarities.

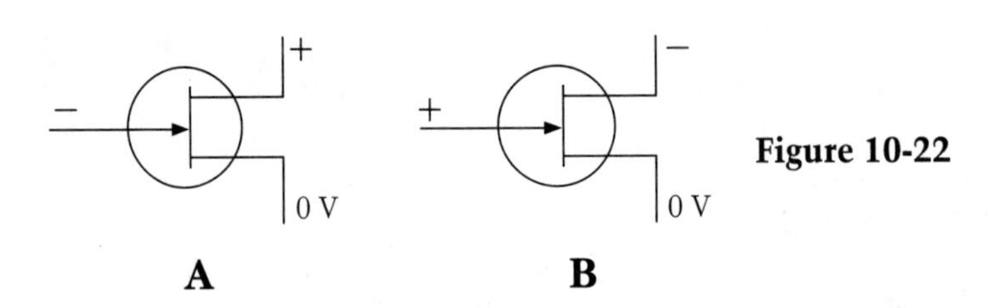

Figure 10-22

16. Which of the following is a diode that can be used as an indicating lamp?

 A. Tunnel diode. B. BSR. C. LED. D. VDR.

17. Which of the following is a diode that can be used as a very fast switch?

 A. Tunnel diode. B. BSR. C. LED. D. VDR.

18. A certain neon lamp is made with two identical electrodes. Except for the voltage values, its characteristic curve would look most nearly like the characteristic curve of a:

 A. tetrode. C. zener diode.
 B. diac. D. phototransistor.

19. In an NPN transistor being operated as a-class-A amplifier:

 A. the base is negative with respect to the collector.
 B. the base is positive with respect to the collector.

20. The dc voltage drop across a silicon rectifier diode is usually about:

 A. 0.2 V. B. 0.8 V. C. 0.08 V. D. 0.02 V.

Electronic components and circuits

1. In a certain vacuum-tube circuit, the signal delivered to the grid is lost. This, in turn, causes the plate current to become so excessively high that the plate glows red. Which of the following is correct?

   A. The tube must be gassy.
   B. The cathode resistor has changed value.
   C. The plate load resistor is shorted.
   D. Grid-leak bias is being used.

2. The filter shown in Fig. 10-23 is:

   A. a band-pass type.
   B. a band rejection (notch) type.
   C. a low-pass type.
   D. a high-pass type.

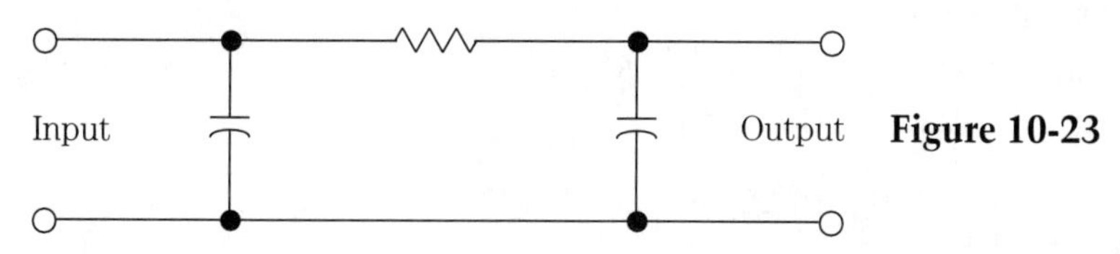

Figure 10-23

3. What is the ripple frequency of the half-wave rectifier circuit in Fig. 10-24?

   A. 60 Hz. B. 120 Hz.

4. In the circuit of Fig. 10-24, capacitor C:

   A. must have a tolerance no greater than ±5%.
   B. is a mica type.
   C. is a semiconductor type.
   D. is an electrolytic type.

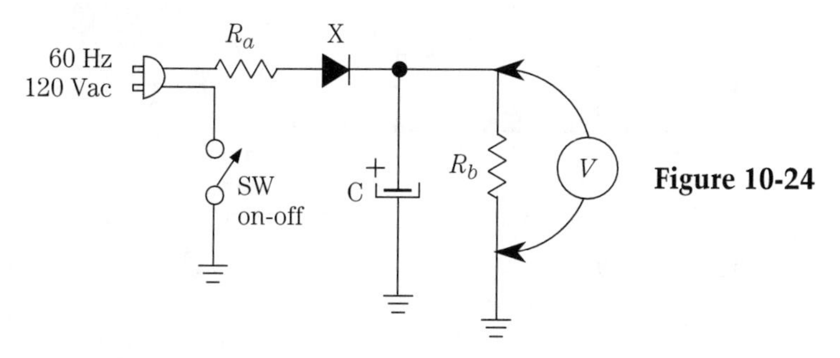

Figure 10-24

5. In the circuit of Fig. 10-24, the purpose of R_a is to:

   A. protect X from an excessive current.
   B. protect C from an excessive voltage.
   C. protect R_b from an excessive power dissipation.
   D. reduce arcing at the ON/OFF switch contacts.

6. In the circuit of Fig. 10-24, the capacitor becomes open circuited. The voltmeter reading will:

   A. increase. B. decrease.

7. Which of the following audio amplifier circuits would have a 180° phase shift between the input and output signals?

   A. Common collector. C. Common emitter.
   B. Common base. D. None of these.

8. In an emitter-follower circuit, you would expect the output signal voltage amplitude to be:

   A. greater than the input signal voltage amplitude.
   B. less than the input signal voltage amplitude.

9. Which of the following bias voltages cannot be measured accurately with a 20,000-ΩV voltmeter?

   A. Grid-leak bias. C. Power-supply bias.
   B. Cathode bias. D. Contact bias.

10. In the circuit of Fig. 10-25, assume that there is no input signal. The collector current will be:

    A. excessively high.
    B. equal to β times the resistance of the secondary winding.
    C. slightly below average.
    D. zero milliamperes.

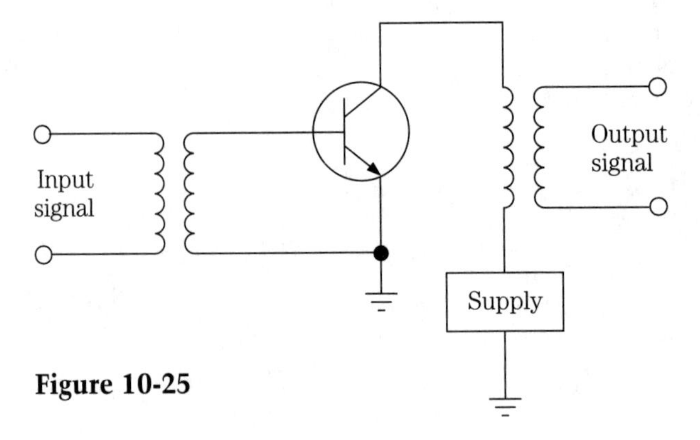

Figure 10-25

11. Which of the following amplifier configurations can be used for matching a high impedance to a low impedance?

    A. Common collector. C. Common emitter.
    B. Common base. D. Grounded grid.

12. A disadvantage of a pentode tube compared to a triode tube is that:

    A. the pentode has a lower gain.
    B. the pentode requires signal power for its operation.
    C. the pentode has a negative resistance characteristic and the triode does not.
    D. the pentode is noisier.

13. The audio amplifiers in the circuits of Fig. 10-26 use the same transistor type. Which will have the broader frequency response?

    A. The one shown in A. B. The one shown in B.

14. Regarding the circuits of Fig. 10-26, point x should be:

    A. positive with respect to ground. B. negative with respect to ground.

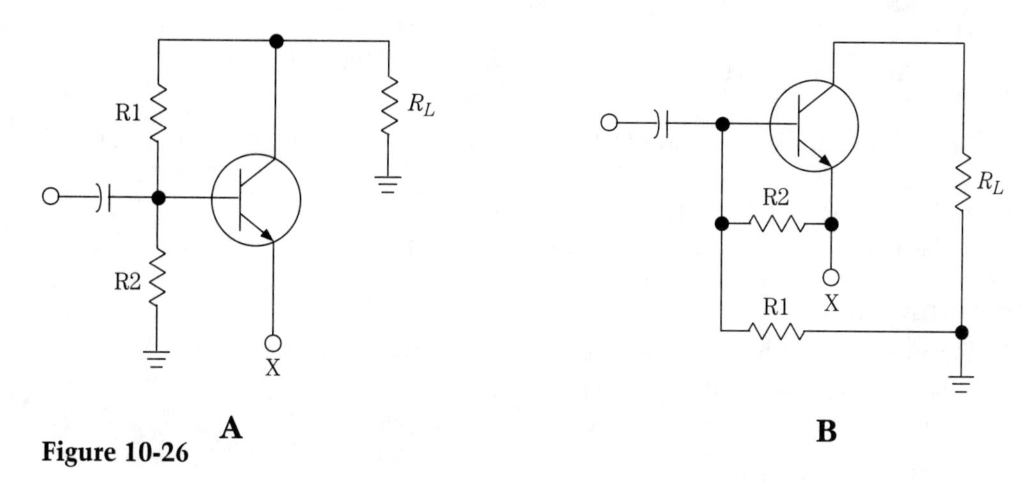

Figure 10-26

15. In which of the following amplifiers would you most likely find a low-frequency compensating network?

    A. RC coupled amplifiers.
    B. Transformer-coupled amplifiers.
    C. Direct-coupled amplifiers.
    D. Impedance-coupled amplifiers.

16. Which of the following circuits would use an NPN transistor and a PNP transistor?

    A. Push-pull amplifiers.
    B. A Darlington circuit.
    C. Parallel-connected transistors.
    D. A complementary symmetry circuit.

17. Peaking coils are used for:

    A. increasing amplifier bandwidth.
    B. increasing amplifier high-frequency gain.
    C. decreasing amplifier high-frequency gain.
    D. obtaining a regenerative feedback.

18. The input impedance of a bipolar transistor amplifier is increased by:

    A. increasing the power-supply voltage.
    B. decreasing the power-supply voltage.
    C. bootstrapping.
    D. direct coupling.

19. Which of the semiconductor packages shown in Fig. 10-27 would be used for a programmable unijunction transistor?

    A. The one marked A.
    B. The one marked B.
    C. The one marked C.
    D. The one marked D.

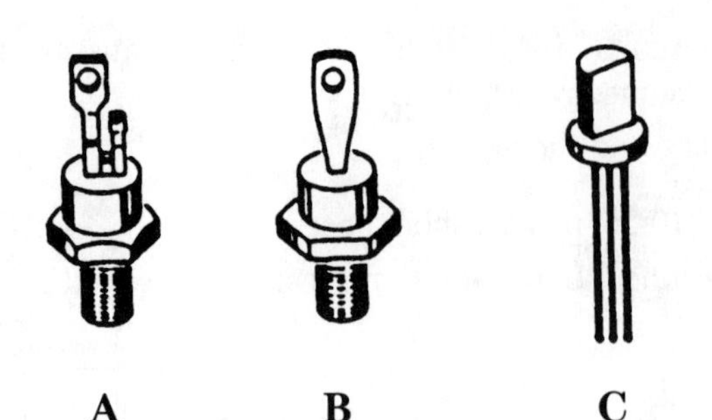

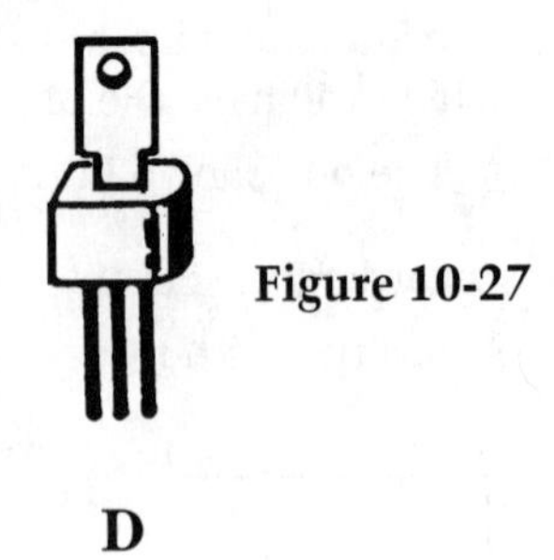

Figure 10-27

20. Which of the packages shown in Fig. 10-27 would be used for a high-current diode?

    A. The one marked A. C. The one marked C.
    B. The one marked B. D. The one marked D.

Instruments

1. To convert a sensitive meter movement to a voltmeter:

   A. place a shunt across the meter movement.
   B. place a multiplier in series with the meter movement.

2. An electronic switch is used for:

   A. switching back and forth between a voltmeter and an ammeter to make power measurements.
   B. selecting the proper ammeter scale.
   C. testing fuses.
   D. simultaneously observing two waveforms on an oscilloscope.

3. Which of the following is used for accurately measuring resistance?

   A. A wheatstone bridge. C. A grid-dip meter.
   B. A marker generator. D. A lecher line.

4. In the circuit of Fig. 10-28, the product of the voltage and current will give the:

   A. vars. C. apparent power.
   B. true power. D. phase angle.

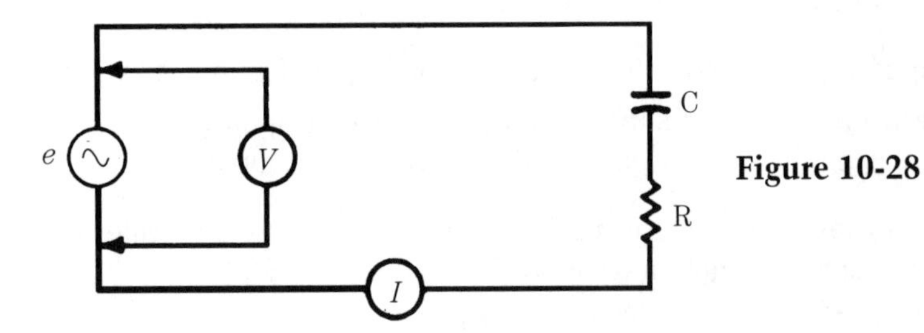

Figure 10-28

5. The pattern obtained by using a transistor curve tracer for an in-circuit test is called:

   A. a bode plot. C. a rise-time plot.
   B. a roll off. D. a signature.

6. Figure 10-29 shows the response curve of an amplifier displayed on an oscilloscope screen. This pattern is obtained with:

   A. the modulated output of an RF signal generator. C. a sweep generator.
   B. the unmodulated output of an RF signal generator. D. a tone generator.

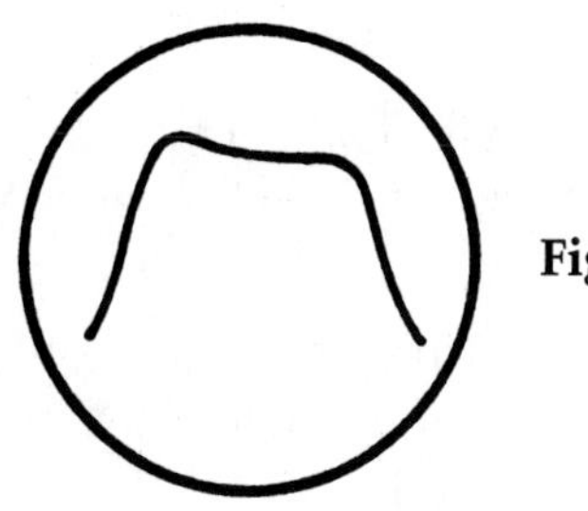

Figure 10-29

7. Important frequency points on response curves like the one shown in Fig. 10-29 are marked with:

   A. a marker generator. C. a slope detector.
   B. a spotter. D. a tracer.

8. It is possible for a voltmeter probe to have a static charge. This might destroy:

   A. a triode. C. a MOSFET.
   B. a CRT. D. an SCR.

9. The square-wave output of a sine/square-wave generator may be used for:

   A. obtaining elliptical Lissajous patterns.
   B. determining capacitor breakdown voltage.
   C. calibrating thermistor probes.
   D. evaluating amplifier frequency response.

10. Insulation can be tested for voltage breakdown with:

    A. an ohmmeter. C. a tangent galvanometer.
    B. a megger. D. an electromyograph.

11. Parallax is an error in reading an analog-type meter caused by:

    A. improper shunt value across the meter movement.
    B. reading the meter at an angle.
    C. connecting the meter across the voltage being measured.
    D. friction in the meter movement bearings.

12. Parallax error is prevented by:
    A. using a very accurate value of shunt resistance.
    B. using a meter with mirrored meter scale.
    C. using a very large value of meter multiplier resistance and a very sensitive meter movement.
    D. using a taut-baud movement.

13. Which of the following is not an advantage of a taut-baud meter movement over the jeweled type?
    A. It is less expensive and easier to repair.
    B. It can be made more sensitive.
    C. It is more rugged.
    D. It permits measurements to be made with better repeatability. In other words, it will deflect to exactly the same point when making the same measurement a number of times.

14. The meter shown in Figs. 10-30A and 10-30B has a jeweled movement. When making a measurement, greater accuracy is obtained when:
    A. the meter is standing as shown in Fig. 10-30A.
    B. the meter is lying flat as shown in Fig. 10-30B.

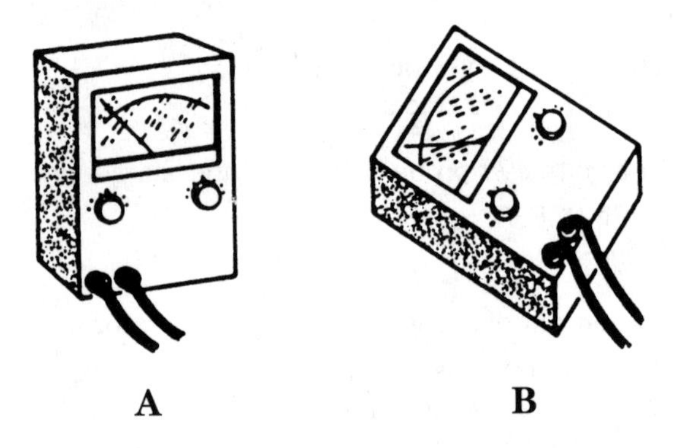

Figure 10-30

15. An oscilloscope is calibrated for 6 V per inch of vertical deflection. Figure 10-31 shows a waveform being measured. If you used a VOM to measure this voltage it would read:
    A. about 42.3 V. C. about 38.2 V.
    B. about 21.2 V. D. none of these answers is correct.

16. Reading a voltage with a VOM that has a jeweled meter movement will be more accurate if:
    A. the reading is taken with the meter range selector set so that the needle is in the upper half of the scale.

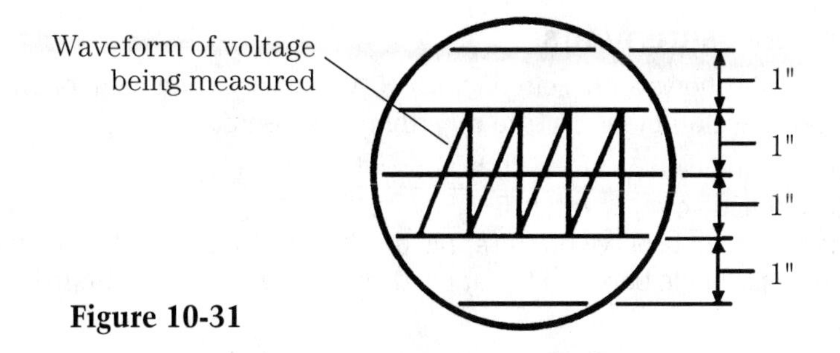

Figure 10-31

B. the reading is taken with the meter range selector set so that the needle is in the lower half of the scale.

17. Which of the following is capable of measuring a voltage with the highest accuracy?

A. A potentiometer.
B. A VOM.
C. A vtvm or FET meter with very high input impedance.
D. An ammeter and an accurately calibrated resistor.

18. A wattmeter actually measures:

A. voltage and resistance.
B. current and resistance.
C. voltage and current.
D. temperature and resistance.

19. The reading on a watt-hour meter is actually a measure of:

A. power. B. energy. C. time. D. temperature.

20. In the circuit of Fig. 10-32, you are to find a value of $R2$ that will result in no current flow through the galvanometer. Which of the following is true?

A. The value of $R2$ cannot be determined unless the values of E_a and R_a are known.
B. $R2 = 100\ \Omega$.
C. $R2 = 300\ \Omega$.
D. $R2 = 900\ \Omega$.

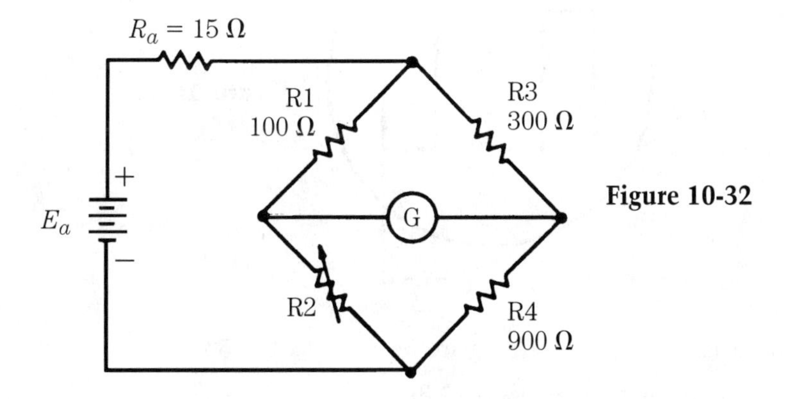

Figure 10-32

Tests and measurements

1. Figure 10-33 shows an RL circuit with 15 Vac across each component. Which of the following statements is true regarding this circuit?

   A. The applied voltage (*e*) is 30 V.
   B. The applied voltage (*e*) is 15 V.
   C. The phase angle between the applied voltage (*e*) and circuit current (*i*) is 45°.
   D. The phase angle between the applied voltage (*e*) and circuit current (*i*) is 90°.

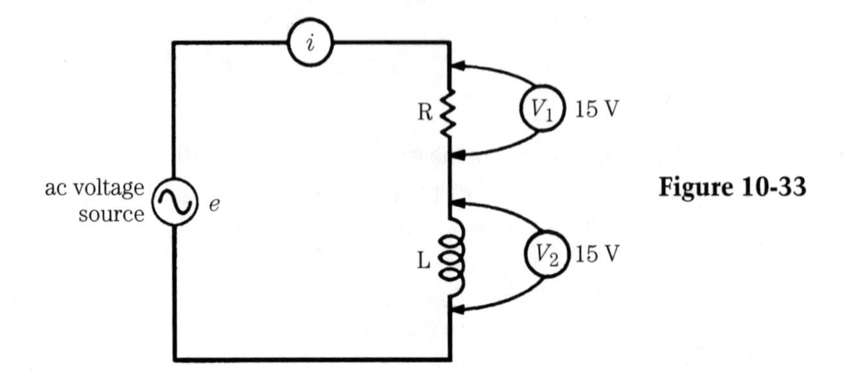

Figure 10-33

2. The oscilloscope display in Fig. 10-34 is obtained when the sweep time is exactly 5 ms. The frequency of the square wave is:

   A. 400 Hz. B. 4 kHz. C. 40 kHz. D. 2500 kHz.

3. Regarding the oscilloscope display of Fig. 10-34, in order to observe one cycle of waveform you should:

   A. increase the sweep time. B. decrease the sweep time.

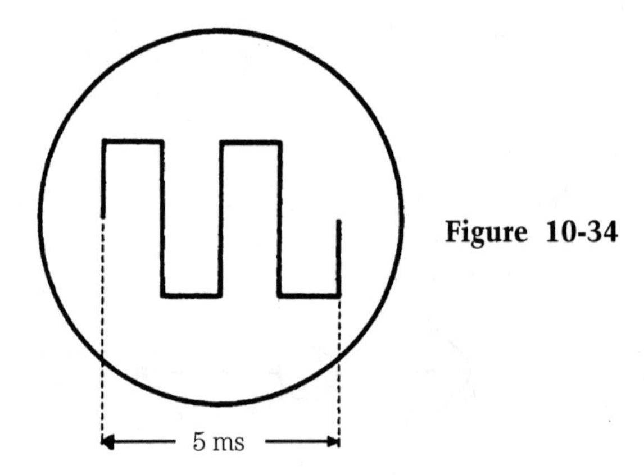

Figure 10-34

4. What is the rise time of the square wave in Fig. 10-35?

   A. 3 ms. B. 4 ms. C. 6 ms. D. 8 ms.

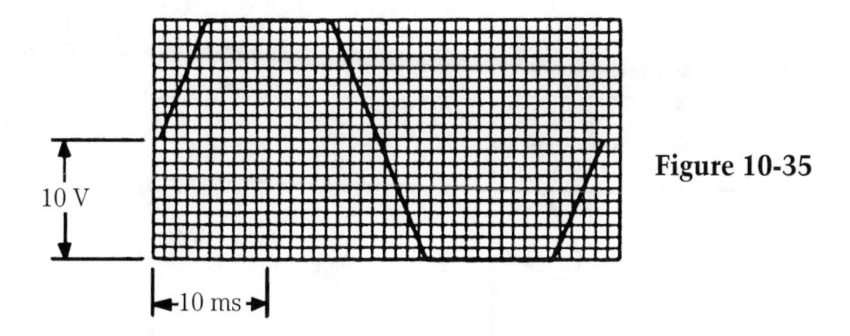

Figure 10-35

5. What is the frequency of the square wave in Fig. 10-35?

   A. 26.8 Hz. B. 250 Hz. C. 166.6 Hz. D. 125 Hz. E. None of these.

6. The Z-axis of an oscilloscope:

   A. permits it to be used for making three-dimensional computer drawings.
   B. is another name for the sweep (or trace).
   C. is used for making Lissajous patterns.
   D. is used for blanking the trace.

7. Two frequencies are being compared on an oscilloscope screen. The Lissajous method is being used, and the pattern is a perfect circle. Which of the following is true?

   A. The waveforms must be hyperbolas equal in frequency and amplitude.
   B. The waveforms are sinusoidal, equal in frequency, and 90° out of phase.
   C. The waveforms are sinusoidal, equal in frequency, and in phase.
   D. The frequency of one waveform must be twice the frequency of the other.

8. Which of the following is correct regarding currents in the bipolar transistor?

   A. The sum of the emitter and base currents should equal the collector current.
   B. The sum of the emitter and collector currents should equal the base current.
   C. The sum of the collector and base currents should equal the emitter current.
   D. Electron current flows from base to collector in an ideal PNP transistor.

9. Figure 10-36 shows a Lissajous pattern obtained by comparing two signal voltages. Which of the following is correct regarding the phase angle (ϕ) between the voltages?

   A. $\sin \phi = A/B$. C. $\tan \phi = B/A$.
   B. $\tan \phi = A/B$. D. $\cos \phi = B/A$.

10. Figure 10-37 shows the Lissajous pattern when:

    A. the vertical frequency equals the horizontal frequency.
    B. the vertical frequency is twice the horizontal frequency.
    C. the horizontal frequency is 4/3 the vertical frequency.
    D. the vertical frequency is 4/3 the horizontal frequency.

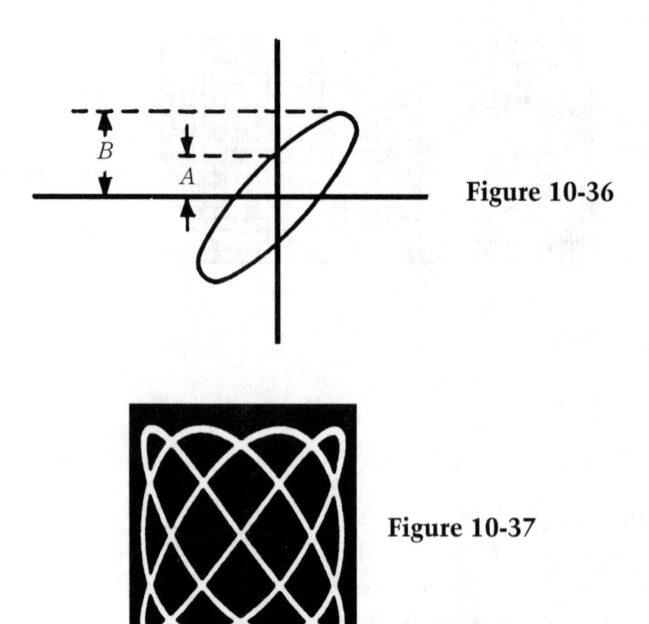

Figure 10-36

Figure 10-37

11. In the circuit of Fig. 10-38, the voltmeter has a resistance of 100,000 Ω. What is the voltage across R2 when the voltmeter is connected across it?

    A. 6 V. C. 1.8 V.
    B. 4.5 V. D. Less than 0.5 V.

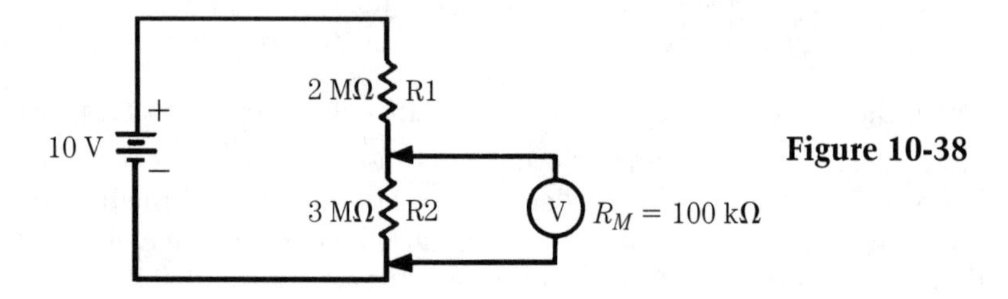

Figure 10-38

12. Which of the waveforms in Fig. 10-39 indicates crossover distortion in transistor push-pull amplifiers?

    A. The one shown in A of Fig. 10-39. C. The one shown in C of Fig. 10-39.
    B. The one shown in B of Fig. 10-39. D. The one shown in D of Fig. 10-39.

13. The problem of crossover distortion in push-pull amplifiers is eliminated by:

    A. using larger heatsinks for the power transistors.
    B. forward biasing the transistors so they do not go to 0 V emitter-base bias at the crossover point.
    C. using a better grade of transformer as a phase splitter.
    D. increasing the collector voltage on the transistors.

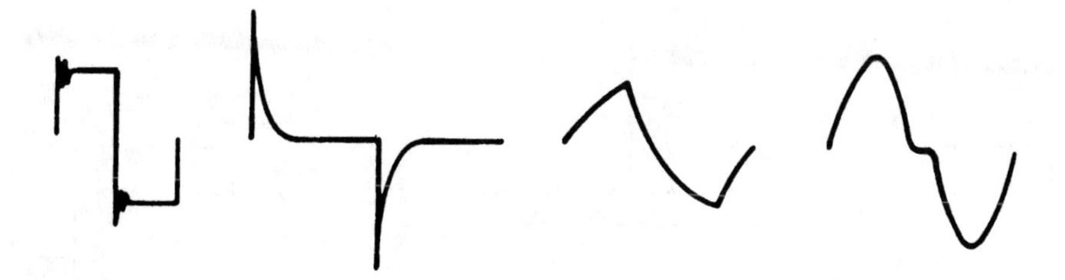

Figure 10-39

14. Which of the waveforms in Fig. 10-39 is obtained by delivering a square wave to the circuit of Fig. 10-40. Assume that the time constant of the circuit in Fig. 10-40 is very short in comparison to the time for one-half cycle of square wave.

    A. The one shown in A of Fig. 10-39. C. The one shown in C of Fig. 10-39.
    B. The one shown in B of Fig. 10-39. D. The one shown in D of Fig. 10-39.

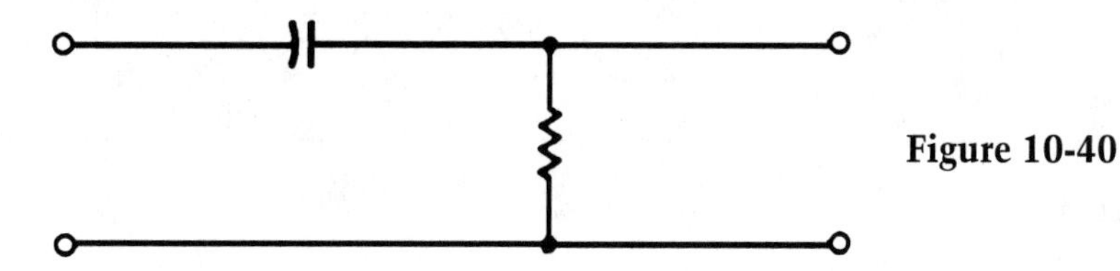

Figure 10-40

15. Which of the following is correct for finding Q or a coil?

    A. $Q = \Delta f/L$. B. $Q = X_L \times R$. C. $Q = R/X_L$. D. $Q = X_L/R$.

16. Which of the following is correct for finding the Q of a tuned circuit?

    A. *Resonant frequency* × *Bandwidth*. C. *Bandwidth* ÷ *Resonant frequency*.
    B. *Resonant frequency* ÷ *Bandwidth*. D. 1/*Bandwidth*.

17. Bandwidth is usually defined as the range of frequencies between the half power points on a response curve. If the response curve is for voltage vs. frequency, the bandwidth is measured between the points where:

    A. the voltage is one-half maximum. C. the voltage is 0.636 times the maximum.
    B. the voltage is one-third maximum. D. the voltage is 0.707 times the maximum.

18. Figure 10-41 shows a phono amplifier that uses an N-channel JFET (Q_1). The power supply should be:

    A. Positive B. Negative.

19. In Figure 10-41, gate bias is obtained with

    A. A power-supply voltage divider.
    B. A source resistor.
    C. A resistor between the gate and drain.
    D. None of these answers is correct.

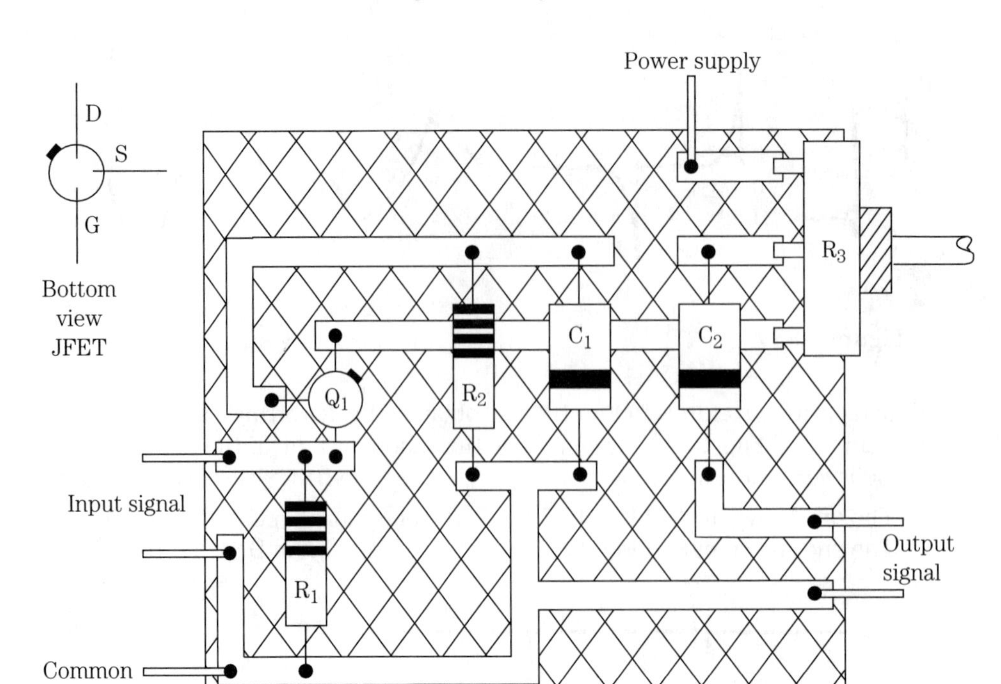

Figure 10-41

20. Regarding the circuit in Figure 10-41, if C1 is open, the result will be:
   A. Complete loss of output signal. B. An increase in gain.
   C. No change in gain. D. A decrease in gain.

21. One FEMTO ampere equals:
   A. 10^{-3} A. B. 10^{-12} A. C. 10^{-15} A. D. 10^{-24} A.

22. A ringing test is used to determine:
   A. if an inductor is good. C. if a resistor is good.
   B. if a capacitor is good. D. if an operational amplifier is good.

23. Some ohmmeters have a high-power and low-power setting. What is the reason for this?
   A. So the ohmmeter can be used for checking both germanium and silicon transistors.
   B. So the ohmmeter can be used to turn a circuit on in the high-power setting.
   C. So the ohmmeter can be used in both voltage and power amplifier circuits.
   D. So the ohmmeter (when in the low-power setting) can make measurements in transistor circuits without forward biasing PN junctions.

24. Transistor dc beta is also known as:
   A. I_{CEO}. B. h_{FE}. C. h_{fe}. D. V_{CB}.

25. You are going to use the 6.3 V transformer secondary voltage to calibrate an oscilloscope. Figure 10-42 shows the waveform. The scope is calibrated to read:
    A. 6.3 V per inch (RMS) for pure sine-wave voltage measurements.
    B. 8.9 V per inch (peak-to-peak) for pure sine-wave voltage measurements.
    C. 12.6 V per inch (RMS) for pure sine-wave voltage measurements.
    D. none of these answers is correct.

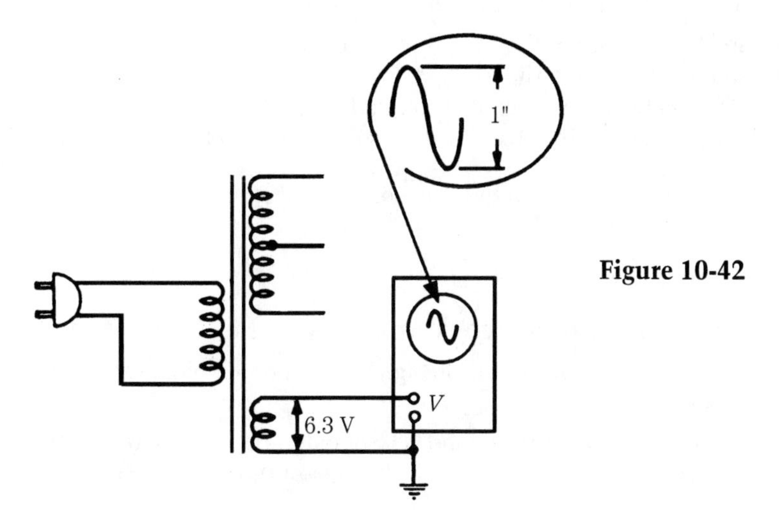

Figure 10-42

26. In the circuit of Fig. 10-43, meter M1 measures:
    A. collector current.
    B. base current.
    C. collector-base voltage.
    D. collector-emitter voltage.

27. In the circuit of Fig. 10-43, transistor Q1 is a:
    A. NPN type. B. PNP type.

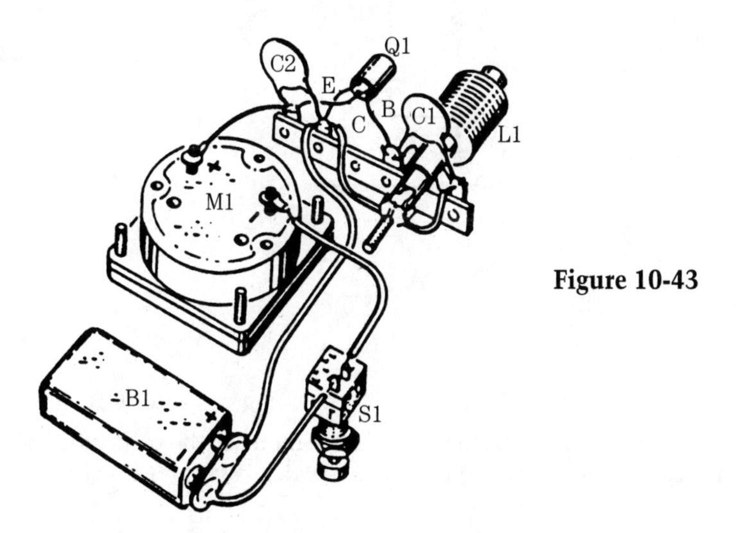

Figure 10-43

28. In the circuit of Fig. 10-43, the tuned circuit comprised of C1 and L1 is connected:
    A. between the collector and power supply.
    B. between the emitter and power supply.
    C. between the collector and the base.
    D. between the emitter and the base.

29. In the circuit of Fig. 10-43, the transistor is:
    A. biased by connecting the base directly to B+.
    B. biased by connecting the emitter directly to B+.
    C. cut off except when a signal is coupled to L1.
    D. destroyed because the base is shorted to the collector through L1.

30. Which of the following is a measure of the leakage current between base and collector?
    A. $\propto$. B. β. C. I_{CBO}. D. I_{CEO}.

Troubleshooting and network analysis

1. In the class-A amplifier circuit of Fig. 10-44, the dc voltage measured by voltmeter V should be:
    A. negative with respect to ground.
    B. equal to voltage *E*.
    C. equal to the voltage drop across R1.
    D. none of these answers is correct.

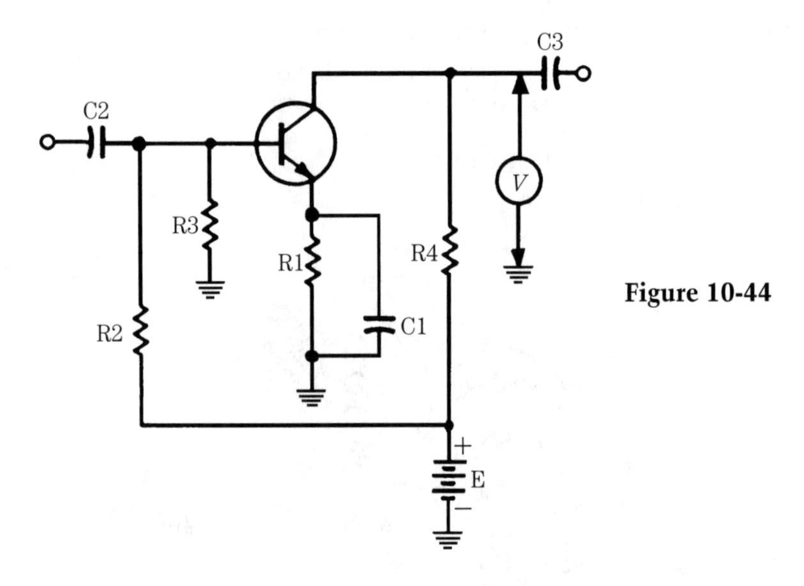

Figure 10-44

2. Refer again to the circuit of Fig. 10-44, removing capacitor C1 will:
    A. increase the gain of the stage.
    B. make the collector voltage increase.
    C. increase the bandwidth of the circuit.
    D. destroy the transistor.

3. Refer again to the circuit of Fig. 10-44, a short circuit is connected between the emitter and base. Which of the following is true?

   A. The transistor will be destroyed. C. Resistor R3 will overheat.
   B. Resistor R2 will be destroyed. D. The reading of voltmeter V will be E.

4. To check the ON/OFF switch of an electronic circuit, a milliammeter is connected across the switch. The meter should read 0 mA when the switch is:

   A. open. B. closed.

5. When troubleshooting the circuit of Fig. 10-45, a technician makes the following dc voltage readings with respect to ground:

$$E = 0\text{ V} \quad B = 0\text{ V} \quad C = -10\text{ V}$$

   Based on these readings, which of the following statements is correct?

   A. The transistor collector lead is open. C. Either R1 or R2 is open.
   B. Emitter resistor R2 is open. D. The readings are normal.

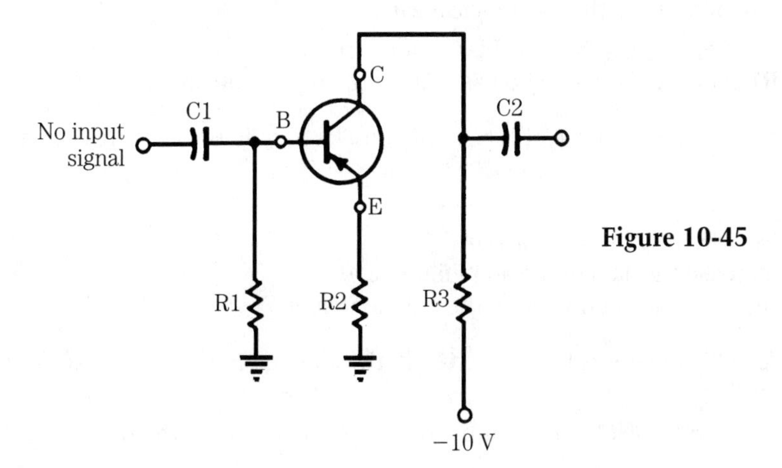

Figure 10-45

6. Assume that you are troubleshooting the triode circuit of Fig. 10-46. Which of the following is not an acceptable method of locating a circuit fault?

   A. Measure the plate voltage to see if the tube is conducting.
   B. Measure the cathode voltage to see if the bias is correct.
   C. Short the grid to the cathode while measuring the plate voltage.
   D. Measure the grid voltage to make sure that C1 is not leaky.

7. In the class-A triode amplifier of Fig. 10-46, the cathode voltage is measured and found to be 90 V. There is no output at point (b) when a signal is delivered to point (a) Which of the following is correct?

   A. R1 is open. C. C1 is open.
   B. R2 is open. D. The tube is open.

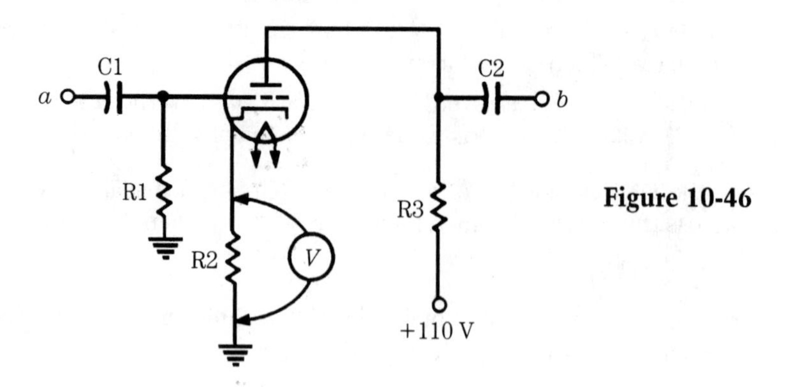

Figure 10-46

8. Refer again to the class-A audio amplifier circuit of Fig. 10-46. A 1000 Hz sinusoidal voltage is applied to point (a) and the output waveform is observed at point (b). The positive peaks of the output waveform are being clipped. Which of the following is the most logical cause?

   A. R3 is open. C. The tube bias is not correct.
   B. R1 is open. D. The tube has a filament-to-cathode short.

9. For the audio amplifier of Fig. 10-46, an electrolytic bypass capacitor across R2 will:

   A. introduce parasitic oscillations.
   B. cause a drop in plate current.
   C. increase the gain of the amplifier stage.
   D. increase the bandwidth of the amplifier stage.

10. A dc milliammeter is to be inserted in the plate circuit of Fig. 10-46. You should connect:

    A. the positive lead to the plate. B. the negative lead to the plate.

11. With no signal applied to the class-A amplifier circuit in Fig. 10-46, a dc milliammeter in the plate circuit registers 40 mA. When a 2000 Hz sine-wave signal is applied to point (a), (assuming the sine-wave signal does not overdrive the amplifier):

    A. the meter reading should increase.
    B. the meter reading should decrease.
    C. there should be no change in the meter reading.
    D. the meter will be destroyed.

12. Most T^2L (TTL) logic in a dual-inline package (DIP) have a positive power supply voltage of +5 V applied to pin #14, and the ground point is at pin #7. Figure 10-47 shows the top view of a DIP package. Which of the following is true?

    A. Pin 1 is marked with an A. C. Pin 1 is marked with a C.
    B. Pin 1 is marked with a B. D. Pin 1 is marked with a D.

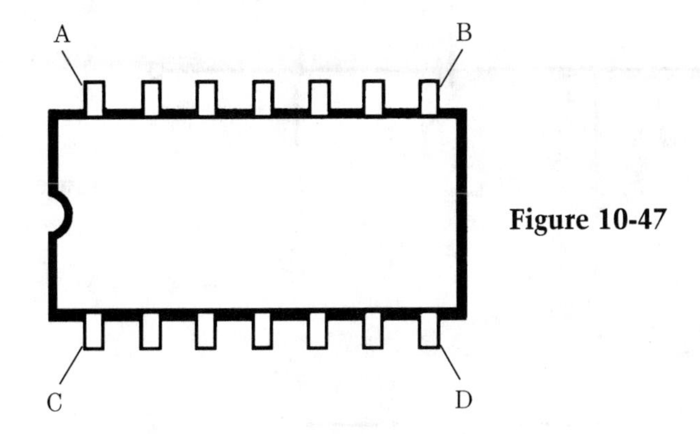

Figure 10-47

13. To make the time constants equal for the two RC circuits in Fig. 10-48, adjust x to equal:

    A. 100 MΩ. B. 10 MΩ. C. 0.01 MΩ. D. 100 kΩ.

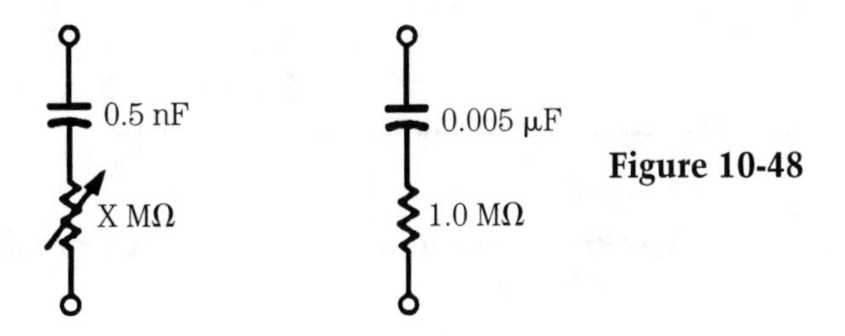

Figure 10-48

14. In an oscilloscope an electronic switch would be used for:

    A. turning the scope on and off.
    B. holding the z-axis in standby until needed.
    C. automatically shutting off the vertical amplifier when the input signal voltage is excessive.
    D. producing a dual trace for simultaneously observing two waveforms.

15. A reel contains 500 feet of 300 Ω twin-lead transmission line. Using an ohmmeter:

    A. a resistance check should show 300 Ω provided the end opposite the point of measurement is shorted.
    B. the resistance measurement will be infinity when the end opposite the point of measurement is open.

16. Figure 10-49 shows a coupling circuit between two amplifiers. When the signal at point A is on the positive half cycle, the signal at point B will be on the:

    A. positive half cycle. B. negative half cycle.

17. In the circuit of Fig. 10-50, one of the three resistors is open. Which of the following is correct?

    A. The voltage across the open resistor will be 0 V.
    B. The voltage across the open resistor will be 100 V.

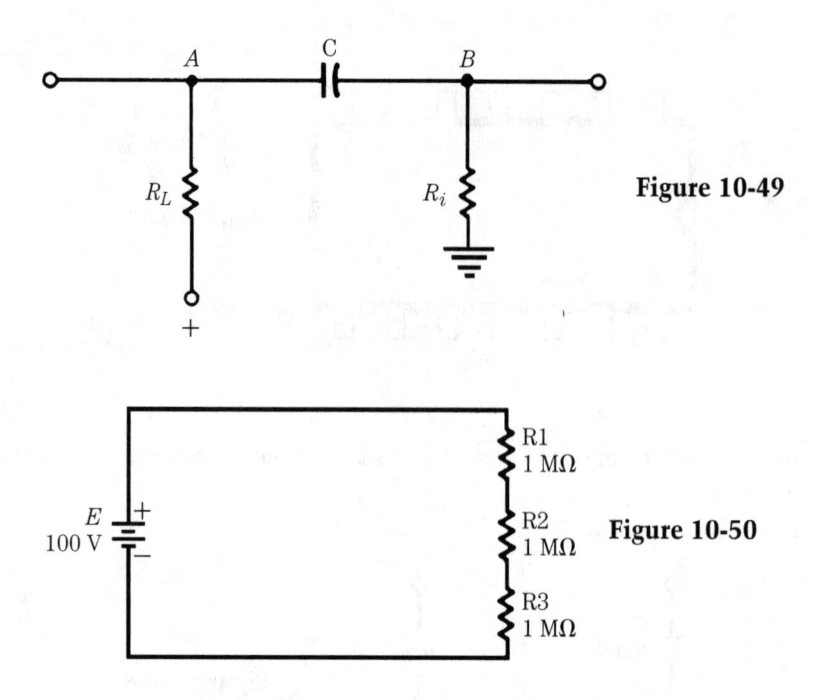

Figure 10-49

Figure 10-50

18. When measuring the voltage drop across a 2.2-MΩ resistor, you should use a voltmeter that has an Ω/V rating:

    A. that is very low. C. that is 1/2.2 Ω/V.
    B. that is 2.2 Ω/V. D. that is very high.

19. Regarding good soldering practices for repairing electronic equipment, which of the following is true?

    A. Always use acid-core solder.
    B. Use 60/40 rosin-core solder for most applications.
    C. Figure 10-51 shows the accepted mechanical connection that should be made before soldering.
    D. The tip of the soldering iron must always be filed after it has been allowed to cool.

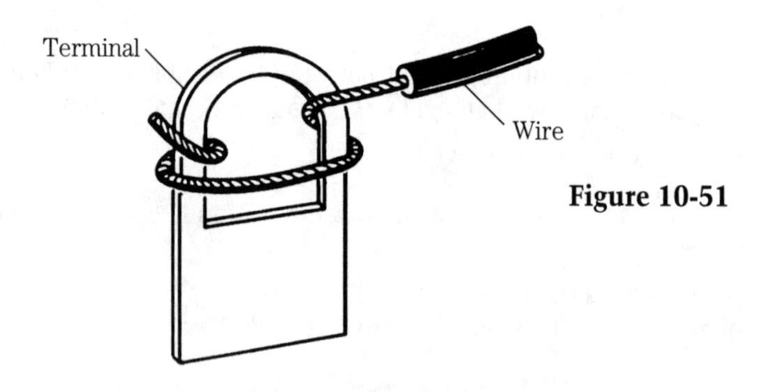

Figure 10-51

20. The brightness of a CRT display in an oscilloscope should increase if you make the CRT:

    A. grid less negative. B. cathode more negative.

21. Which of the following is not true regarding capacitors?

    A. If you charge a capacitor to a voltage then move its plates further apart, the voltage across the capacitor will increase.
    B. The black band around one side of a capacitor indicates the lead connected to the outside foil. This lead is normally connected to the ground side of a circuit.
    C. When an ac voltage (e) is applied to one lead of a capacitor the voltage at the other lead will be 90° out of phase with e.
    D. A varactor is actually a semiconductor diode, but it is used as a capacitor.

22. When mounting a power transistor on a heatsink it is important to get good thermal conductivity. This can be accomplished by the use of:

    A. an aquadag coating between the transistor and the heatsink.
    B. a thin film of machine oil between the surfaces.
    C. paint with a lead base applied to both surfaces (surfaces must be mated before paint is dry).
    D. silicone grease applied to the mating surfaces.

23. You are going to change the resonant frequency of a parallel LC circuit by bending the capacitor plates. This procedure is sometimes called knifing the plates. The resonant frequency will be increased by:

    A. moving the plates closer together. B. moving the plates farther apart.

24. Transistor junctions can be checked with an ohmmeter. For some types of ohmmeters, it is not advisable to do this with the ohmmeter set on the:

    A. $R \times 1$ scale. B. $R \times 100$ scale. C. $R \times 10$ scale. D. $R \times 1000$ scale.

25. Which of the following explains why measuring the terminal voltage of a dry cell with a high-impedance voltmeter does not tell anything useful about how it will perform in a circuit?

    A. Even if the dry cell is not capable of delivering its rated voltage under load, its terminal voltage will be equal to its rated voltage when measured with a high-impedance load.
    B. At almost the instant the voltmeter is connected the dry cell will surely be destroyed. This is because the capacitance of the probe causes a high surge current and destroys the chemical balance.

26. In the circuit of Fig. 10-52, transistors Q1 and Q2 are:

    A. stacked. C. connected as a differential amplifier.
    B. direct coupled. D. in a Darlington configuration.

27. With a voltmeter connected between the collector of Q1 and ground in the circuit of Fig. 10-52, you short the emitter to the base of Q1. Which of the following statements is correct?

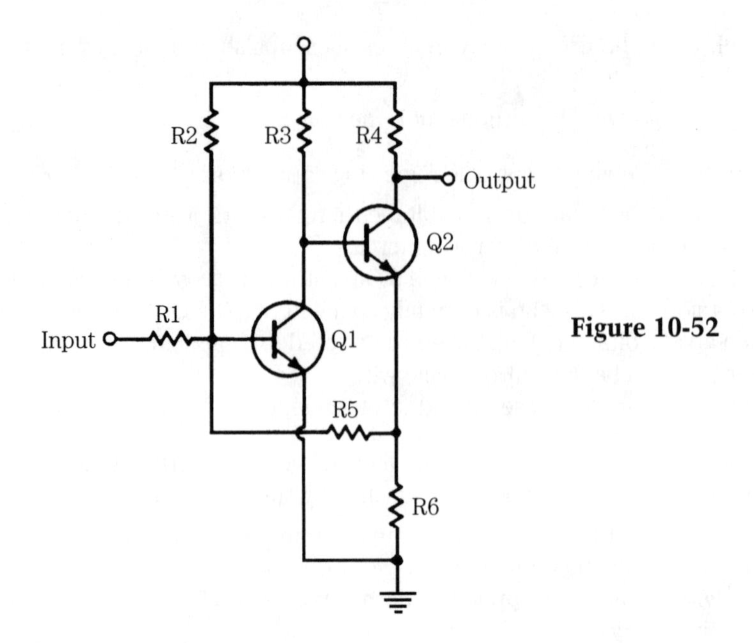

Figure 10-52

A. You have very likely destroyed Q1.
B. You have very likely destroyed Q2.
C. This is an acceptable procedure for troubleshooting in this circuit.
D. You have very likely destroyed R3.

28. In the circuit of Fig. 10-52:

A. no feedback is employed.
B. regenerative feedback occurs from the circuit of Q2 to the circuit of Q1.
C. degenerative feedback occurs from the circuit of Q2 to the circuit of Q1.
D. Q1 is an emitter-follower circuit.

29. Which of the waveforms shown in Fig. 10-53 would be correct for voltage on the deflection plates of an electrostatic deflection system if it is assumed that a linear trace is desired?

A. The one shown in A. C. The one shown in C.
B. The one shown in B. D. The one shown in D.

Figure 10-53

30. Which of the waveforms shown in Fig. 10-53 would be correct for current in the deflection coils of an electromagnetic deflection system if it is assumed that a linear trace is desired?

    A. The one shown in A. C. The one shown in C.
    B. The one shown in B. D. The one shown in D.

Answer sheet
CET practice test Associate-level

Mathematics

1. C	2. A	3. B	4. D	5. C
6. B	7. B	8. D	9. C	10. A

dc circuits

1. C	2. A	3. A	4. A	5. B
6. A	7. B	8. B	9. D	10. A

ac circuits

1. A	2. D	3. C	4. B	5. A
6. B	7. C	8. B	9. A	10. C

Transistors and semiconductors

1. D	2. D	3. B	4. C	5. E
6. D	7. A	8. B	9. D	10. D
11. A	12. D	13. C	14. C	15. C
16. C	17. A	18. B	19. A	20. B

Electronic components and circuits

1. D	2. C	3. A	4. D	5. A
6. B	7. C	8. B	9. D	10. D
11. A	12. D	13. A	14. B	15. A
16. D	17. B	18. C	19. C	20. B

Instruments

1. B	2. D	3. A	4. C	5. D
6. C	7. A	8. C	9. D	10. B
11. B	12. B	13. C	14. B	15. D
16. A	17. A	18. C	19. B	20. C

Tests and measurements

1. C	2. A	3. B	4. A	5. E
6. D	7. B	8. C	9. A	10. D

11. D	12. D	13. B	14. B	15. D
16. B	17. D	18. A	19. B	20. D
21. C	22. A	23. D	24. B	25. A
26. A	27. B	28. C	29. C	30. C

Troubleshooting and network analysis

1. D	2. C	3. D	4. B	5. D
6. C	7. B	8. C	9. C	10. B
11. C	12. C	13. B	14. D	15. B
16. A	17. B	18. D	19. B	20. A
21. C	22. D	23. B	24. A	25. A
26. B	27. B	28. C	29. A	30. A

Journeyman CET practice test communications option Please read this first!

This is not the actual Communications CET test that you will take to become a Certified Electronics Technician, but it covers questions that are similar in nature and are prepared by some of the same people who prepared actual tests.

In this practice test, helpful instructions precede each section of the test. But, titles of sections here might differ from the actual test that you receive. Refer to Appendix A for more information on ETA and ISCET Journeyman CET exams.

In recent communications tests, you might not get questions on transmitters, transmission lines, etc. But expect questions on tests, measurements, and troubleshooting and the polarities of voltages needed to get transistors and other semiconductors into operation.

Example In an NPN transistor, should the base be positive or negative with respect to the collector?

Answer Negative

Circuits using the above devices are included. Know the advantages and disadvantages of each basic configuration.

Example Which FET configuration has a high input impedance and low output impedance?)

Answer Common drain, which is also known as a *source follower*.

Be able to identify amplifier configurations. For example, know the types of power amplifiers, such as totem poles, quasi-complementary, and parallel-operated. In questions on amplifiers, it is important to be able to follow signal paths. Know when a signal is inverted and when it is not inverted.

Be able to identify the different oscillator circuits, and whether an oscillator is series or shunt fed. Crystal oscillators are especially important in this type of test.

Questions on filters, tuned circuits, and series-parallel combinations of components are usually given in the Associate-level test. However, it is not unusual to have a few questions on basics here. You will find examples in this practice test.

Practice test questions

1. Regarding the device illustrated in Fig. 10-54, and with the dc voltages shown, which of the following statements is correct?

   A. The device is conducting the maximum possible drain current, and may be destroyed.
   B. The device is a P-channel enhancement MOSFET.
   C. The input signal is normally delivered to the drain.
   D. None of these choices is correct.

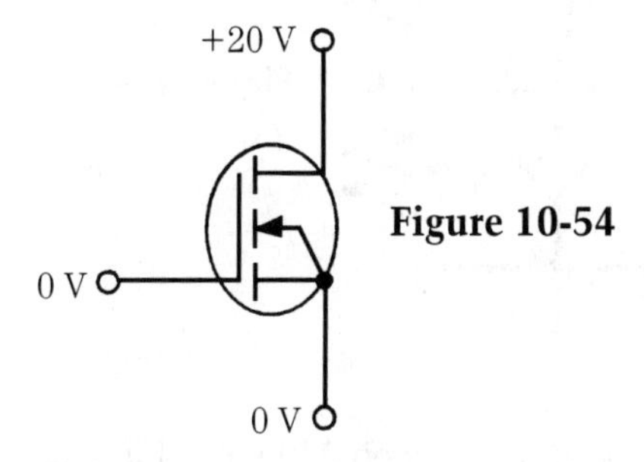

Figure 10-54

2. Which of the following equations can be used for finding the wavelength of a radio wave?

   A. *Wavelength in meters* = 300,000,000/*Frequency in Hz*
   B. *Wavelength in miles* = 186,000/*Frequency in Hz*
   C. Both equations are correct.
   D. Neither equation is correct.

3. The circuit in Fig. 10-55 can be used as a/an:

   A. wavemeter. B. Maxwell bridge. C. RF wattmeter. D. local oscillator.

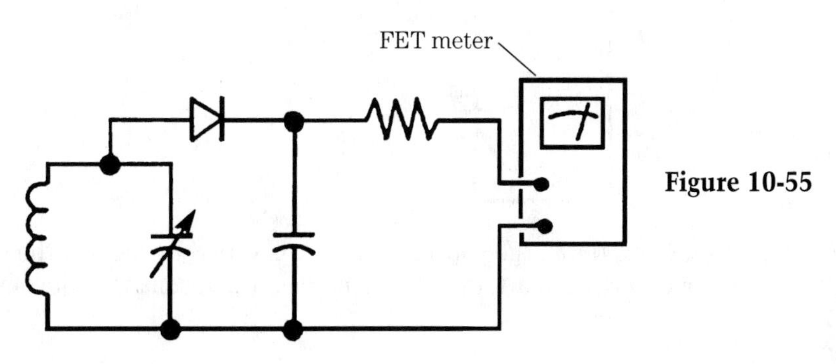

Figure 10-55

4. Figure 10-56 shows two rectifier diodes in series. The purpose of the resistors is to:
   A. limit current surges due to a charging filter capacitor.
   B. equalize voltage drops across the diodes.
   C. provide additional filtering for low-frequency ripple.
   D. none of these answers is correct.

5. The purpose of the bypass capacitors in Fig. 10-56 is to:
   A. prevent inductive kickback damage from the ripple filter.
   B. increase the junction capacitance of the diodes.
   C. prevent damage to the diodes from transient spikes.
   D. filter low-frequency ripple.

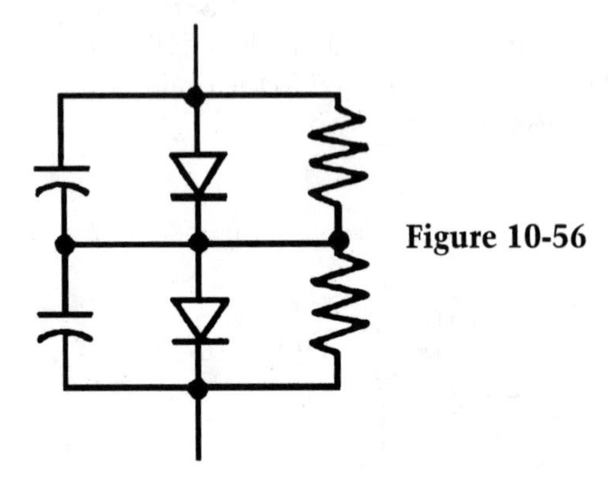

Figure 10-56

6. Which of the following components could be used in place of C in the circuit of Fig. 10-57?
   A. Forward biased hot-carrier diode.
   B. Reverse biased silicon junction diode.
   C. Forward biased PIN diode.
   D. Reverse biased germanium point-contact diode.

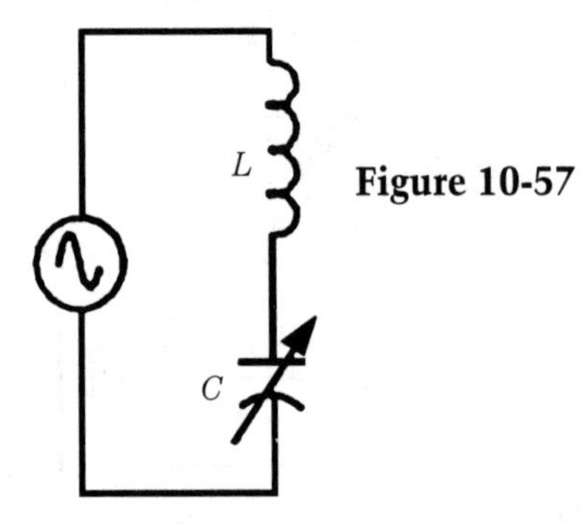

Figure 10-57

7. In the circuit of Fig. 10-58, the capacitance of C is varied by moving the plates closer together or further apart. In order to increase the resonant frequency, you should:

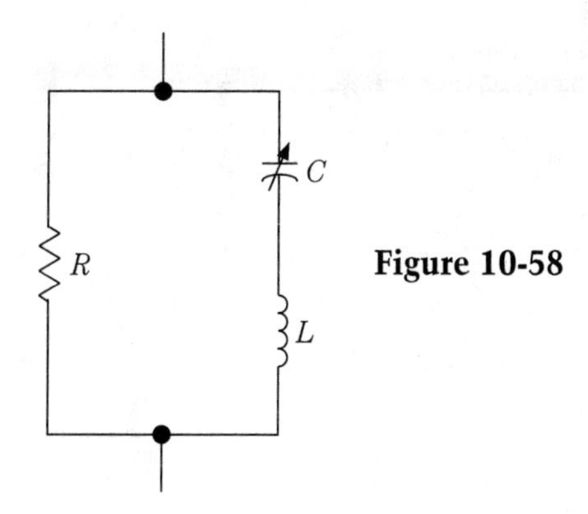

Figure 10-58

A. move the capacitor plates further apart.
B. move the capacitor plates closer together.

8. Undesired, high-frequency oscillations in an amplifier are called:

   A. parasitics. B. transients. C. VCOs. D. flutters.

9. A certain resistor is color coded Brown, Black, Black, Gold. The highest resistance this resistor can have and still be in tolerance is:

   A. 105 Ω. B. 1.05 Ω. C. 10.5 Ω. D. none of these answers is correct.

10. For the op-amp circuit of Fig. 10-59:

    A. the output signal is 180° out of phase with the input signal.
    B. the gain is $R_3/R_1 + R_2$.
    C. the terminal marked with a plus sign is never used for input signal.
    D. the terminal marked with a negative sign shows where the negative side of the power supply is connected.

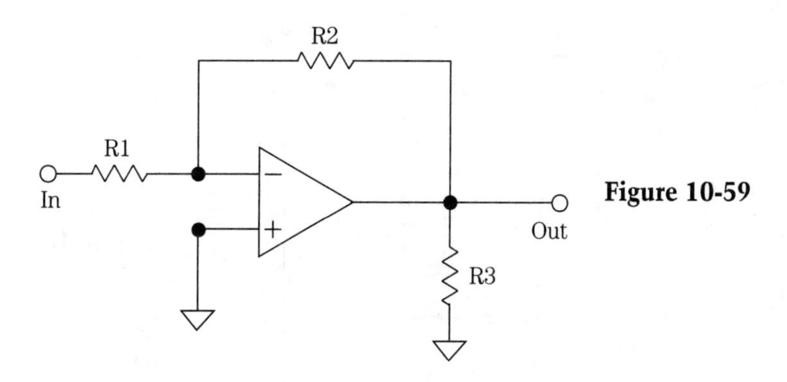

Figure 10-59

11. The power factor of the circuit in Fig. 10-60 will have a value of 1.0 when the switch is in position:

    A. x. B. y. C. z. D. none of these answers is correct.

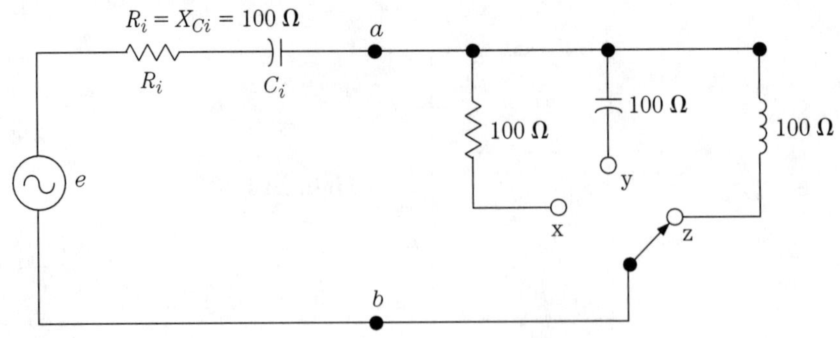

Figure 10-60

12. In the circuit of Fig. 10-60, the ac generator terminals are a and b. R_i and C_i are the internal resistance and reactance of the generator. The generator will deliver the most power to the load when the switch is in position:

    A. x. B. y. C. z. D. none of these answers is correct.

13. The transistor in the oscillator circuit of Fig. 10-61 is operated:

    A. Class A. B. Class B. C. Class C. D. cannot determine from information given.

14. The power supply voltage in the circuit of Fig. 10-61 should be:

    A. positive. B. negative.

15. The transistor oscillator in the circuit of Fig. 10-61 is:

    A. series fed. B. shunt fed.

16. The transistor in the circuit of Fig. 10-61 is connected in the:

    A. common-base configuration. B. common-emitter configuration.

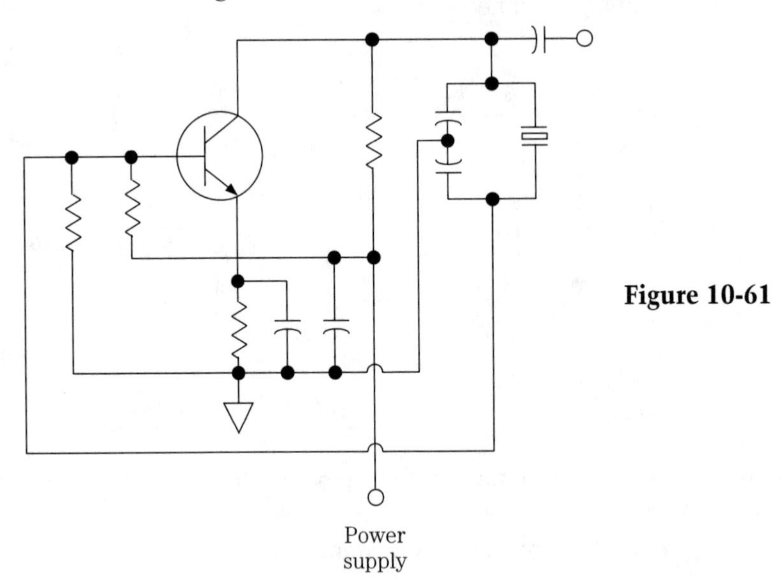

Figure 10-61

17. In Fig. 10-62, the resonant frequency of the parallel L-C circuit is 100 kHz. As seen by the 110-kHz ac generator, the circuit is:
    A. open. B. resistive. C. capacitive. D. inductive.

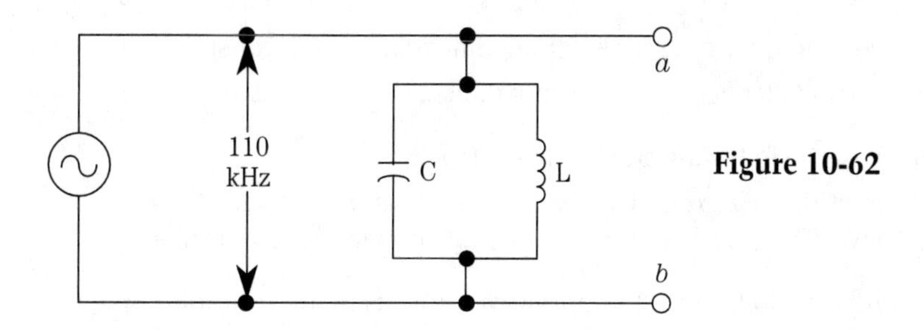

Figure 10-62

18. Adding a resistor between terminals a and b in the circuit of Fig. 10-62 will:
    A. make the circuit tune more like y in Fig. 10-63.
    B. make the circuit tune more like x in Fig. 10-63.

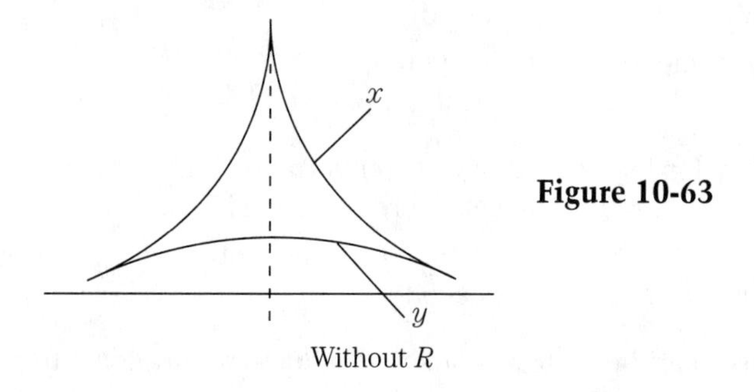

Figure 10-63

19. Greater "talk power" is obtained in a transmitter by using:
    A. a speech clipper. C. a carbon microphone with dc bias.
    B. an equalizer. D. slightly greater than 100% modulation.

20. Regarding receiver operation, the initials BFO stand for:
    A. Bare Foot Operation. C. Break Front Operation.
    B. Beat Frequency Oscillator. D. Backward-Forward Oscillation.

Antennas and transmission lines

In addition to transporting a signal from one point to another, transmission line segments are used for impedance matching, tuning, mechanical support, and frequency measurement.

It would be a good idea to know the characteristics of both open and shorted lines, at various lengths up to and including half wave. Standing waves, and measurement of standing wave ratios, should be clearly understood. Impedances of transmission lines will also be the subject of some questions in this section.

The methods of matching transmission lines to antennas are important. Also, the types of antennas, especially the quarter-wave and half-wave types, must be known. The use of loading capacitors and loading coils to change the electrical length of antennas will be useful information for this part of the test.

21. Another name for the characteristic impedance of a transmission line is:

    A. bode impedance. B. surge impedance. C. I^R impedance. D. E-I impedance.

22. Which of the following statements is true?

    A. Above 10-MHz ground-wave coverage is greatly reduced.
    B. Above 10-MHz ground-wave coverage increases with frequency.

23. Which of the following standing-wave ratios is more desirable?

    A. 1:1. B. 10:1.

24. The base of a vertical quarter-wave antenna that is shunt fed:

    A. must never be grounded. B. can be grounded.

25. A quarter-wave stub, with a shorted end, has:

    A. nearly zero impedance at the open end.
    B. an extremely high impedance at the open end.

26. The tendency for high-frequency currents to flow near the surface of conductors, rather than through the entire cross section is called:

    A. skin effect.
    B. the outer flow ratio.
    C. inverse resistance ratio.
    D. peel effect.

27. You could use a grid-dip meter (or semiconductor equivalent) to measure:

    A. undesired radiated power from a transmission line.
    B. IC voltages on Dual Inline Packages.
    C. Grid modulation percentage.
    D. the resonant frequency of an L-C tank circuit.

28. Lecher lines are used for measuring:

    A. frequency. B. radiation resistance. C. power. D. efficiency.

29. If you cut a 100-foot section of a 72 Ω coaxial cable into two 50-foot lengths, each section will have an impedance of:

    A. 36 Ω. B. 72 Ω. C. 144 Ω. D. 300 Ω.

30. In the transformer symbol of Fig. 10-64, the purpose of the shield (marked X) is to:

    A. reduce the primary current.
    B. reduce hysteresis losses.
    C. reduce eddy current losses.
    D. prevent electrostatic coupling between the primary and secondary.

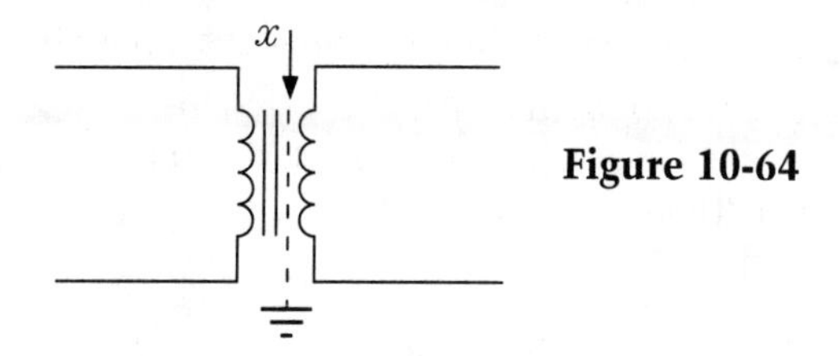

Figure 10-64

31. Which of the antennas in Fig. 10-65 will radiate a wave with a horizontal magnetic field?

    A. The one marked B. B. The one marked A.

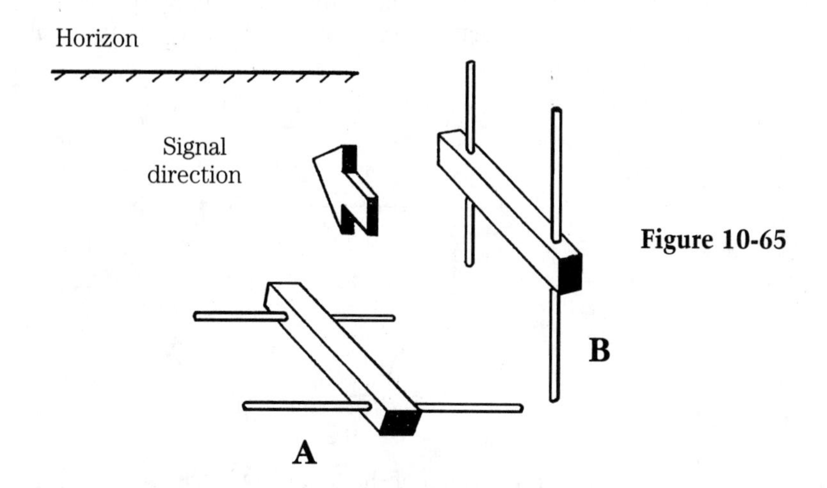

Figure 10-65

32. If a piece of wire is too short to be used as a quarter-wave antenna, it can be made electrically longer by adding:

    A. inductive loading. B. capacitive loading.

33. Devices for matching a balanced transmission line and load to an unbalanced transmission line and generator, with maximum transfer of power between the two, are called:

    A. reeds. B. diacs. C. baluns. D. Z-angle matches.

34. Which of the following is not an example of transmission line loss?

    A. Hysteresis loss. B. Dielectric loss. C. Copper loss. D. Radiation loss.

35. Which of the following antennas most nearly approaches an isotropic radiator?

    A. Loop. B. Rhombic. C. Marconi. D. Hertz.

36. A director in an antenna system is an example of:

    A. a shield. B. a parasitic element. C. a driven element. D. an array.

37. With a diversity antenna system, the receiving antennas should be:

    A. not less than one wavelength apart.
    B. more than one-half wavelength, but less than a full wavelength apart.
    C. one-half wavelength apart.
    D. one-fourth wavelength apart.

38. Which of the following is not a method of matching a transmission line to an antenna?

    A. Transmission line stub. B. Gamma match. C. Delta match. D. Kappa match.

39. As shown in Fig. 10-66, a 20-foot length of 72 Ω coaxial cable is terminated with a pure resistance (no inductance or capacitance). Assuming that the generator is matched to the line, the resistor will dissipate the maximum possible heat when the resistance value is:

    A. 20 × 72 = 1440 Ω. B. 72 Ω. C. 36 Ω. D. 0 Ω.

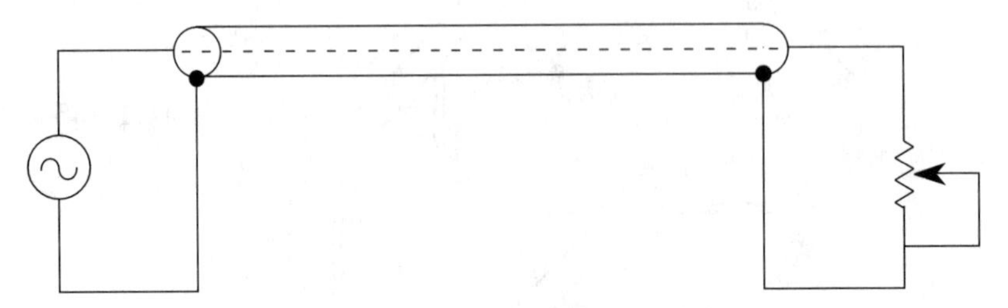

Figure 10-66

40. Regarding a Marconi antenna, which of the following is true?

    A. It is no longer being used.
    B. Voltage is maximum at the base, minimum at the top.
    C. Current is maximum at the base, minimum at the top.
    D. It is a half-wave, center-fed antenna.

Modulation and demodulation

Questions in this section are mostly related to AM, FM, and single-sideband operation. There might be an occasional question on pulse position modulation. However, unlike the FCC test, there are no questions on television modulation or demodulation. Know the Hazeltine, Rice, and low-level vs. high-level modulation circuits.

It is very important to know the measurements related to modulation. This includes such factors as percent modulation, modulation index, deviation, and sideband frequencies.

All types of demodulators are included in questions. Know the difference between ratio detectors, discriminators, and other FM detectors.

Phase-locked loops (PLLs) are becoming increasingly important because of their use as detectors, and also because of their use in frequency synthesizers. Part of this popularity is due to the availability of integrated circuit PLLs, such as the 565. You should know the basics of PLLs for this section and for other sections of the CET test.

Superheterodyne receivers are most popular, but other types should also be understood. Remember that the converter in a superhet is sometimes called the *first detector*, and this means that you might get questions on image frequencies, frequency conversion, and double conversion.

41. When a pure sine wave with a frequency of 1200 Hz is used to modulate a 27.5 MHz signal in an AM transmitter, the lower sideband frequency is:

    A. 26.3 MHz. B. 26.12 MHz. C. 27.5 MHz. D. none of these answers is correct.

42. A downward carrier shift with modulation in an AM transmitter:

    A. means that the negative peaks are greater than the positive peaks.
    B. means that the frequency of the carrier goes down when modulation is applied.

43. A certain superheterodyne receiver has a 455-kHz IF frequency. When tuned to a 1500-kHz signal, one of the image frequencies will be:

    A. 3000 kHz. B. 2410 kHz. C. 1545.5 kHz. D. 900 kHz.

44. The circuit in Fig. 10-67 is a:

    A. high-pass filter. C. low-pass filter.
    B. band-rejection filter. D. band-pass filter.

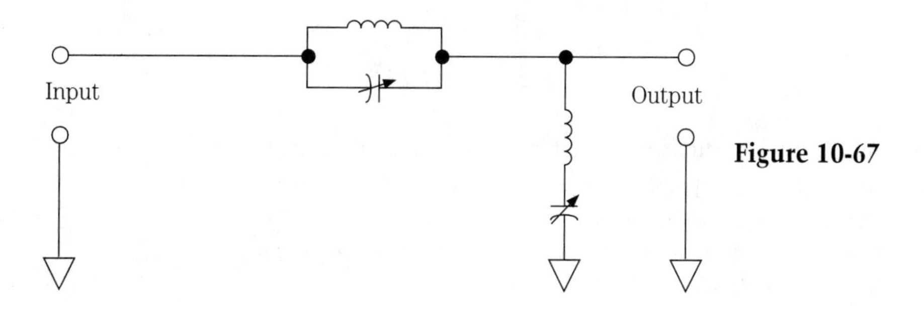

Figure 10-67

45. Which of the following might be used as a local oscillator in the TRF receiver?

    A. Armstrong. C. Both of these may be used.
    B. Shunt-fed Hartley. D. Neither of these could be used.

46. To eliminate unnecessary noise from a receiver when it is not receiving a signal, the receiver may use a:

    A. squelch circuit. B. reducer. C. noise gate circuit. D. noise limiter circuit.

47. In a broadcast FM signal, the bandwidth increases with:

    A. an increase in the amplitude of the audio modulating signal.
    B. an increase in the frequency of the audio modulating frequency.

48. A voltage-controlled oscillator:

 A. could not be used in a direct FM transmitter to generate the carrier.
 B. could be used in a direct FM transmitter to generate the carrier.

49. In comparing typical Hartley oscillators:

 A. both tube and bipolar transistor circuits are operated class B.
 B. both tube and bipolar transistor circuits are operated class A.
 C. both tube and bipolar transistor circuits are operated class C.
 D. those using tubes are operated class C, while bipolar transistor types are forward biased for class AB operation.

50. You would expect to find a de-emphasis circuit in:

 A. an FM transmitter. B. an FM receiver. C. a single-sideband transmitter.

51. In the modulation display of Fig. 10-68, $A = 2.5$ cm and $B = 3.5$ cm. The percent modulation is at:

 A. 41.66%. B. 28.57%. C. 71.4%. D. 16.66%.

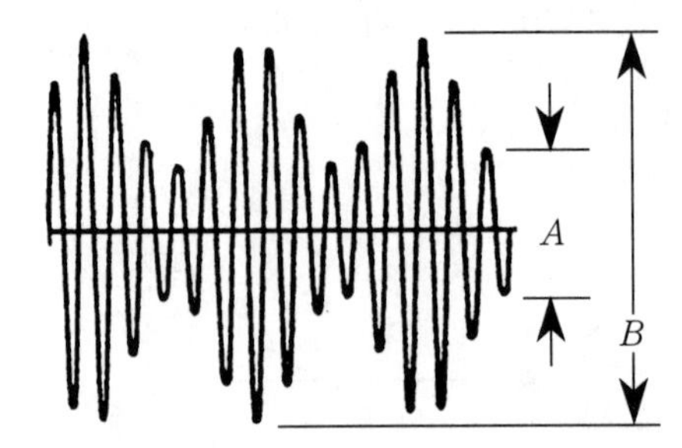

Figure 10-68

52. A balanced modulator is used in single-sideband transmitters because:

 A. it automatically eliminates the carrier.
 B. it has no phase distortion.
 C. it can be made with one PNP and one NPN transistor.
 D. it automatically eliminates one sideband.

53. The oscillator shown in Fig. 10-69 is:

 A. an Armstrong type. C. a Colpitts type.
 B. a Pierce type. D. an R-C phase shift type.

54. If the input signal of the frequency multiplier in Fig. 10-70 is off frequency by 0.05%, the error in output frequency will be:

 A. 23.25 kHz. B. 19.375 kHz. C. 0.0316 MHz. D. 0.018 MHz.

55. Consider this equation for an FM signal:

 Deviation ratio = Deviation/Modulating frequency

 A. The equation is not correct. B. The equation is correct.

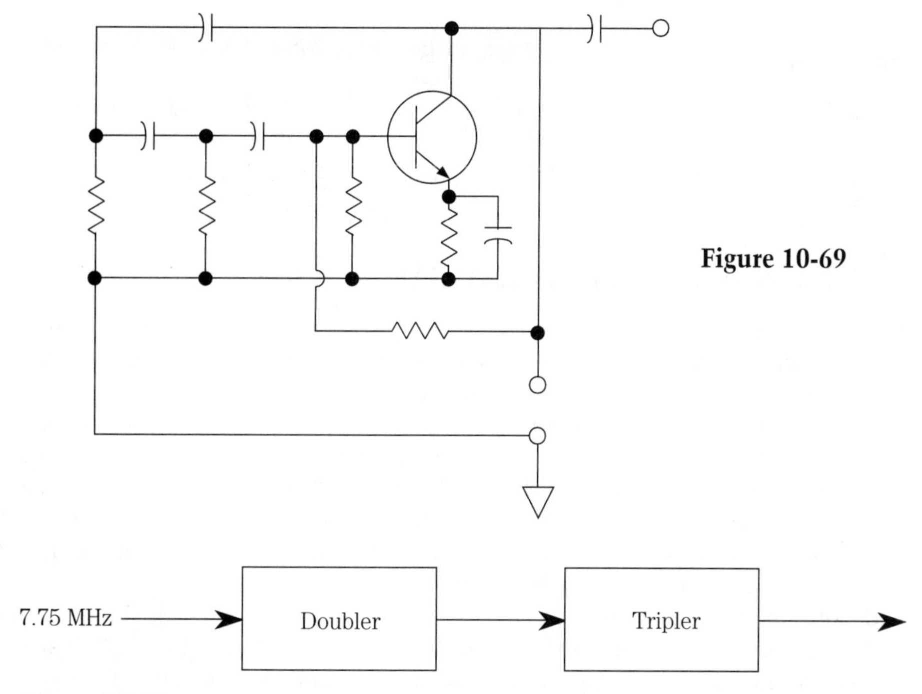

Figure 10-69

Figure 10-70

56. The output frequency for the frequency multiplier system in Fig. 10-70 should be:
    A. 12.75 MHz. B. 46.5 MHz. C. 38.75 MHz. D. none of these answers is correct.

57. Consider this equation for an FM signal:

    Frequency deviation × *Modulation frequency* = *Modulation index*

    A. This equation is not correct. B. This equation is correct.

58. Which of the following FM demodulators is preceded by a limiter stage?
    A. Gated beam detector. C. Phase-locked loop.
    B. Discriminator. D. Ratio detector.

59. Splattering is caused by:
    A. broken parasitic elements on an antenna. C. excessive modulation.
    B. oscillator pulling. D. insufficient modulation.

60. Which of the following is true regarding the trapezoid pattern of Fig. 10-71?
    A. It is used only for determining the FM modulation index.
    B. It can be used for calculating percent modulation of an AM signal using the following equation: *% Modulation* = $x/y \times 100$
    C. It shows that the output signal is greatly distorted.
    D. None of these statements is correct.

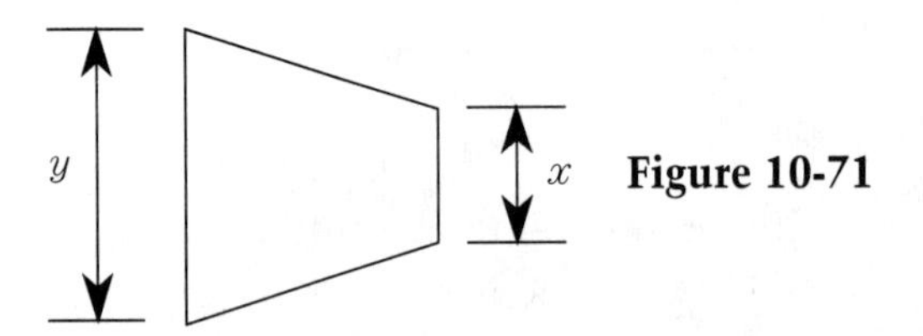

Figure 10-71

Circuits and components

Technicians usually do well with this section. It includes circuits and components used in transmitters, receivers, and transceivers. As in other sections, troubleshooting and measurements are emphasized.

An important part of this section is the test of your ability to work directly in circuits. There are examples in this practice test (Questions 61 through 67). It is best to study the complete circuit board, first, to get an idea of the circuitry. This approach is better than starting to answer questions about parts of the circuit without having an overall picture. Get the input and output terminals clear in your mind, and locate various points for making measurements.

Be sure to read each question carefully, and be sure you answer the question being asked. Interviews with technicians after they have taken the CET test reveal that carelessness is an important factor in the number of questions answered incorrectly.

Figure 10-72 shows the component layout for a regulated power supply. Questions 61 through 67 test your ability to analyze a circuit from components rather than from a schematic diagram. In this practice test, the parts layout for the circuit is shown in Fig. 10-72. The schematic would *not* normally be given in the actual CET test!

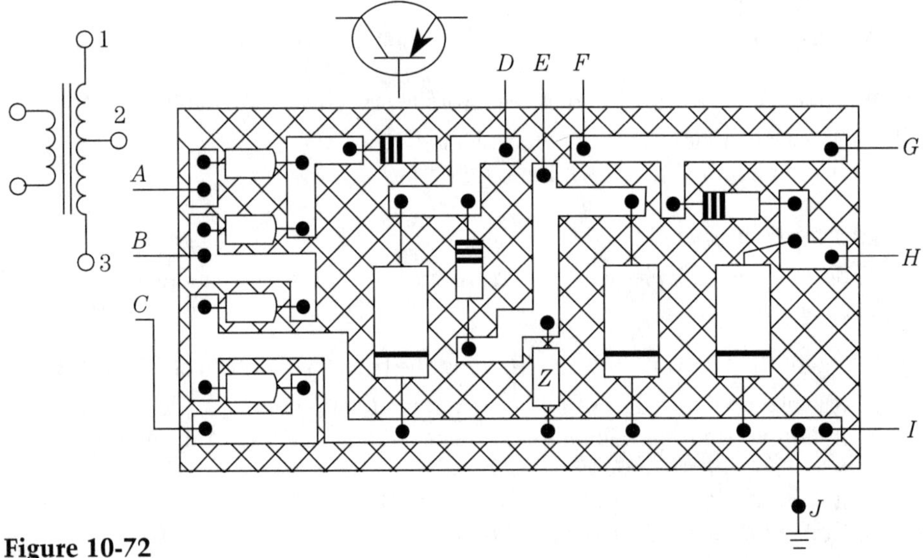

Figure 10-72

61. To make a bridge rectifier in the circuit of Fig. 10-72:
    A. connect leads A and B to transformer terminal #2 and lead C to transformer terminal #1.
    B. connect lead A to transformer terminal #1, lead B to transformer terminal #2, and lead C to transformer terminal #3.
    C. connect leads A and B to transformer terminal #1, and lead C to transformer terminal #3.
    D. connect leads A and C to transformer terminal #1, and lead B to transformer terminal #3.

62. Regarding the circuit of Fig. 10-72, the PNP power transistor must:
    A. be connected with the collector at lead F.
    B. be connected with the collector at lead E.
    C. be connected with the collector at lead D.
    D. not be used in this positive power supply.

63. Regarding the circuit of Fig. 10-72, the cathode of the zener diode (Z) should be connected to:
    A. the copper strip that leads I and J are connected to.
    B. the copper strip that lead E is connected to.

64. There is a resistor between the foils connected to leads G and H in Fig. 10-72. It is used:
    A. as a surge limiter. C. as a bleeder.
    B. for obtaining bias. D. none of these answers is correct.

65. There is a resistor connected between the foil connected to lead D and the rectifier in Fig. 10-72. This resistor is used as a:
    A. surge-limiting resistor. C. voltage-dropping resistor.
    B. bias resistor. D. none of these answers is correct.

66. Assume that a power transistor is mounted outboard in the circuit of Fig. 10-72. To increase the current rating of the regulated power supply, an additional transistor can be added in:
    A. an AND configuration. C. a stacked amplifier configuration.
    B. a Darlington configuration. D. a totem pole configuration.

67. A ferrite bead is to be added to the circuit of Fig. 10-72. Its purpose is to prevent parasitic oscillations. It would normally be added to:
    A. Lead I. B. Lead E. C. Lead B. D. Lead A.

68. Which of the following stages might be used to follow the oscillator that generates the carrier in an AM transmitter?
    A. Low-pass filter. B. Inverter-converter. C. AGC-AFC. D. Buffer-multiplier.

69. A number of different frequencies for a transmitter carrier can be obtained with two crystals and:

A. a synthesizer. C. a digital/analog converter.
B. a Loftin-White amplifier. D. an op amp.

70. Which of the following is a circuit used for neutralizing an amplifier?

A. Rice. B. Parametric. C. Phase lock. D. Snubber.

71. Which of the following amplifiers has a high input impedance and low output impedance?

A. Bootstrap amplifier. C. Common-base amplifier.
B. Common-collector amplifier. D. Common-emitter amplifier.

72. Figure 10-73 shows the response curve of a tuned circuit. The bandwidth for this curve is:

A. measured at the half power points.
B. equal to the center frequency divided by the circuit Q.
C. measured at the points where the amplitude is down to 0.707 of the maximum.
D. all of these answers are correct.

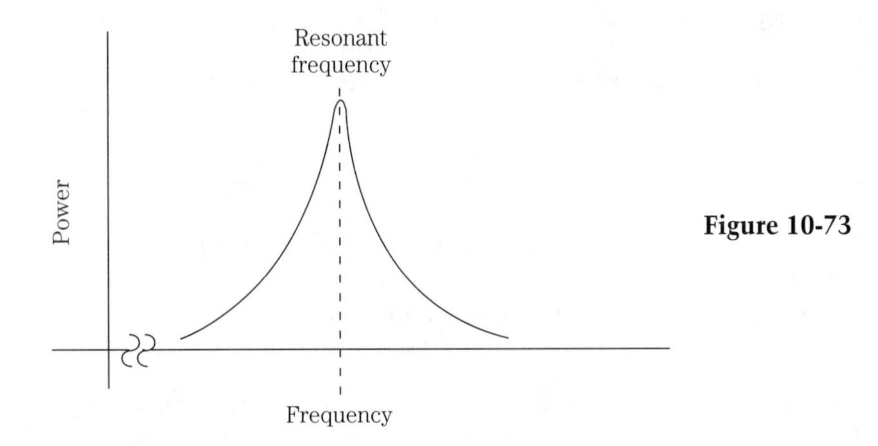

Figure 10-73

73. In the circuit of Fig. 10-74 the phase angle between the voltage and current would:

A. increase if the plates of capacitor C were moved further apart.
B. increase if the plates of capacitor C were moved closer together.

74. Which of the following statements is true regarding the circuit in Fig. 10-74?

A. The value of e cannot be obtained by adding V_1 and V_2.
B. $e = V_1 + V_2 = 12$ V.

75. A dc voltage is not normally required for the operation of a:

A. carbon microphone. C. "condenser" microphone.
B. ribbon microphone. D. all of these answers are correct.

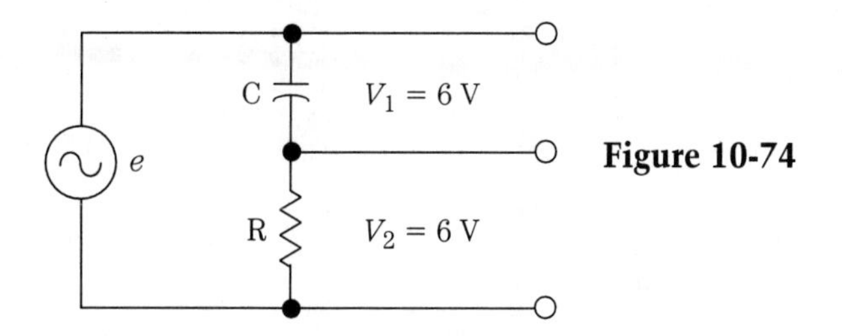

Figure 10-74

Answer sheet for Journeyman question test

1. D	2. C	3. A	4. B	5. C
6. B	7. A	8. A	9. C	10. A
11. C	12. C	13. A	14. A	15. B
16. B	17. C	18. A	19. A	20. B
21. B	22. A	23. A	24. B	25. B
26. A	27. D	28. A	29. B	30. D
31. A	32. A	33. C	34. A	35. C
36. B	37. A	38. D	39. B	40. C
41. D	42. A	43. B	44. B	45. D
46. A	47. A	48. B	49. D	50. B
51. D	52. A	53. D	54. A	55. A
56. B	57. A	58. B	59. C	60. D
61. D	62. A	63. B	64. D	65. A
66. B	67. B	68. D	69. A	70. A
71. B	72. A	73. B	74. A	75. B

11
CHAPTER
GMDSS

Element 9—Maintenance practices and procedures

The initials GMDSS stand for Global Maritime Distress and Safety System. This chapter has all possible questions and answers in the Element 9 Question Pool.

Examiners must use questions from the latest pool made available to the public when preparing a question set for a written examination element. The pool of questions in this chapter was effective as of February 1994.

Before taking a GMDSS exam, you should check with the FCC for a possible new version of the test. However, you can use this version to determine if you have sufficient training to answer 75% of the questions on a similar test. The instructions from the FCC say that you must answer at least 38 questions out of a possible 50. There are two GMDSS licenses:

GMDSS/0 Radio Operating practices and procedures license
GMDSS/M Radio Maintenance practice and procedures license

The GMDSS/0 license is not covered here. This book covers only the GMDSS/M Radio Maintenance practices and procedures license. In addition to passing the required test, you must:

- be able to receive and transmit in spoken English; and
- be a legal resident, or be eligible for employment, in the United States.

The requirements for the test are:
GMDSS/M—Elements 1, 3, and 9.
The elements are described in the following paragraphs:
Element 1: 24 questions on radio law and operating practices (reviewed in chapter 10 of this book).

Element 3: 76 questions on electronic fundamentals and techniques required to adjust, repair, and maintain transmitters and receivers licensed by the FCC. The equipment may be operated in aviation, maritime and/or fixed public radio service (reviewed in chapters 1 through 7 of this book).

Element 9: 50 questions on GMDSS/M maintenance practices and procedures (reviewed in chapters 1 through 7 and in this chapter).

FCC review questions for the GMDSS Element 9

(Answers are at the end of this chapter)

1. In a radio receiver, the AGC voltage must always be:
   A. present before adjustments can be made.
   B. dc, but may be either polarity.
   C. positive dc.
   D. negative dc.

2. What would be indicated by a measurement of little or no voltage across an emitter resistor in the IF stage of a receiver?
   A. The resistor is probably open.
   B. The transistor is probably open.
   C. There is no input signal.
   D. The bypass capacitor is open.

3. Which of the following would be indicated by a high-voltage reading across an emitter resistor of a receiver's transistor IF stage?
   A. The transistor is open.
   B. The bypass capacitor is shorted.
   C. The transistor is shorted.
   D. The resistor is shorted.

4. Which of the following is least likely to be the cause of intermittent trouble in a solid-state receiver?
   A. An electrolytic capacitor.
   B. A break in printed circuit wiring.
   C. A cold solder joint.
   D. Incorrect value replacement parts installed.

5. Of the following, which is not true of Gunn effect diodes?
   A. They require a resonant circuit to oscillate.
   B. They employ a metal to semiconductor junction.
   C. They produce current pulses at the anode.
   D. They can generate microwaves when applied to cavities.

6. An audio amplifier stage which exhibits frequency response roll-off at low frequencies:
   A. should be checked to see if it is oscillating.
   B. is probably direct coupled.
   C. may have a bypass capacitor in the emitter circuit.
   D. should be grounded at the input.

7. In testing a receiver, you find that an SAW filter output is 6 dB down from the input.
   A. This is normal.
   B. The device is faulty.
   C. The filter is improperly terminated.
   D. The filter is not resonating.

8. When choosing a communications receiver, the least important of the following is:
   A. selectivity.
   B. intermodulation performance.
   C. sensitivity.
   D. dynamic range.

9. While coastal sailing, your operator reports that his receiver seems to go dead sporadically in a certain geographic area. The most probable cause is:
   A. overheating finals.
   B. intermittent connection.
   C. blocking.
   D. VSWR mismatch.

10. An FM receiver active IF limiter:
    A. is operated as an amplifier on very weak signals.
    B. is used to cause the IF amplifier carrier to be a constant, fairly low value, regardless of input strength of the signal.
    C. is used to limit the deviation to the FM detector stage.
    D. B and C above.

11. An FM receive signal will:
    A. vary in amplitude with modulation.
    B. vary in frequency with modulation.
    C. vary in frequency and amplitude with wideband modulation.
    D. B and C above.

12. A superheterodyne receiver:
    A. is super to the heterodyne receiver.
    B. uses a local oscillator feeding the mixer stage.
    C. is only used with FM receivers.
    D. A and C above.

13. On maritime superheterodyne receivers that utilize triple conversion, a wave trap:

A. is used in the antenna circuit to eliminate an unwanted received signal.
B. is used in the antenna input circuit to reduce high impulse interference such as lightning static.
C. is used to reduce the corrosive effect of sea waves.
D. is peaked to resonance of the desired received signal.

14. In a triple conversion superheterodyne receiver, diode mixer stages are:

A. operated in the linear region.
B. operated in the nonlinear region.
C. operated as class-A amplifiers.
D. operated as class-B amplifiers.

15. The bandwidth of an FM signal is:

A. not proportional to the modulating frequency.
B. proportional to the modulating frequency.
C. inversely proportional to the modulating frequency.
D. A and C above.

16. The advantage of using a superheterodyne receiver over a TRF receiver is:

A. greater selectivity at higher frequencies.
B. reduces or eliminates ghosts.
C. greater sensitivity at higher frequencies.
D. A and C above.

17. When two FM signals of different strengths are received on the same frequency:

A. one steady heterodyne will appear in the AF output.
B. both will appear in the AF output.
C. only the stronger will appear in the AF output.
D. neither signal will be intelligible unless the weaker signal is at least 10 times weaker.

18. In an FM receiver the discriminator should be:

A. adjusted for maximum voltage across the two series resistors at center frequency.
B. adjusted for zero voltage across the two series resistors at center frequency.
C. adjusted with a very weak signal.
D. A and C above.

19. Which of the following types of modulation produces sidebands?

A. AM. B. SSB. C. FM. D. All of the above.

20. For a given carrier, the minimum baseband frequency is:

A. half the carrier frequency.
B. equal to the carrier frequency.
C. 0 Hz.
D. one tenth the carrier frequency.

21. Which of the following types of modulation does not produce sidebands?

    A. AM. B. SSB. C. FM. D. None of the above.

22. When a transmitter is operated with a modulation index of 2.4, the carrier level will:

    A. be a Bessel null.
    B. be maximum for any modulation frequency.
    C. not change because it is FM.
    D. increase by a factor of 2.4.

23. Of the following parameters, which should be checked when servicing an FM transmitter?

    A. Clipping.
    B. Noise figure.
    C. Linearity, frequency, and deviation.
    D. Pre-emphasis, clipping, and noise figure.

24. A power amplifier in an FM transmitter that is not linear:

    A. may not be operated legally on maritime frequencies.
    B. causes severe distortion in the audio signal.
    C. may be desirable because of its efficiency.
    D. is usually caused by improper bias.

25. The peak amplitude of the audio signal applied to a phase modulator:

    A. is limited to a value less than the deviation.
    B. should be limited.
    C. is unlimited.
    D. is of no concern with phase modulation.

26. Of the following, which is not characteristic of both AM and FM receivers?

    A. Oscillator/mixer. C. IF amplifier.
    B. Discriminator. D. RF amplifier.

27. Which of the following is not normally included in a phase-locked loop?

    A. Voltage-controlled oscillator. C. Phase detector.
    B. Low-pass filter. D. High-pass filter.

28. A ground-mounted vertical antenna should be:

    A. no longer than ½ wavelength of the highest frequency used.
    B. as long as possible, regardless of operating frequency.
    C. no longer than ¾ wavelength of the highest frequency used.
    D. fed with balanced feedline.

29. A trap antenna is operated on a frequency below the resonant frequency of the trap. The trap operates as:

A. an inductor. C. an antenna gain multiplier.
B. a capacitor. D. a high-pass filter.

30. Regarding the driven element of a ground plane antenna used for VHF:

A. the shorter the more selective.
B. length determines operating frequency.
C. the longer the better.
D. must be top loaded.

31. The plane elements of a ground plane antenna are sometimes referred to as a:

A. lightning protector. C. counterpoise.
B. parasitic element. D. dummy element.

32. In a one-half wavelength Hertz antenna, the:

A. voltage and current are 90 degrees out of phase.
B. voltage and current are in phase.
C. voltage and current are 180 degrees out of phase.
D. feed points are unbalanced and maximum impedance.

33. An antenna parasitic element:

A. increases directivity.
B. is unnecessary to improve directivity.
C. is required for all vertical antennas.
D. has to be the same physical length as the active element.

34. Matching the impedance of an antenna to the transmission line:

A. produces lowest SWR.
B. produces maximum power transfer.
C. causes reduced harmonic radiation.
D. A and B above.

35. On a transmitter, transmission line, and antenna system, high SWR will:

A. provide maximum power transfer from transmitter to antenna.
B. reflect a portion of the RF power from the antenna back to the transmitter.
C. distort modulation.
D. cause a power gain.

36. ERP takes into account:

A. transmitter output. C. transmission line loss.
B. antenna gain. D. all of the above.

37. SWR stands for:

A. shortwave reception. C. standing wave radio.
B. single wave radius. D. safety with radio.

38. Increasing the physical length of a longwire antenna:

    A. increases its resonant wavelength.
    B. decreases its resonant wavelength.
    C. increases its Q.
    D. increases its power handling capability.

39. For best SWR, R_l should be: (Please refer to Fig. 11-1.)

    A. 20 Ω. B. 50 Ω. C. 30 Ω. D. 10 Ω.

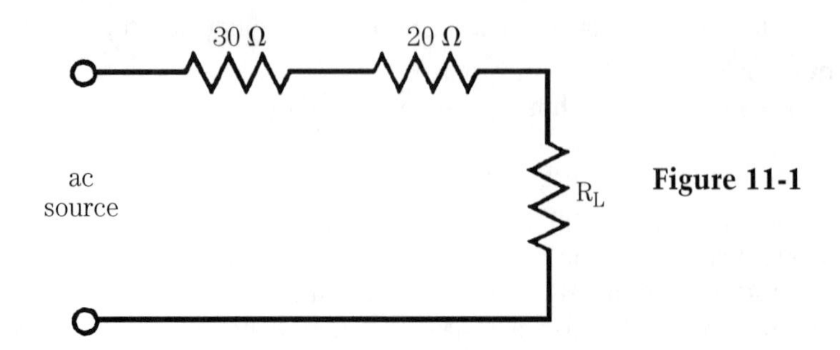

Figure 11-1

40. An antenna with a gain of 3 dB with 100 W entering the antenna feed point will have an ERP of:

    A. 200 W. B. 100 W. C. 50 W. D. 300 W.

41. A 200-foot transmission line that has a loss of 6 dB per 100 ft. has 100 W inserted from a transmitter at 4 MHz. How much power will actually be at the antenna feed point?

    A. 6.250 W. B. 33.3 W. C. 166 W. D. 12 W.

42. SWR is not important or significant at:

    A. microwave frequencies. C. HF/MF frequencies.
    B. VHF/UHF frequencies. D. none of the above.

43. SWR can be expressed and calculated in terms of:

    A. power. B. voltage. C. current. D. all of the above.

44. Theoretically, a perfect impedance match between the transmitter and the antenna would produce an SWR of:

    A. 1:1. B. 1:1.5. C. 1:2. D. 1:4.

45. An operator reports that calls can be made with the HF/MF communications equipment on 2182 kHz, but that greatly reduced power is indicated on all higher frequencies. After verifying that the transmitter functions properly into a dummy load, the antenna tuner of the system could be investigated by:

    A. observing changes in variable component position or relay selections as other channels are tried.

B. observing the coupler output with an oscilloscope of suitable bandwidth.
C. observing the reflected power between the transmitter and the coupler.
D. both A and C above.

46. The criteria for oscillation is:

A. 360° phase shift; gain greater than 1.
B. 180° phase shift; unity gain.
C. 360° phase shift; unity gain.
D. 180° phase shift; gain greater than 1.

47. A PLL is locked onto an incoming signal with a frequency of 1 MHz at a phase angle of 50°. The VCO signal feeds the phase detector at a phase angle of 20°. The peak amplitude of the incoming signal is 0.5 V and that of the VCO output signal is 0.7 V. What is the VCO frequency?

A. 200 kHz. B. 700 kHz. C .1 MHz. D. 1.2 MHz.

48. The output frequency of a certain VCO changes from 50 kHz to 65 kHz when the control voltage increases from 0.5 V to 1 V. What is the conversion gain?

A. 30 kHz/V. B. 15 kHz. C. 65 kHz. D. 15 kHz/V.

49. In troubleshooting a PLL, you find that the signals f_1 and f_2 are the same frequency, but are 45° out of phase (please refer to Fig. 11-2).

A. The VCO is low in gain.
B. Check the low-pass filter for correct cutoff frequency.
C. This is normal.
D. The input signal is slowing faster than the VCO can follow.

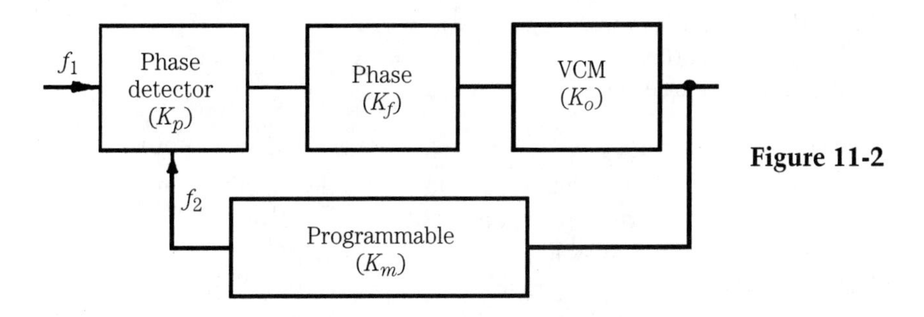

Figure 11-2

50. Measuring the frequencies, you find that the VCO is running at 34.5 MHz. f_1 is at 34.4 MHz (please refer to Fig. 11-3).

A. Replace the low-pass filter.
B. This is within system and measurement accuracies.
C. You need to look further to find the problem within the PL.
D. Replace the gain block, Av.

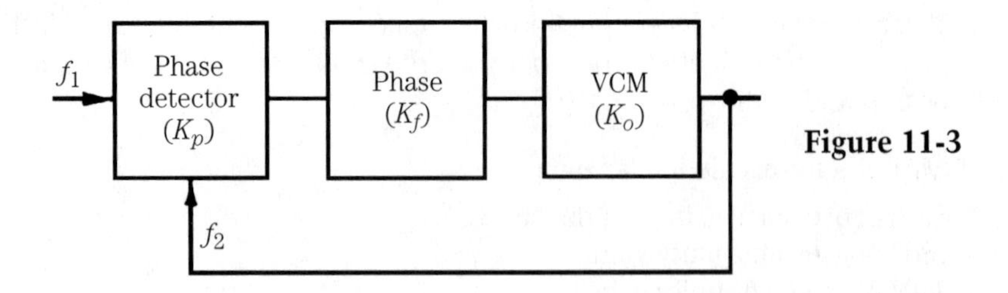

Figure 11-3

51. An RC circuit is configured as in the diagram. (Please refer to Fig. 11-4.) With a 5 V, 10 kHz signal in, the output is:

 A. 2.66 V. B. 1.929 V. C. 2.343 V. D. 1.632 V.

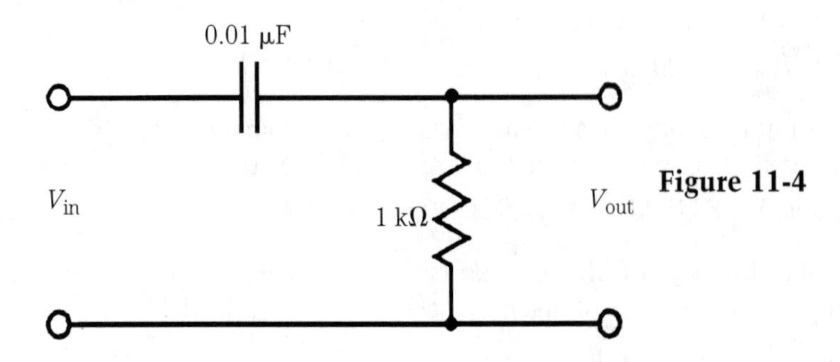

Figure 11-4

52. In longitude 160W, you wish to place a call to a city that is located at 78W. Daylight savings time is not in force. What is the time difference between yourself and that city?

 A. It is 5 hours later in that city.
 B. It is 6 hours later in that city.
 C. It is 7 hours earlier in that city.
 D. It is 6 hours earlier in that city.

53. The advantage of a stepping motor over a conventional dc motor is:

 A. it has higher position accuracy because of the D/A conversion.
 B. it can be directly driven by digital inputs.
 C. it has higher speed capability.
 D. it has higher positioning accuracy.

54. The gyros in the steerable antenna system are of the following type:

 A. not precession limited. C. mercury erected.
 B. universally mounted. D. rate gyro.

55. In a gyro, precession is in what direction?

 A. In a direction which will line itself up in the opposite direction that the allied force is trying to turn it.
 B. Right angles to the applied force.

C. At a 90° angle formed by the center of the earth and the vector of the applied force.
D. In a direction where the gyro wheel is spinning toward the applied force.

56. Choose the most correct statement:
    A. In frequency modulation, the deviation is proportional to the voltage of the modulating signal.
    B. In frequency modulation, the deviation is proportional to both the voltage and the frequency of the modulating signal.
    C. Frequency modulation is the same as phase modulation.
    D. Frequency modulation requires less bandwidth than amplitude modulation.

57. Choose the most correct answer:
    A. *Deviation ratio = Phase deviation/Maximum frequency.*
    B. *Deviation ratio = Maximum frequency/Deviation frequency.*
    C. *Deviation ratio = Frequency deviation × Maximum frequency.*
    D. *Deviation ratio = Frequency deviation/Maximum frequency.*

58. The sensitivity of an FM modulator is measured in:
    A. modulation index. C. kHz/V.
    B. rad/sec. D. V/radian.

59. Direct FM modulation is not normally used in today's systems because:
    A. of the cost of Schottky diodes.
    B. it consumes too much space.
    C. inherent stability is insufficient for modern standards.
    D. of the inherent nonlinearity of direct modulation.

60. Which of the following cannot be used as an FM detector?
    A. Quadrature detector. C. Ratio detector.
    B. Phase discriminator. D. All answers can be used.

61. The most correct definition of a transmission line is:
    A. one conductor separated by a dielectric from conductive braid.
    B. a conductor surrounded by conductive braid.
    C. one insulated conductor.
    D. two conductors separated by a dielectric.

62. Which of the following is not a source of RF noise?
    A. All answers are noise sources. C. Radio receiver.
    B. Resistor. D. Neon lights.

63. In an Inmarsat C system, the main source of noise is:
    A. rain. C. the receiver.
    B. sky noise. D. atmospheric noise.

64. Antenna impedance mismatch will have the following effect on the noise figure:

    A. increase in noise figure.
    B. no effect.
    C. decrease in noise figure.
    D. increase in noise factor.

65. What is meant by companding?

    A. Detecting and demodulating a single-sideband signal by converting it to a pulse modulated signal.
    B. Using an audio frequency signal to produce pulse length modulation.
    C. Combining amplitude and frequency modulation to produce a single-sideband signal.
    D. Compressing speech at the transmitter and expanding it at the receiver.

66. The SITOR operator cannot make contact with a shore station. Your forecast shows that he or she is operating at a frequency 12 times MUF. Your solution is:

    A. change baud rate to correspond to frequency.
    B. go to a higher band.
    C. decrease frequency to 0.8 MUF.
    D. change frequency to the predicted MUF.

67. Low receive amplitude in an Inmarsat C station might be caused by:

    A. all of the answers are correct.
    B. longitude of the vessel.
    C. latitude of the vessel.
    D. obstruction by superstructure.

68. Choose the most correct answer regarding the ionosphere.

    A. The F2 layer is the principal reflector for long-distance HF communications.
    B. The E layer is important for HF daytime propagation and MF night propagation.
    C. The F1 layer exists only during daylight.
    D. All are correct.

69. In an FM system, pre-emphasis/de-emphasis is used to:

    A. improve output signal-to-noise ratio.
    B. decrease system bandwidth.
    C. precompress high frequencies.
    D. increase fidelity.

70. A red LED on the control panel has failed. You find a spare that has the same electrical characteristics, but is marked 1.06 micrometers:

    A. it can be used with a filter.
    B. this can be used but is the wrong color.
    C. this is not an LED.
    D. this cannot be used as it is not visible.

71. The solar cell in the schematic is being illuminated and is producing a current of 250 μA (please refer to Fig. 11-5). What is the voltage out of the circuit?

    A. –0.25 V. B. 2.5 V. C. –2.5 V. D. 5 V.

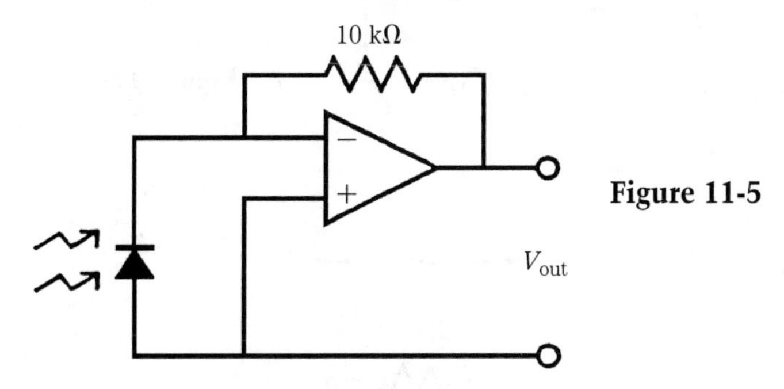

Figure 11-5

72. What is the characteristic difference that is responsible for the emission of light from light emitting diodes, but not from ordinary rectifying diodes?

    A. Rectifying diodes are usually black.
    B. LEDs are constructed of opaque materials.
    C. Gallium and arsenide are necessary to produce photons.
    D. LEDs are constructed of semitransparent materials.

73. Which of the following statements is not true of liquid crystal displays?

    A. LCDs must be externally illuminated.
    B. LCD performance is enhanced at lower temperatures.
    C. LCDs operate at microampere current levels.
    D. The nematic liquid is used to shift light polarity.

74. With regard to methods of joining optical fibers, which of the following statements is incorrect?

    A. Splices are inferior to any detachable connector.
    B. Refractive indices should match at coupling interfaces.
    C. Couplers should be clean and of high quality.
    D. Precision tools should be used when making splices.

75. With a 1 Vdc input, the output of the operational amplifier shown in Fig. 11-6 is:

    A. –3 Vdc. B. –2 Vdc. C. 3 Vdc. D. –2 Vdc.

76. With an input voltage of +5 V, the voltage measured at the inverting (–) input of the op amp shown in Fig. 11-7 will be:

    A. 0 V. B. +5 V. C. 1.3 V. D. 0.7 V.

77. In the op-amp circuit shown in Fig. 11-8, the output voltage will be:

    A. $2 \times (V_1 - V_2)$. C. $V_2 - V_1$.
    B. $-2 \times (V_1 + V_2)$. D. $V_1 + V_2$.

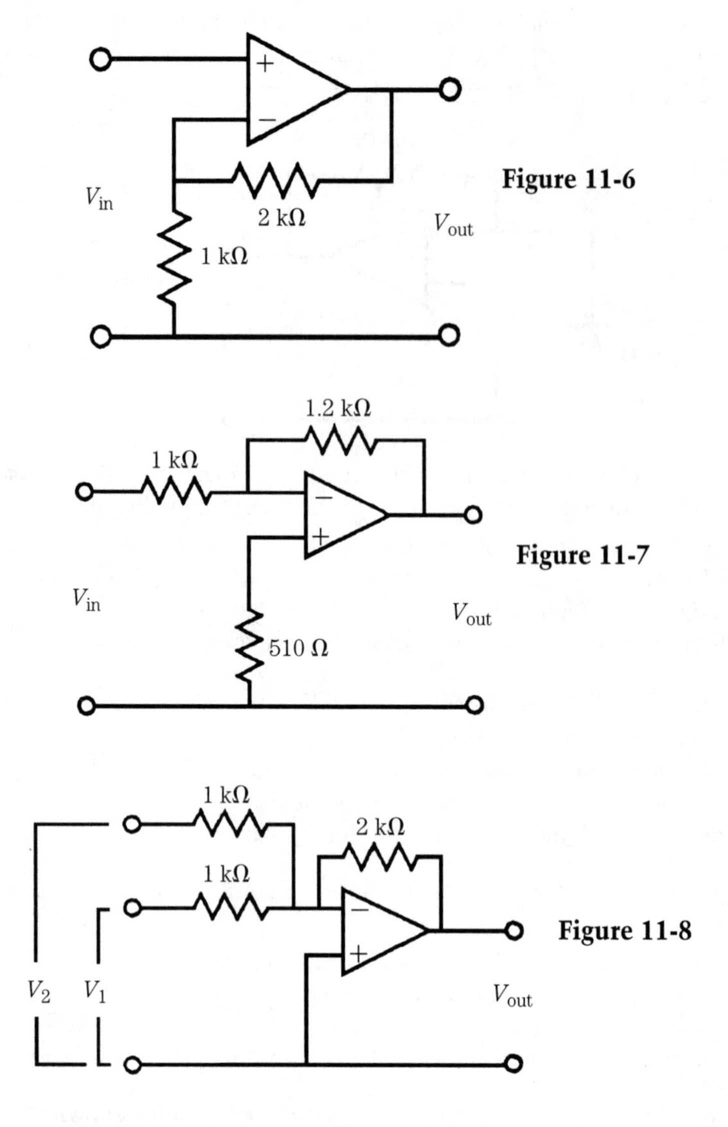

Figure 11-6

Figure 11-7

Figure 11-8

78. In the comparator circuit shown in Fig. 11-9, the nominal output voltage when the output is high is 4.5 V. The upper trip point will be:
    A. 0.7 V. B. 0 V. C. 4.5 V. D. 450 mV.

79. In the op-amp circuit in Fig. 11-10, the input voltage is –100 mVdc. The output voltage you would expect to measure is:
    A. 5 Vdc. B. –5 Vdc. C. 7.5 Vdc. D. 100 mVdc.

80. In the circuit shown in Fig. 11-11, resistor R3 has opened because of vibration of the vessel. The input is a 1-V p-p sine wave centered about 0. The output waveform will:

A. change in frequency.
B. be 0.
C. increase in peak-peak amplitude.
D. be an oscillation.

81. In the circuit shown in Fig. 11-12, both zener diodes have the same zener breakdown voltage. The output will be limited by:

A. the zener voltage plus one diode drop.
B. the power supplies.
C. the zener voltage.
D. twice the zener voltage.

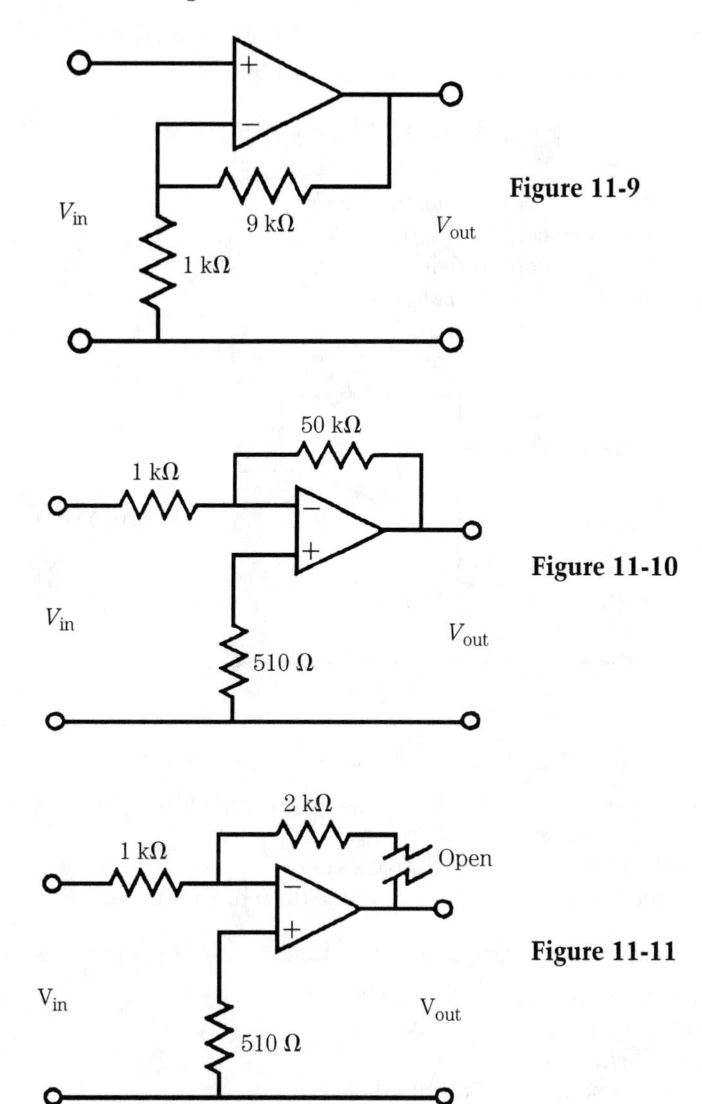

Figure 11-9

Figure 11-10

Figure 11-11

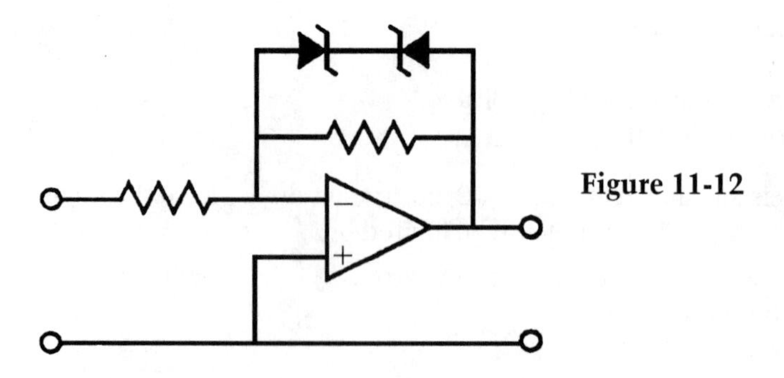

Figure 11-12

82. With a 2 Vdc input, the output of the operational amplifier shown in Fig. 11-6 is:
    A. +4 Vdc. B. –6 Vdc. C. +6 Vdc. D. –4 Vdc.

83. The input to the op amp in Fig. 11-13 is a 1-V sine wave centered about +2 V. The output will be:
    A. a 1-V sine wave centered about +2 V.
    B. a 2-V sine wave centered about –4 V.
    C. a 3-V sine wave centered about 1.5 V.
    D. a 3-V sine wave centered about –1.5 V.

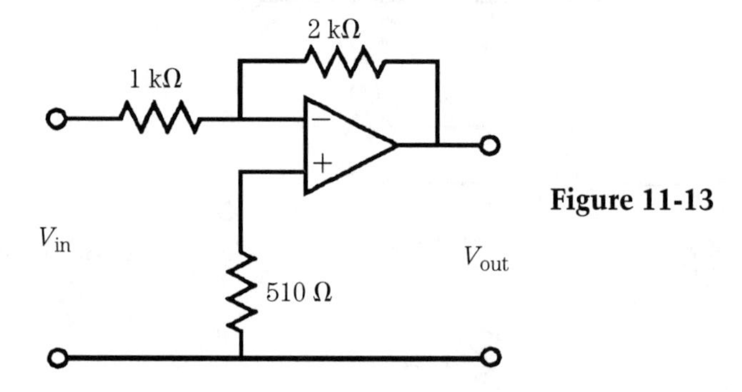

Figure 11-13

84. In the circuit of Fig. 11-14, the purpose of the diode is to:
    A. slow the rise time during the negative portion of the input signal swing.
    B. speed up the rise time of the output waveform.
    C. bound the circuit in the positive swing.
    D. limit the negative swing to one diode drop below ground.

85. An advantage of high-frequency operation of a dc/dc converter is:
    A. the output does not need to be filtered.
    B. a small filter capacitor can be used.
    C. less RFI is generated.
    D. fewer components are required.

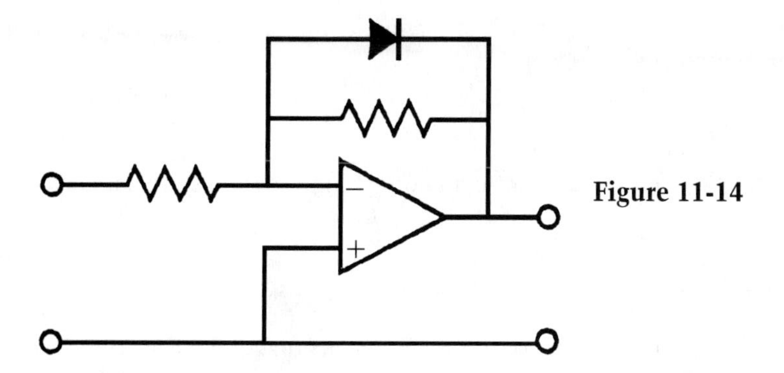

Figure 11-14

86. What happens if the gate voltage on an SCR falls below the trigger point with anode and cathode voltage applied?

    A. The SCR turns on. C. The SCR turns off.
    B. The SCR stays on. D. A pulse is generated.

87. A power supply crowbar circuit might consist of a:

    A. rheostat, switch, and indicator lamp.
    B. transformer, capacitor, and diode.
    C. transformer, rectifier, and filter.
    D. zener diode and SCR.

88. What would be the effect of an open filter capacitor in a dc power supply under load?

    A. No power would be transferred to the load.
    B. The transformer would overheat if the fuse didn't blow.
    C. The dc output voltage would drop.
    D. The rectifier could be damaged due to excessive PIV.

89. Most switching power supplies do not use a line isolation transformer. This implies the following:

    A. This is a transformerless system, which reduces parts cost and makes it a desired unit.
    B. No line isolation will improve the no-load/full-load response.
    C. Low voltages cannot be achieved due to the high power required for voltage conversion.
    D. There is no line isolation. Using a grounded scope in this system will destroy the supply.

90. In a switching power supply, the output voltage is controlled in the following manner:

    A. Using a ferro resonant reactor to control output.
    B. Feeding the output back to either a dc/dc chopper or a pulse width modulated inverter.

C. Feeding the output back to a magnetic amplifier, which in turn controls the output voltage.
D. The error voltage is used to control SCR gates, which accomplish the regulation.

91. Which of the following is not commonly found in a three-terminal voltage regulator?

A. Reference.
B. Error amplifier.
C. Thermal shutdown.
D. None of the above.

92. In a voltage regulator incorporating foldback current limiting, output current decreases with:

A. excessive input/output voltage differential.
B. excessive output voltage.
C. decreasing input voltage.
D. device temperature.

93. When it has been determined that a zener reference diode in a regulated power supply has avalanche breakdown, the best thing to do is:

A. replace the diode with one with higher ratings.
B. replace the diode with one having the same specifications.
C. remove the diode from the circuit and retest it.
D. none of the above.

94. If a zener breakdown occurs at 24 V, in which mode does it operate?

A. Avalanche.
B. Diode.
C. Catastrophic failure.
D. Rectifier.

95. Which of the following is true regarding the circuit in Fig. 11-15?

A. Normally, D1 is forward biased by the battery.
B. Normally, D1 is reverse biased by the voltage on C1.
C. The battery will charge to the peak voltage across C1.
D. Charging stops when ac power to T1 is removed.

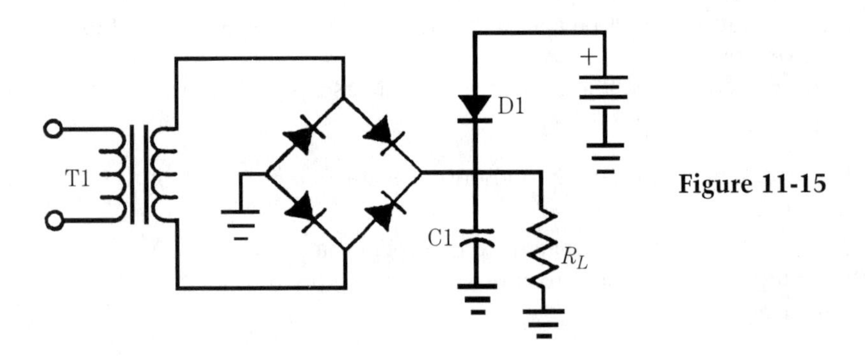

Figure 11-15

96. After the emergency generator was tested on-line, the GMDSS VHF transceiver failed. The blown power supply fuse was replaced, but blew again. Which of the following corrective actions could safely get the power supply working again?

    A. Connect the receiver power leads in parallel with those of another transceiver that is still working.
    B. Install a fuse rated for 10× the current draw in transmit.
    C. Replace the shorted SCR in the crowbar.
    D. Connect a 12-V lamp in series with the power supply output.

97. Edison-type storage-cell battery-charge condition can be determined by:

    A. observing the color of the electrolyte.
    B. measuring voltage with cell under load.
    C. measuring the specific gravity of the electrolyte.
    D. measuring maximum voltage with cell not loaded.

98. Lead-acid type storage-cell battery-charge condition can best be determined by:

    A. measuring maximum voltage with cell not loaded.
    B. observing the color of the electrolyte.
    C. measuring voltage with cell under load.
    D. measuring the specific gravity of the electrolyte.

99. Internal resistance of a battery is:

    A. a fixed resistor placed physically inside battery to limit current flow.
    B. a fixed resistor placed physically inside battery to provide a trickle flow of-electrons during unused periods.
    C. resistance measured when plates short circuit internally.
    D. none of the above.

100. The most common acid used in lead-acid batteries is:

    A. nitric acid. C. sulphuric acid.
    B. muric acid. D. phosphoric acid.

101. The approximate voltage of a fully charged lead-acid battery cell is:

    A. 2.1. B. 6.3. C. 12.6. D. 1.27.

102. The common dry-cell battery contains:

    A. zinc and lead. C. carbon and zinc.
    B. lead and zinc. D. lead and lead sponge.

103. Energy stored in a lead-acid battery is:

    A. piezoelectric. B. kinetic. C. dynamic. D. chemical.

104. A fully charged lead-acid battery rated at 500 amp-hours could power a:

    A. 10-A receiver for 50 hours. C. 5-A lamp for 100 hours.
    B. 50-A load for 10 hours. D. all of the above.

105. The schematic in Fig. 11-16 shows:

A. a digital line driver.
B. a low-noise preamplifier.
C. a pi network output circuit.
D. a power factor correction circuit.

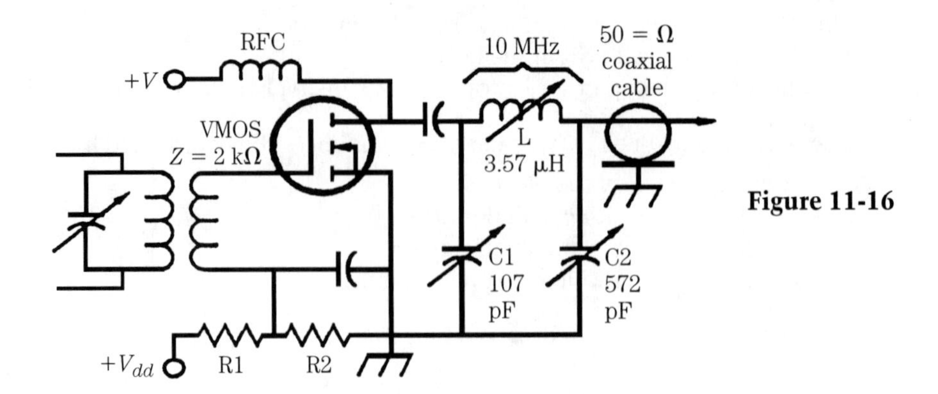

Figure 11-16

106. Crossover distortion in a push-pull amplifier occurs when:

A. the high-frequency tweeters are shorted.
B. high- and low-frequency signals are not separating in crossover network.
C. dc base voltage is zero and input signal voltage has not exceeded V_{be}.
D. base current is not equal in both transistors.

107. A single stage transistor RF amplifier operated class B:

A. operated in the linear range for 180 degrees of the input signal cycle and is cutoff for the remaining 180 degrees.
B. is more efficient than a class-A amplifier.
C. requires an output tank circuit to produce an undistorted output.
D. all of the above.

108. The maximum efficiency for a class-B solid-state power amplifier is:

A. 60%. B. 78.5%. C. 40%. D. 25%.

109. The maximum efficiency for a class-A solid-state power amplifier is:

A. 25%. B. 60%. C. 40%. D. 78.5%.

110. A class-C RF power amplifier has 2000 V on the plate at 300 mA under operating conditions. The output power of the amplifier is between:

A. 600 to 700 W.
B. 420 to 510 W.
C. 500000 to 600000 W.
D. lower than 400 W.

111. Parasitic oscillations are undesirable in RF amplifier stages and:

A. can be clamped.
B. must be tolerated because they are close to the operating frequency.

C. can be shifted far from the desired frequency to cause no interference.
D. can be reduced or eliminated with a parasitic choke.

112. A class-B RF amplifier with an output tuned resonant tank requires:
A. a two-transistor configuration known as push-pull.
B. bias to be twice the cutoff value.
C. only one amplifier transistor.
D. neutralization.

113. The schematic in Fig. 11-17 is an example of:
A. a varactor tuning stage. C. a JFET RF amplifier.
B. a MOSFET RF amplifier. D. an IGFET RF amplifier.

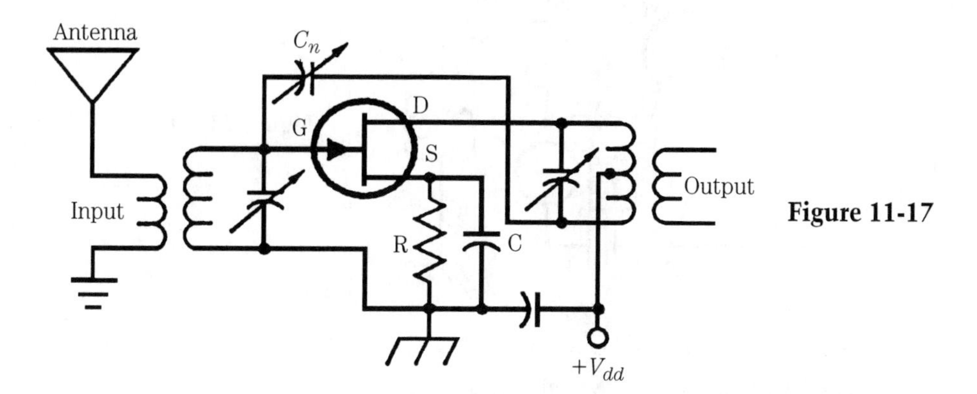

Figure 11-17

114. The schematic in Fig. 11-18 is best described as:
A. a dual gate MOSFET or IF amplifier. C. a JFET audio amplifier.
B. a BJT RF or IF amplifier. D. a JFET RF or IF amplifier.

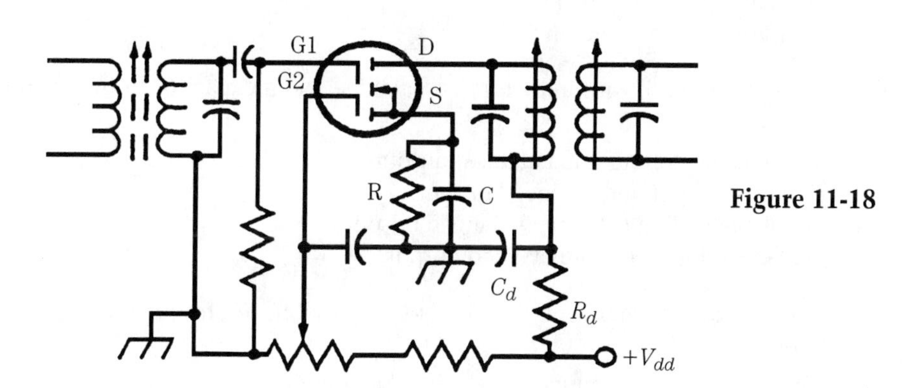

Figure 11-18

115. When operating a VMOS (power FET) in a high-frequency class-C amplifier, the following is true:
A. The VMOS FET is interchangeable with a BJT.

B. The FET must be forward biased in a quiescent state.
C. The stage might require neutralization if operated in a common-source configuration.
D. If not loaded heavily enough, the stage might break into oscillation.

116. The schematic in Fig. 11-19 illustrates:

A. a series-fed RF amplifier output circuit.
B. a shunt-fed RF amplifier output circuit.
C. a high-impedance input circuit.
D. a common-base RF amplifier.

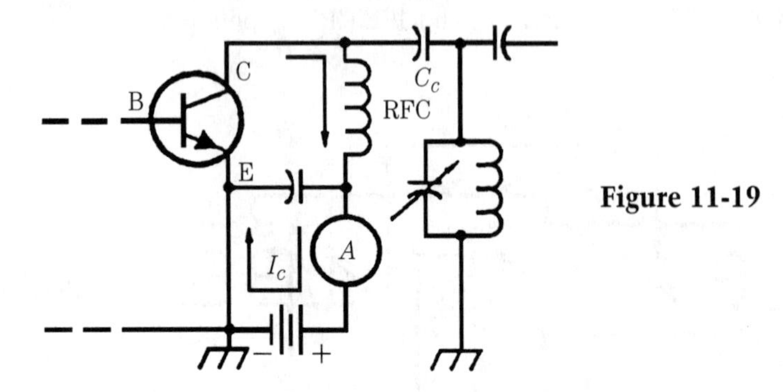

Figure 11-19

117. Audio filters are commonly used for:

A. bandpass filters.
B. high-pass filters.
C. low-pass filters.
D. all of the above.

118. A simple pi network with inductors on each leg and a capacitor in the center is a:

A. high-pass filter.
B. bandpass filter.
C. low-pass filter.
D. notch filter.

119. A simple pi network with inductors on each leg and a capacitor in the center has:

A. an unbalanced input and balanced output.
B. a balanced input and balanced output.
C. an unbalanced input and unbalanced output.
D. a balanced input and unbalanced output.

120. Communication transmitters with microphones that have a frequency response to 10 kHz:

A. provide better clarity under weak-signal conditions.
B. usually contain a high-pass filter to limit the frequencies that modulate the transmitter.
C. are more desirable than microphones with a lower frequency response.
D. none of the above.

121. To obtain maximum power level of an SSB transmitter, the audio circuit should have a:

    A. low-pass filter.
    B. a speech clipper after modulation.
    C. compression of the audio wave.
    D. all of the above.

122. Linear amplification can be achieved at audio frequencies by operating class-B. With regard to this, which of the following is true?

    A. A single-stage class-B amplifier must be proceeded by a class-A amplifier.
    B. A push-pull configuration is necessary.
    C. This can be achieved using a single transistor stage.
    D. It is not possible to obtain linear amplification.

123. You determine on the oscilloscope that time t is 500 μs (please refer to Fig. 11-20). The frequency is:

    A. 2 kHz. B. 500 μs. C. 500 kHz. D. 20 kHz.

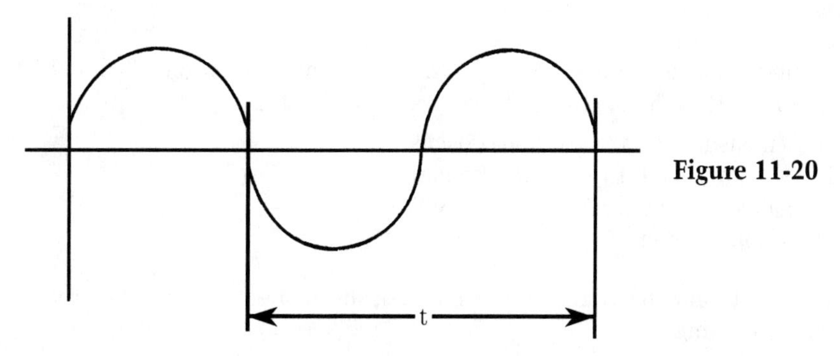

Figure 11-20

124. Low-frequency phase measurements should be made in the following mode:

    A. high-frequency reject. C. chopped.
    B. alternate. D. dual trace.

125. Rise-time measurements are most commonly made from the following amplitudes:

    A. 0 to 100%. C. 20 to 80%.
    B. 10 to 90%. D. 30 to 70%.

126. Pulse-width measurements on the oscilloscope are normally made at:

    A. 50% amplitude. C. 80% amplitude.
    B. 90% amplitude. D. 70% amplitude.

127. An oscilloscope has a rise time of 7 ns. The probe has a rise time of 7 ns. The system BW is:

    A. 40.36 MHz. C. 35.36 MHz.
    B. 50.36 MHz. D. 30.36 MHz.

128. A frequency counter has the following specs: Accuracy ±1 count ± time base accuracy; Time base accuracy: Aging < 0.3 ppm/month, Temp ± 10 ppm 0–40C. The counter was calibrated 10 months ago. To measure a 2182 kHz oscillator, the max error of the counter will be:

    A. 31 Hz. C. 21 Hz.
    B. 28 Hz. D. 18 Hz.

129. Using the 10-second gate of a frequency counter rather than 1 second will improve:

    A. accuracy. C. long term stability.
    B. resolution. D. consistency.

130. When measuring the power output of a VHF transmitter, the power output does not change when modulated:

    A. this test is invalid on dummy load.
    B. there is a probable modulator problem.
    C. there is a strong parasitic oscillation.
    D. this is normal.

131. During troubleshooting, you are in doubt on the calibration of your RF signal generator. You decide that the best way to check it is:

    A. to measure it with a 100-MHz oscilloscope.
    B. by comparing it to a crystal oscillator.
    C. against the carrier frequency of WWV.
    D. by comparing it to your receiver synthesizer.

132. A discontinuity in a transmission line (i.e., short, open, etc.) can be precisely located by using:

    A. an impedance bridge. C. a Weinbridge.
    B. a TDR. D. a precision ohmmeter.

133. You are testing a copper constantan thermocouple that is part of an over-temperature alarm. The resistance is specified as 250 Ω at 25 C 1% with a t_c of 0.003%/C. At 30 C is should read:

    A. 251.22 ± 1%. B. 252.37 ± 1%. C. 253.77 ± 1%. D. 254.22 ± 1%.

134. The printers found in GMDSS can interface in the following way or ways:

    A. serial RS 232C, TT1-Level, 20- or 60-mA current loop.
    B. any method at all could be used.
    C. parallel Centronics.
    D. serial RS-232C.

135. Which of the following is not a recommended maintenance action for printers?

    A. Vacuum regularly to remove dust and paper debris.
    B. Wash with water and mild soap when needed.

C. Wipe with fabric softener to prevent static buildup.
D. Spray inside with a thin film of oil to prevent rust.

136. In troubleshooting a dot-matrix printer, the print head moves, but it does not print. The ink ribbon is new and correctly positioned. A logical next item to check would be:

A. the print-head driver signals.
B. enable, strobe, or busy signals, if used.
C. the paper supply.
D. print-head coil resistances.

137. In troubleshooting a dot-matrix printer, the printer prints, but the print head does not move. A logical first area to check would be:

A. the print head drive motor.
B. the drive head position sensor and belt.
C. the ribbon.
D. the select, busy, or enable signals, if used.

138. Which of the following detrimental factors is the most serious for a typical modern printer in the GMDSS application?

A. Cables not neat or connector mating screws not secured.
B. Located in an RFI environment.
C. Room temperatures in excess of 70° F.
D. Fans or vents blocked.

139. A color CRT that is part of the required GMDSS SITOR terminal will not turn on. The bridge rectifier in the unit is working. Where is the problem most likely to be found?

A. The degaussing coil.
B. The horizontal sweep circuit.
C. The ac line fuse.
D. The luminance circuits.

140. The GMDSS operator has been unable to use the satellite communication system because it has a video display monitor that does not work. A substitute monitor is not available. What action(s) might be appropriate to localize the problem?

A. Use an oscilloscope to verify that there is a video signal.
B. Spray nonlubricating contact cleaner on each switch and connector while observing the monitor.
C. Resolder each connector of the video cable while observing the monitor.
D. Perform a complete alignment of the video monitor in accordance with the manufacturer's specifications.

141. The suppressor grid in a vacuum tube RF amplifier circuit:

A. normally increases and decreases the signal gain by control bias.
B. helps to suppress interference.
C. catches/attracts loose electrons that happen to bounce off the plate.
D. accelerates electron flow to the plate.

142. The presence of a blue glow in a vacuum tube operated as an audio amplifier:
    A. could indicate RF energy is being generated to cause the blue glow.
    B. could indicate air or gas in the tube.
    C. has no effect on the efficiency of the tube.
    D. can be eliminated by reducing the frequency of the input signal.

143. A beam power tetrode has beam forming plates between the:
    A. cathode and control grid. C. control grid and screen grid.
    B. screen grid and the plate. D. suppressor grid and the plate.

144. The main purpose of a screen grid in a vacuum amplifier tube is to:
    A. permit higher amplification.
    B. decrease secondary emission.
    C. absorb some heat from the plate.
    D. provide isolation from signal input to control grid signal input.

145. The gray code minimized the possibility of ambiguity when changing state by:
    A. changing only bit at a time.
    B. using a common clock to synchronize inputs.
    C. changing state on leading or trailing pulse edges.
    D. requiring coincidence between two or more samples.

146. Convert the decimal number 164 to hex:
    A. 5B. B. 104. C. 10110100. D. A4.

147. A computer memory location is designated F09Fh. The decimal equivalent is:
    A. 41231. B. 1010101. C. 10110100. D. 4367.

148. What is the maximum number of characters or commands available in the ASCII code?
    A. 256. B. 128. C. 500. D. 4096.

149. Which of the following items is not an ASCII item?
    A. BS. B. END. C. 8. D. A.

150. If the input lead to an operating TTL inverter became grounded, what would the output lead measure?
    A. More than +5 Vdc.
    B. Ground.
    C. Between +2.5 Vdc and +5, depending on the load.
    D. +0 to +0.7 Vdc.

151. What is the range of supply voltage (V_{DD}) to a CMOS logic IC?
    A. +4.5 to +5.5 Vdc. C. +3 Vdc to +15 Vdc.
    B. –3 Vdc to +10 Vdc. D. +5 to +25 Vdc.

152. A TTL IC will recognize the following voltage levels as valid 1 and 0 levels in operational logic circuits:

A. binary 0 = +0.4 Vdc, binary 1 = +3.6 Vdc.
B. binary 0 = 0.0 Vdc, binary 1 = +5.0 Vdc.
C. binary 0 = +0.3 Vdc, binary 1 = +4.7 Vdc.
D. all of the above.

153. The output of a TTL gate measures 2.0 Vdc:

A. there is a problem in the gate.
B. this is a normal high.
C. this is a normal low.
D. there is a problem either in the gate or the loading.

154. A digital logic chip has a supply voltage of –5.2 V. This chip belongs to which family?

A. RTL. B. TTL. C. ECL. D. CMOS.

155. Which of the following integrated circuit or semiconductor devices normally require special handling to avoid damage by static electricity?

A. MOV. B. ECL. C. CMOS. D. TTL.

156. In Fig. 11-21, the function D is described as:

A. C = (~A)B. B. AB(~C). C. C + AB. D. A + BC.

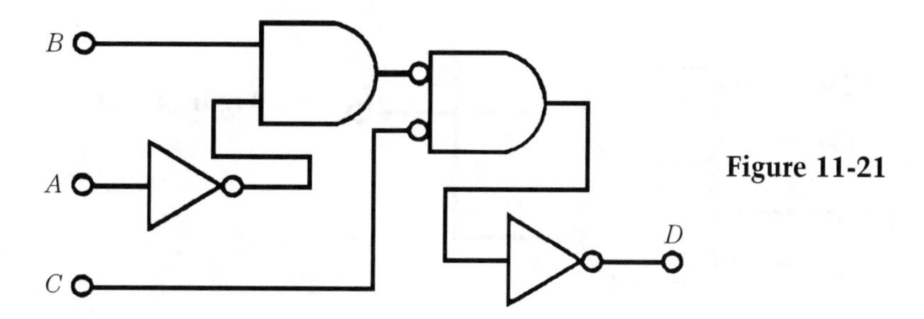

Figure 11-21

157. ECL achieves high speed because of:

A. operation in the low-noise negative-supply region.
B. construction in small geometries.
C. the use of gallium arsenide conductors.
D. the operating transistors being unsaturated.

158. The logic family that typically has the largest fanout is:

A. CMOS. B. ECL. C. TTL. D. RTL.

159. In order of highest to lowest speed, the logic families are ranked:

A. Schottky, Standard TTL, ECL, CMOS.

B. ECL, Schottky TTL, Standard TTL, CMOS.
C. CMOS, Standard TTL, Schottky TTL, ECL.
D. ECL, CMOS, Schottky TTL, Standard TTL.

160. In a 3-bit binary ripple counter, the state following 111 will be:

A. 110. B. 000. C. 001. D. 111.

161. In a 4-bit BCD ripple counter the state following 1001 will be:

A. 0000. B. 1001. C. 1111. D. 1000.

162. Synchronous counters are distinguished from ripple in that:

A. there is no ripple on synchronous counters.
B. counter feedback is synchronous.
C. logic inputs are applied in parallel.
D. the clock is applied to all flip-flops simultaneously.

163. The circuit shown in Fig. 11-22:

A. is a wired OR.
B. uses open collector gates.
C. is used for high-speed operation.
D. is a wired EXCLUSIVE OR.

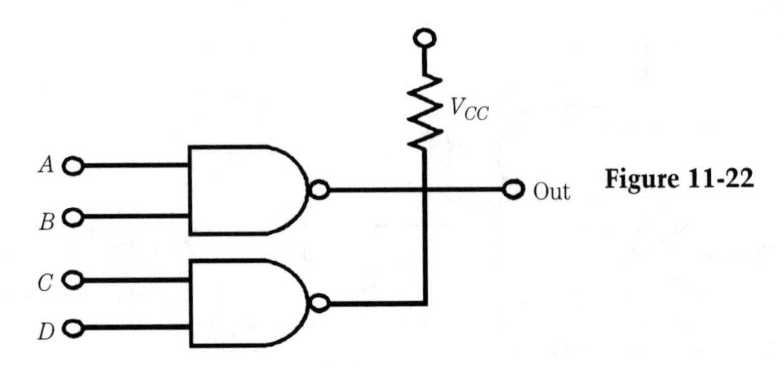

Figure 11-22

164. What is the result of adding binary 110111 and 1001?

A. 100000. B. 11111. C. 1000000. D. 111111.

165. What is the binary result of multiplying hexadecimal 1C by 7?

A. 11001100. B. 11101100. C. 11000100. D. 11100100.

166. The value 123456 is based upon a number system that has a minimum radix of:

A. 6. B. 7. C. 10. D. 2.

167. A 2s complement number is formed by the following method:

A. add the number to all ones; then add 1.
B. complement individual bits; then add 1.

C. subtract the number from all ones; then add 1.
D. add individual bits; then add 1.

168. These numbers are in 2s complement form: 0101 + 0010 = Choose the correct solution:

A. 0111. B. 1000. C. 1011. D. 01000.

169. With x and y as inputs to an AND gate, what is the output waveform? Please refer to Fig. 11-23.

A. Waveform A. C. Waveform C.
B. Waveform B. D. Waveform D.

170. With x and y as inputs to an OR gate, what is the output waveform? Please refer to Fig. 11-23.

A. Waveform A. C. Waveform C.
B. Waveform B. D. Waveform D.

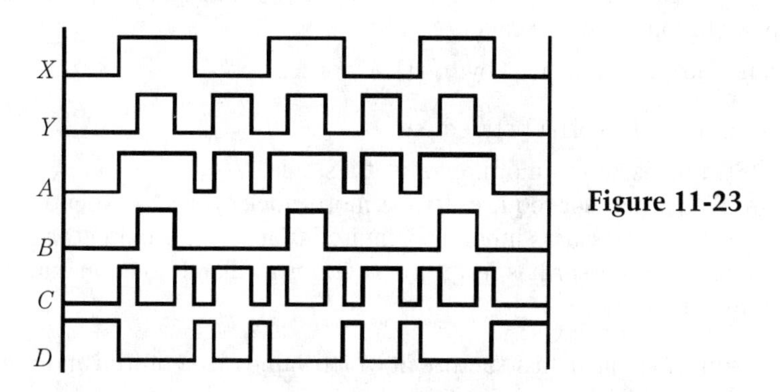

Figure 11-23

171. With x and y as inputs to an EXCLUSIVE OR gate, what is the output waveform? Please refer to Fig. 11-23.

A. Waveform A. C. Waveform C.
B. Waveform B. D. Waveform D.

172. The standard serial output of a PC conforms to the following specification:

A. Centronics. C. NMEA0180.
B. 20-mA loop. D. RS232.

173. Which of the following logic levels are in a normal range for TxD and RxD signals on an RS232 interface line?

A. –10 V and 0 V. C. –2.5 V and +2.5 V.
B. –10 V and +10V. D. +0.20 V and +4. 5 V.

174. Depressing a key on a computer or message input keyboard:

A. generates a coordinate or scan code output.
B. sends a character address to a DMA line.

C. directs access to a ROM where the associated character is stored.
D. both A and C.

175. What happens in a microprocessor system if electrical power input is interrupted?

A. Data stored in ROM is lost forever.
B. Data stored in ROM is lost, but can be restored by rebooting.
C. Data stored in RAM is lost.
D. Both B and C above.

176. In a microprocessor, the program counter contains:

A. a sequential instruction set.
B. data.
C. the address of the next instruction to be executed.
D. an instruction set.

177. Complete the following sentence: A bit string manipulated by a computer in one operation is usually called:

A. a nibble. B. a bit. C. a byte. D. a word.

178. Choose the most correct statement:

A. RISC processors cannot implement a stack.
B. An RISC processor requires two or more clock cycles to execute a command.
C. An RISC processor is inherently limited to a 32-bit architecture.
D. An RISC processor has fewer instructions available than an equivalent non-RISC processor.

179. A one-dimensional data structure in which values are entered and removed one item at a time at one end is called what?

A. A FIFO. C. A stack pointer.
B. A pushdown stack. D. A ring counter.

180. Choose the most correct statement:

A. The output of an assembler is an object module.
B. Machine language and assembly language are the same.
C. Assembly language can be directly run on a machine.
D. In assembly language, a label is mandatory.

181. In the program excerpt in Fig. 11-24, which address is the first that does not contain a processor command?

A. 1000. B. 1001. C. 1002. D. 1003.

182. In the program in Fig. 11-24, which address contains the command that would cause an exit from the loop?

A. 1004. B. 1005. C. 1006. D. 1009.

183. In the program excerpt in Fig. 11-24, which pair of addresses contains data and no commands?

A. 1006, 1007. C. 1002, 1003.
B. 1000, 1001. D. 1004, 1005.

```
                   . D0 WILL ACCUMULATE PROD
00001000  4240  MOLT CLR.W D0
                   . D1 HOLDS LOOP COUNT
00001002  3238       MOVE. W   MPY,D1
00001004  101C
                   . DONE IF COUNT DOWN TO 0
00001006  670A       BEQ       DONE
                   . ELSE ADD MCND TO PROD
00001008  D078  LOOP ADD.W MCND,D0
```

Figure 11-24

184. Compilers are used with which type of code?

A. High-level languages. C. Machine languages.
B. Assembly languages. D. Object codes.

185. Choose the correct answer:

A. a nibble is 4 bits. C. a byte is 8 bits.
B. all answers are correct. D. a bit is one binary digit.

186. An interpreter is used with which type of code?

A. High-level languages. C. Object codes.
B. Assembly languages. D. Machine codes.

187. If you check a junction field-effect transistor out of circuit with an ohmmeter:

A. you will destroy the FET.
B. gate to source and gate to drain resistance should be nearly infinite.
C. you must first determine if it is a P-channel or N-channel.
D. the reading will be the same either way between drain and source.

188. An internal short between the base and collector of a bipolar transistor might be indicated by which of the following?

A. dc supply voltage on the collector.
B. No signal at the output.
C. A weak signal at the collector, in phase with the input.
D. Little or no signal at the collector and a reversal of phase.

189. When attempting to test a bipolar silicon or germanium transistor, which of the following is likely to be correct if the test is conducted after the device has been removed from its circuit?

A. If a PNP transistor is being tested, forward junction resistance will be greater than reverse resistance.

B. Forward junction resistance should be less than reverse junction resistance.
C. If an ohmmeter is used without an external limiting resistor, excessive base current will destroy the device.
D. Circuit effects that cannot be accounted for, preclude the use of resistance measurements.

190. When replacing a diac which is used with a silicon-controlled rectifier, which way must the diac be installed?

A. The cathode must be connected to the gate.
B. The anode must be connected to the gate.
C. Either way, depending on the polarity of the SCR.
D. Either way will work.

191. Which of the following statements about diac bidirectional trigger diodes is incorrect?

A. Lamps and battery chargers are typical loads for a diac.
B. A diac switch functions in either direction.
C. The breakover voltage is usually between 28 and 36 V.
D. A diac is a three-layer device with two terminals.

192. Which of the following statements correctly describes photo diode operation?

A. Dark current flows in response to black light.
B. Photodiodes generate light in response to incident photons.
C. Efficiency is a measure of photons per electron.
D. Reverse biased photodiodes are photoresistive.

193. What happens if the ambient light reaching a photodiode is reduced to zero and the reverse bias is increased?

A. PN junction leakage current flows.
B. No current flows.
C. Conductivity is at minimum.
D. A and C above.

194. Which of the following statements distinguish phototransistors from photodiodes?

A. Phototransistors are faster than photodiodes.
B. Photodiodes are more sensitive than phototransistors.
C. Photodarlingtons are the fastest photoconductors.
D. None of the above.

195. Photodiode reverse current is called:

A. dark current in very low illumination.
B. Zener current if the diode is reversed biased.
C. dark current in the absence of light.
D. photocurrent if the diode is forward biased.

196. What is the life expectancy of a light-emitting diode operated continuously under normal operating conditions?

   A. Light-emitting diodes can last up to 20 years.
   B. Unlimited life expectancy.
   C. Up to 100 years.
   D. Approximately 87,660 hours.

197. How can the polarity of the lead of an LED be identified prior to installation in a circuit?

   A. If the cathode can be seen, it is usually smaller.
   B. There might be a flat edge on the body near the anode.
   C. The cathode lead is usually longer than the anode lead.
   D. The ground lead is usually lighter in color.

198. In testing a GaAs light-emitting diode within the manufacturer's operating parameters of voltage and current, it appears that no light is emitted. What explanation could be given?

   A. The LED might be emitting invisible light.
   B. The LED might have run out of GaAs.
   C. The LED is connected backwards.
   D. A and C above.

199. What is the typical minimum bias voltage required for normal operation of a GaAs light-emitting diode?

   A. The LED will operate at 2 V reverse bias.
   B. The LED will operate at 0.6 V forward bias.
   C. The LED will operate at 1.2 V forward bias.
   D. The LED will operate at 1.2 V reverse bias.

200. Which of the following statements is not true of light-emitting diodes?

   A. They can be made to emit nearly pure white light.
   B. They can operate at very high speed.
   C. They can be manufactured to emit various wavelengths.
   D. They are vulnerable to failure because of over-current.

201. What is the usual method of protecting a light-emitting diode from damage that would result if the operating voltage became too high?

   A. A series current-limiting resistor is used.
   B. A zener voltage regulator is used.
   C. A fast-blow fuse is used in series with the LED.
   D. The LED is attached to a heatsink.

202. Assuming that a light-emitting diode has an internal resistance of 5 Ω, what value of series current-limiting resistor should be used if the power supply voltage is 6 V and the diode current is to be 0.05 A at 1.6 Vdc?

   A. 32 Ω. B. 120 Ω. C. 88 Ω. D. 83 Ω.

203. With regard to a 7-segment LED display, which statement is not true?

   A. Only one segment at a time can be illuminated.
   B. An external decoder/driver is usually used.
   C. They are often used as simple on-off indicators.
   D. Both A and C above.

204. Identify the statement below that is incorrect with respect to infrared light-emitting diodes:

   A. Photodiodes can be used with IR LEDs.
   B. IR LEDs are often used in optical couplers, but not in optoisolators.
   C. Phototransistors can be used with IR LEDs.
   D. Light from an IR LED is invisible.

205. When a power MOSFET has 0 V from gate to source the following is true:

   A. It is drawing gate current.
   B. It is saturated.
   C. It is in a conducting region.
   D. It is in pinch off.

206. The purpose of the diode in the circuit shown in Fig. 11-25 is to:

   A. increase current capacity.
   B. speed switching time.
   C. protect the transistor.
   D. compensate for thermal variations.

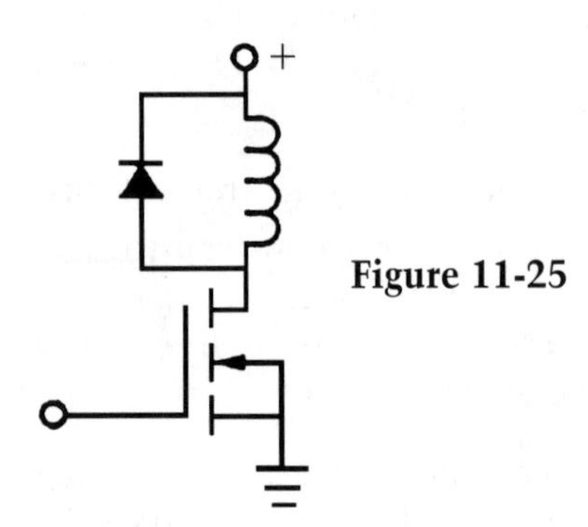

Figure 11-25

207. In the silicon transistor circuit shown in Fig. 11-26, you test the input and find that it is a 0.5-V P-P sine wave centered about 4 V. The output is a 1.5-V P-P sine wave centered about 7 V. Neglect drop across 100 Ω resistor as negligible. V_{be} assumed as 0.7 V (silicon junction). V_{re} = 3.3 V, I_c = 1.65mA, with no load V_o = 7.05 V. The most likely fact is that:

   A. the circuit is operating normally and is driving a high impedance.
   B. the circuit is operating normally and is heavily loaded.
   C. the transistor is bad.
   D. one of the circuit components is bad.

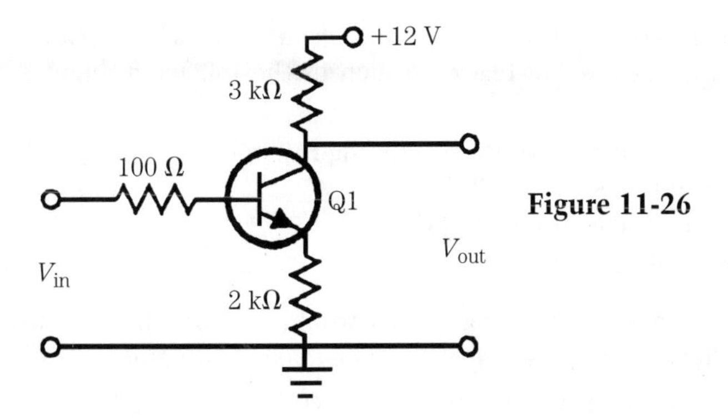

Figure 11-26

208. Relay K1 contact do not close. Q1 base measures 0.7 V. Q1 collector measures 12 V. The proper course of action is (please refer to Fig. 11-27):

A. change Q1. C. change K1.
B. clean the K1 contacts. D. change base resistor.

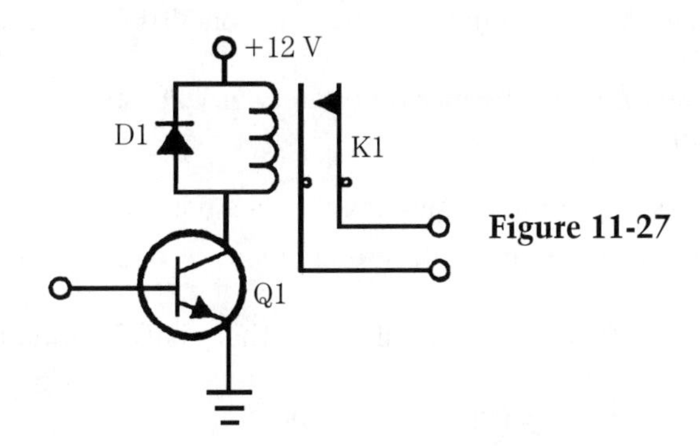

Figure 11-27

209. V_{in} measures 3.0 V, the Q1 emitter measures 2.3 V, and the Q1 collector measures 2.3 V. Proper course of action is to (please refer to Fig. 11-28):

A. change Q1. C. change the 5 kΩ resistor.
B. check 10 kΩ for an open. D. check the next stage.

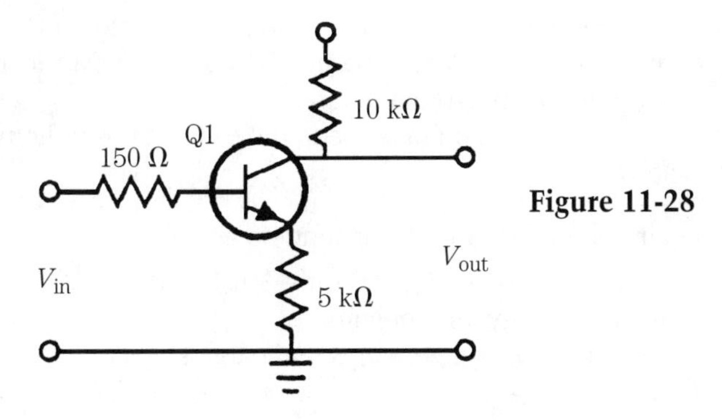

Figure 11-28

210. It has been reported that a VHF transceiver is unusable because the LCD display characters are "all black." Which of the following might get it working again?

    A. Replace the display's current-limiting resistor.
    B. Drain and replace the LCD liquid.
    C. Warm the display up to 20° C.
    D. Replace the crystal.

211. Which of the following procedures would be appropriate with metal-oxide semiconductor integrated circuits while making equipment repairs?

    A. Keep spare ICs neatly organized in Styrofoam.
    B. Use a nonconductive mat on the work surface.
    C. Touch each lead to equalize electrostatic potentials.
    D. Wear a conductive wrist strap with a 1 million-Ω resistor to ground.

212. One licensed GMDSS operator aboard ship must be:

    A. dedicated to receive and transmit distress traffic.
    B. responsible for maintaining the watches on VHF Channel 70, 2187.5 kHz, and VHF Channel 16.
    C. responsible for receiving maritime safety broadcasts.
    D. all the above.

213. Under GMDSS regulations, a ship may leave a port only if:

    A. all GMDSS distress and safety equipment is operational except for VHF Channel 70.
    B. excluding VHF Channel 16, only two other GMDSS equipments are out of order.
    C. all safety and distress equipment is operational.
    D. only the Navtex receiver has failed.

214. While docked, Federal Communications Commission regulations require that the capacity of the emergency batteries be checked at intervals of:

    A. 7 days. B. 12 months. C. 30 days. D. 24 hours.

215. Ship requirements of VHF Channel 70 are:

    A. to maintain a continuous watch with a dedicated radio on Channel 70.
    B. to maintain a continuous VHF Channel 70 watch with a receiver that scans with other required VHF frequencies.
    C. that a distress transmission must be initiated from the radio room.
    D. A and C above.

216. GMDSS requirements for radar transponders are:

    A. one for cargo ships exceeding 300 tons, but less than 500 tons.
    B. at least one unit on any kind of ship.
    C. one on each side of passenger ships and cargo ships of 500 tons upwards.
    D. A and C above.

217. GMDSS requirements for radar transponders are:
   A. they must be stowed in locations so that they can be rapidly placed in any survival craft.
   B. they must be permanently installed in survival craft required by the SOLAS Convention.
   C. they must have a keyboard for loading ship or survival craft navigation position.
   D. A and C above.

218. GMDSS requirements for portable VHF radio apparatus are:
   A. one for cargo ships exceeding 300 tons, but less than 500 tons .
   B. two for cargo ships exceeding 300 tons, but less than 500 tons.
   C. three on each passenger ship and cargo ships of 500 tons upwards.
   D. B and C above.

219. Navtex broadcasts are sent:
   A. in categories of messages.
   B. immediately following traffic lists.
   C. on request of maritime mobile stations.
   D. regularly after the radiotelephone silent periods.

220. Navtex uses _______________ for broadcasts:
   A. 518 kHz and selected HF frequencies.
   B. 0.581 MHz.
   C. 2182 kHz.
   D. VHF Channel 16.

221. Poor copy of Navtex broadcasts can be caused by:
   A. weak signals.
   C. interference.
   B. phase distortion.
   D. all of the above.

222. Controls on a dedicated GMDSS Navtex must be able to select for printout:
   A. one category of messages.
   B. no less than two categories of messages.
   C. no less than three categories of messages.
   D. no less than four categories of messages.

223. Navtex broadcasts depend on transmission by:
   A. ground waves.
   C. sky waves.
   B. line of sight waves.
   D. interrupted continuous waves.

224. The following is not true of geo-synchronous satellites:
   A. they are not effective from polar regions.
   B. they orbit at the same altitude.
   C. they are always at the same elevation.
   D. their orbit is elliptical.

225. The ship INMARSAT antenna is aligned for maximum signal from the Atlantic West Satellite. It is desired to down link to the Italy CES, which has been properly selected:

A. The ship INMARSAT system will not accept the command and fail before accessing the INMARSAT telex system.
B. The INMARSAT telex system will accept the command, process the call and give a "GA+" back to the operator.
C. The ship radio alignment system will realign to the Italian Satellite and process the call successfully.
D. The INMARSAT telex system will accept the command, process the call and disconnect.

226. To keep the INMARSAT antenna pointing at the desired satellite regardless of the ship nautical position or changes in ship's course, it has an input from the vessel's:

A. radar with Automated Radar Plotting Aid (ARPA).
B. gyro.
C. operational radar.
D. steering control.

227. Over the INMARSAT system, telephone and telex communications channel usage is:

A. one ship per telephone channel and many ships per telex analog channel frequency.
B. one ship per analog telephone channel and one ship per telex analog channel frequency.
C. many ships on same telex analog channel frequency and many ships per same TDM telephone channel.
D. one ship per analog channel frequency whether telephone or telex.

228. When a telex call is initiated over the INMARSAT system, the channel used is:

A. one of three channels a ship is assigned by COMSAT and dedicated only for calling.
B. an idle channel chosen by the scanning device of the ship INMARSAT radio.
C. a common channel shared by many ship stations.
D. time division multiplexed on any busy channel with an idle time-division channel selected by ship's equipment.

229. The INMARSAT system uses the following method for voice transmission and reception:

A. AM. B. TDM. C. SSB. D. FM.

230. Over the INMARSAT telex system communication, the automatic answer back (AAB) request is first used by the CES to:

A. identify the SES making the request.
B. let the CES operator know the printer is functioning correctly.

C. get the identity of the ship station and start the channel assignment process.
D. both A and B.

231. Over the INMARSAT system, the SF signaling tones used by the SES and CES are:

A. 1688 Hz. B. 2600 Hz. C. 1100 Hz. D. 400 Hz.

232. In a properly functioning GMDSS VHF DSC system, the receiver bandwidth requirements is specified as:

A. as narrow as possible, but not less than 170 Hz.
B. 5 kHz.
C. GMDSS imposes no requirement.
D. sufficiently wide to pass video pulses.

233. A properly operating VHF DSC transmitter used in the GMDSS system can be expected to have a carrier frequency accuracy of:

A. ±25 Hz coastal and marine.
B. ±10 Hz coastal, ±25 Hz marine.
C. ±10 Hz coastal and marine.
D. ±100 Hz coastal and marine.

234. The properly operating VHF DSC transmitter in the GMDSS system differs from older non-GMDSS VHF voice transmitters in what way?

A. Frequency response is flat rather than 6-dB/octave pre-emphasis.
B. Required carrier stability.
C. Direct FM is used rather than PM.
D. 800-Hz deviation is used rather than 5 kHz.

235. If you were to observe the output power of a properly operating VHF transmitter in the DSC mode using a typical in-line RMS wattmeter as calls were made, what would you expect to see?

A. Normal carrier power would be indicated.
B. The meter would flicker for approximately 0.5 seconds of transmission time, but readings could be inaccurate.
C. The meter would flicker at half-second intervals.
D. Nothing is observed because of the quickness of the communication.

236. In checking the demodulator for a VHF DSC receiving system:

A. the RF generator must be tuned 1.7 kHz below carrier.
B. it must recognize the ships assigned tones.
C. it must recognize 1300 and 2100 Hz within 10 Hz.
D. it must recognize 1300 and 2100 Hz within 10%.

237. The modulation index associated with 1200 baud, 1700 Hz carrier, 800-Hz shift as would be used in VHF DSC mode is:

A. 1.0 ±10%. B. 1.0 ±20%. C. 2.0 ±10%. D. 2.0 ±20%.

238. SImplex Teleprint Over Radio code does not have which of the following properties:

    A. it is asynchronous.
    B. there are seven bits in each character.
    C. each character contains four ones.
    D. each character contains three zeros.

239. The receiving station in a SITOR link uses which of the following methods to check for invalid characters while operating in the ARQ mode?

    A. Odd or even polarity.
    B. A 4/3 mark to space ratio.
    C. Correct number of start and stop bits.
    D. Three marks to four spaces.

240. Of the following, which is true of SITOR ARQ mode direct printing radioteletype transmission?

    A. A continuous data stream is transmitted.
    B. The acceptance code consists of three characters.
    C. Each data block consists of three characters.
    D. Forward error correction reduces the number of errors.

241. Incorrect functioning of an ARQ SITOR transmitting station would be indicated by:

    A. transmitter on time equals off time.
    B. transmitter off time is greater than on time.
    C. transmitter on time is less than off time.
    D. both B and C above.

242. The minimum time for two-way ARQ SITOR communication of data is:

    A. about 5 s.
    B. about 244 ms.
    C. about 0.21 s.
    D. equal to 240 ms.

243. Which of the following is true of SITOR ARQ mode?

    A. This is an interactive mode.
    B. Each character is repeated three times.
    C. Each character is transmitted twice.
    D. Both A and B above.

244. With regard to SITOR, what should happen when an RQ code is received correctly by the transmitting station?

    A. The acknowledge light should illuminate.
    B. The next block will be sent.
    C. The last block will be reversed.
    D. The last block will be resent.

245. If a SITOR transmitting station operating in ARQ mode has no data waiting to be sent:

A. idle characters will be sent.
B. a break character will be sent and the link will drop.
C. the link will reverse.
D. synchronization will be lost.

246. In SITOR Mode A, the time required to transmit a single group of data characters as measured using a calibrated oscilloscope with the horizontal timebase set to 50 ms/div would be indicated by a length of:

A. 3.0 centimeters. C. 4.2 divisions.
B. 4.8 divisions. D. 3.0 divisions.

247. In a properly functioning MF/HF DSC system, the receiver bandwidth should be:

A. as narrow as possible, but not less than 170 Hz.
B. 1.7 kHz.
C. 300 Hz maximum.
D. sufficiently wide to pass video pulses.

248. A properly operating MF/HF DSC transmitter used in the GMDSS system can be expected to have a carrier frequency accuracy of:

A. ±10 Hz coastal and marine.
B. ±10 Hz coastal, ±25 Hz marine.
C. ±25 Hz coastal and marine.
D. ±100 Hz coastal and marine.

249. The properly operating HF DSC transmitter in the GMDSS system differs from older nonGMDSS SSB voice transmitters in what way?

A. Frequency response is flat, rather than 6 dB/octave pre-emphasis.
B. Required carrier stability.
C. Direct FSK is used, rather than AFSK.
D. 170-Hz shift is used rather than 1700 kHz.

250. If you were to observe the output power of a properly operating MF/HF transmitter in the DSC mode using a typical in-line RMS wattmeter as calls were made, what would you expect to see?

A. The meter would flicker for approximately 0.5 second, but readings could be inaccurate.
B. Normal carrier power would be indicated.
C. The meter would flicker at half-second intervals.
D. Nothing is observed because of the quickness of the communication.

251. In checking the demodulator for an MF/HFDSC receiving system:

A. it must recognize audio of 1615 and 1785 Hz.
B. it must recognize the ships assigned tones.

C. the RF generator must be tuned 1.7 kHz below carrier.
D. it must recognize audio of 800 Hz and 1700 Hz within 10 Hz.

252. Federal Communication Commission requirements for GMDSS specify an EPIRB capable of:

A. transmitting distress alerts on the 406-MHz band.
B. transmitting a beacon on the 121.5 MHz.
C. receiving distress alerts on 156.8 MHz.
D. A and B above.

253. Federal Communication Commission regulations for GMDSS specify that certain ships have a 406-MHz EPIRB capable of:

A. acknowledging reception by search and rescue by its flashing light.
B. commencing transmissions after a 60-second delay only after floating free from a sinking ship.
C. floating free of a ship, in the event that it sinks, and automatically transmitting a 121.5-MHz homing beacon and a 406.025-MHz signal, which contains a unique identification code for each EPIRB station.
D. automatically activating when floating away from a sinking ship and transmitting on the 406 MHz and 158.6 MHz.

254. When activated, the GMDSS EPIRB will transmit:

A. only a 406-MHz beacon to polar satellites and 121.5 MHz beacon to search and rescue craft.
B. only a 406-MHz signal to a MARISAT satellite.
C. a 406-MHz beacon to polar satellites and 158.6-MHz homing signal for search and rescue craft.
D. 156.8-MHz and 406-MHz beacons for search and rescue craft.

255. The radiotelephone alarm tone frequencies are:

A. 1100 Hz and 1900 Hz.
B. 700 Hz and 1900 Hz.
C. 1300 Hz and 2200 Hz.
D. 700 Hz and 2200 Hz.

256. The radiotelephone silence period(s) are required:

A. set by the captain of the vessel at the beginning of the voyage.
B. one time each hour for 3 minutes duration commencing at X.00 UTC.
C. 4 times each hour for 3 minutes duration commencing at X.00, X.15, X.30, and X.45 UTC.
D. 2 times each hour for 3 minutes duration commencing at X.00 and X.30 UTC.

257. The radiotelephone silence period is:

A. a 3-minute period that is reserved for distress, urgency or safety messages.
B. a period dedicated for testing so as not to interfere with normal message traffic.

C. used to pass message traffic on a non-interference basis.
D. a time period that the radiotelephone watch receiver can be powered off as no traffic is sent during this period.

Answer sheet

1. B	2. B	3. C	4. D
5. B	6. C	7. A	8. B
9. C	10. A	11. B	12. B
13. A	14. B	15. B	16. D
17. C	18. B	19. D	20. C
21. D	22. A	23. C	24. C
25. B	26. B	27. D	28. C
29. A	30. B	31. C	32. A
33. A	34. D	35. B	36. D
37. C	38. A	39. B	40. A
41. A	42. D	43. D	44. A
45. D	46. C	47. C	48. A
49. C	50. C	51. A	52. B
53. B	54. D	55. B	56. B
57. D	58. C	59. C	60. D
61. D	62. A	63. C	64. B
65. D	66. C	67. A	68. D
69. A	70. D	71. C	72. D
73. B	74. A	75. C	76. A
77. B	78. D	79. A	80. C
81. A	82. C	83. B	84. D
85. B	86. B	87. D	88. C
89. D	90. B	91. D	92. A
93. D	94. A	95. B	96. C
97. B	98. D	99. D	100. C
101. A	102. C	103. D	104. D
105. C	106. C	107. D	108. B
109. A	110. B	111. D	112. C
113. C	114. A	115. C	116. B
117. D	118. A	119. C	120. D
121. D	122. B	123. A	124. C
125. B	126. A	127. C	128. A
129. B	130. D	131. C	132. B
133. C	134. B	135. D	136. A
137. B	138. B	139. B	140. A
141. C	142. B	143. B	144. A
145. A	146. D	147. A	148. B
149. B	150. C	151. C	152. D
153. D	154. C	155. C	156. A
157. D	158. A	159. B	160. B

161. A	162. D	163. B	164. C
165. C	166. B	167. B	168. A
169. B	170. A	171. C	172. D
173. B	174. D	175. D	176. C
177. D	178. D	179. B	180. A
181. B	182. C	183. D	184. A
185. B	186. A	187. D	188. C
189. B	190. D	191. A	192. D
193. D	194. D	195. C	196. B
197. C	198. A	199. C	200. A
201. A	202. D	203. D	204. B
205. D	206. C	207. A	208. A
209. B	210. C	211. D	212. D
213. C	214. B	215. A	216. D
217. A	218. D	219. A	220. B
221. D	222. C	223. A	224. C
225. D	226. B	227. A	228. C
229. D	230. A	231. B	232. C
233. C	234. B	235. B	236. D
237. C	238. A	239. B	240. C
241. A	242. B	243. A	244. D
245. A	246. C	247. C	248. A
249. B	250. A	251. A	252. D
253. C	254. A	255. C	256. D
257. A			

Appendix
Present ETA and ISCET Associate and Journeyman Exams

ETA exams

The present ETA Journeyman Communications CET Exams are available as either a Radio Option or a Telecommunications Option. The Radio Option covers 2-way radio, amateur radio, and business radio. It also covers maritime, aviation, and navigation radio; communications, walkie talkies, short-range radio, military radio, and radar.

The Radio Journeyman Telecommunications Option includes television, telephone and TV distribution, cellular phones and pagers, data, microwaves, and fiberoptic communications.

ISCET Exams

The present ISCET Associate CET exams cover the following sections: basic mathematics, ac circuit, and dc circuits; transistors and semiconductors, and components and circuits (10 questions each); and tests and measurements, and troubleshooting (15 questions each).

The present ISCET Journeyman Communications CET Exam includes four Sections: Basic Communications Electronics (25 questions), AM and FM Transmitters (20 questions); Communications Receivers (20 questions), and Communications Systems (20 questions).

ISCET also has Journeyman CET tests on the Radar Electronics Option, the Video Option, the Audio Option, and FCC Legal.

Index

A

Illustrations are in **boldface.**

D

E

S

About the authors

Sam Wilson

Sam Wilson earned his bachelor's degree from Long Beach State College and his master's degree from Kent State University. He also has diplomas from Capitol Radio Engineering Institute and RCA Institutes.

Wilson is now a full-time technical writer and consultant. In 1983, he was selected as *Technician of the Year* by the International Society of Certified Electronics Technicians (ISCET). He has been the CET Test Consultant for that organization; and, has been the Technical publications Director for the National Electronics Service Dealers Association (NESDA).

His electronics experience includes 18 years as an instructor and professor. He also has 12 years of practical experience as a technician and engineer. He is presently a specialist in training equipment design and preparation of technical training publications. Over the years, Wilson has written over 25 technical electronics books.

Joe Risse

Joseph A. "Joe" Risse has worked as an assembler, technician, engineer, maintenance engineer, transmitter operator, chief broadcast engineer, director of electronics department for a correspondence school, project manager for industrial training programs, and is active on advisory committees for a technical institute and a vo tech school. He has completed courses in the military, college courses in electronics, correspondence school courses, and industrial group/training programs.

He holds the B.A. degree in Natural Science/Mathematics from Thomas Edison College, is a Fellow of the Society of Broadcast engineers and the Radio Club of America, a Life Member of the Electronic Technicians Association, and Member of the International Society of Certified Electronics Technicians. He is a registered Professional Engineer by the Commonwealth of Pennsylvania, and certified as an electronics technician by both ISCET and ETA.

Risse has completed all of the electronics, mathematics, and physics courses required for the Electronics major in Physics degree at the University of Scranton. He has prepared certification and practice exams for ISCET. He was the editor of the *Journal of the Society of Broadcast Engineers* during the early years of the SBE, and held the national office of executive vice president for 2 years.

About the Authors